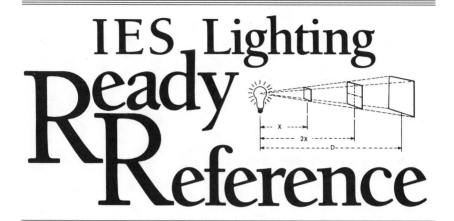

IES Lighting
Ready Reference

A Compendium of:

Definitions

Conversions Factors

Lighting Source Tables

Illuminance Recommendations

Calculation Data

rgy Management Considerations

Cost Analysis Methods

Survey Procedures

Edited by John E. Kaufman, PE, FIES and Jack F. Christensen

Published by the
Illuminating Engineering Society of North America
345 East 47th Street, New York, NY 10017

ISBN 0-87995-030-7
Library of Congress Catalog Card Number: 89-082769

First Edition Published 1985
Second Edition 1989

Printed in the United States of America

Contents

"Its object shall be . . . the advancement of the theory and practice of illuminating engineering and the dissemination of knowledge relating thereto."

Preface

In 1949, the monthly journal of the IES and Society reports moved from a six- by nine-inch format to an eight- by eleven-inch size, and in 1966 the *IES Lighting Handbook* followed suit. As the latter publication became thicker and, in 1981, was split into two volumes, it became less portable, especially for today's lighting specialists when on the go. This volume is an attempt to answer their needs as well as others who want a smaller, handier reference.

This second edition represents a compendium of the most often used material from the IES *Lighting Handbook, 1987 Application Volume* and *1984 Reference Volume*. The title page for each section indicates the source or sources used for that section. This compendium is not meant to replace any of these sources, but only as a "ready reference" for selected material.

John E. Kaufman
Editor

Lighting Terminology and Conversion Factors

(From IES Lighting Handbook—*1984 Reference Volume*)

As the title implies this first section contains terminology directly related to light and lighting practice. All terms are presented in alphabetical order and are followed by their standard symbols and defining equations where applicable, by their definitions and by other related terms of interest. No attempt has been made to provide information on pronunciations or etymologies. Definitions of electrical terms common to lighting and to other fields are available in *American National Standard Dictionary of Electrical and Electronics Terms* (ANSI/IEEE 100-1984).

Any of the radiometric and photometric quantities that follow may be restricted to a narrow wavelength interval $\Delta\lambda$ by the addition of the word spectral and the specification of the wavelength λ. The corresponding symbols are changed by adding a subscript λ, *i.e.*, Q_λ, for a spectral concentration, or a λ in parentheses, *i.e.*, $K(\lambda)$ for a function of wavelength. Fig. 1 is a tabulated summary of standard units, symbols and defining equations for the fundamental photometric and radiometric quantities. Other symbols, abbreviations and conversion factors are given in Figs. 7 to 15 at the end of this section.

DEFINITIONS

A

absolute luminance threshold: luminance threshold for a bright object like a disk on a totally dark background.

absorptance, $\alpha = \Phi_a/\Phi_i$: the ratio of the flux absorbed by a medium to the incident flux. See *absorption.*
  NOTE: The sum of the hemispherical reflectance, the hemispherical transmittance, and the absorptance is one.

absorption: a general term for the process by which incident flux is converted to another form of energy, usually and ultimately to heat.
  NOTE: All of the incident flux is accounted for by the processes of reflection, transmission and absorption.

accent lighting: directional lighting to emphasize a particular object or to draw attention to a part of the field of view. See *directional lighting.*

accommodation: the process by which the eye changes focus from one distance to another.

adaptation: the process by which the visual system, becomes accustomed to more or less light or light of a different color than it was exposed to during an immediately proceding period. It results in a change in the sensitivity of the eye to light. See *scotopic vision, photopic vision, chromatic adaptation.*
  NOTE: Adaptation is also used to refer to the final state of the process, as reaching a condition of dark adaptation or light adaptation.

adaptive color shift: the change in the perceived object color caused solely by change of the state of chromatic adaptation.

adverse weather lamp: See *fog lamp.*

aerodrome beacon: an aeronautical beacon used to indicate the location of an aerodrome.
  NOTE: An aerodrome is any defined area on land or water—including any buildings, installations, and equipment—intended to be used either wholly or in part for the arrival, departure and movement of aircraft.

aeronautical beacon: an aeronautical ground light visible at all azimuths, either continuously or intermittently, to designate a particular location on the surface of the earth. See *aerodrome beacon, airway beacon, hazard or obstruction beacon, landmark beacon.*

aeronautical ground light: any light specially provided as an aid to air navigation, other than a light displayed on an aircraft. See *aeronautical beacon, angle-of-approach lights, approach lights, approach-light beacon, bar (of lights), boundary lights, circling guidance lights, course light, channel lights, obstruction lights, runway alignment indicator, runway-end identification light, perimeter lights, runway lights, taxichannel lights, taxiway lights.*

aeronautical light: any luminous sign or signal specially provided as an aid to air navigation.

Fig. 1 Units, Symbols and Defining Equations for Fundamental Photometric and Radiometric Quantities*

Quantity†	Symbol†	Defining Equation	Unit	Symbol
Radiant energy	$Q, (Q_e)$		erg joule‡ calorie kilowatt-hour	erg J cal kWh
Radiant energy density	$w, (w_e)$	$w = dQ/dV$	joule per cubic meter‡ erg per cubic centimeter	J/m^3 erg/cm^3
Radiant flux	$\Phi, (\Phi_e)$	$\Phi = dQ/dt$	erg per second watt‡	erg/s W
Radiant flux density at a surface 　Radiant exitance (Radiant emittance§) 　Irradiance	 $M, (M_e)$ $E, (E_e)$	 $M = d\Phi/dA$ $E = d\Phi/dA$	watt per square centimeter, watt per square meter,‡ etc.	W/cm^2 W/m^2
Radiant intensity	$I, (I_e)$	$I = d\Phi/d\omega$ (ω = solid angle through which flux from point source is radiated)	watt per steradian‡	W/sr
Radiance	$L, (L_e)$	$L = d^2\Phi/(d\omega dA \cos\theta)$ $= dI/(dA \cos\theta)$ (θ = angle between line of sight and normal to surface considered)	watt per steradian and square centimeter watt per steradian and square meter‡	$W/sr\cdot cm^2$ $W/sr\cdot m^2$
Emissivity, spectral-total hemispherical	ϵ	$\epsilon = M/M_{blackbody}$ (M and $M_{blackbody}$ are respectively the radiant exitance of the measured specimen and that of a blackbody at the same temperature as the specimen)	one (numeric)	
Emissivity, spectral-total directional	$\epsilon(\theta, \phi, T)$	$\epsilon(\theta, \phi, T) = L(T)/L_{blackbody}(T)/[L(T)$ and $L_{blackbody}$ (T) are, respectively, the radiance of the measured specimen and that of a blackbody at the same temperature (that of the specimen)].	one (numeric)	
Emissivity, spectral directional	$\epsilon(\theta, \phi, \lambda, T)$	$\epsilon(\lambda, \theta, \phi, T) = L_\lambda(\lambda, \theta, \phi, T)/L_{\lambda,blackbody}(\lambda, T)$ $(L_\lambda$ and $L_{\lambda blackbody}$ are respectively the spectral radiance of the measured specimen and that of a blackbody at the same temperature of the specimen)	one (numeric)	
Emissivity, spectral hemispherical	$\epsilon(\lambda, T)$	$\epsilon(\lambda, T) = M_\lambda(\lambda, T)/M_{\lambda,blackbody}$ (λ, T) are respectively the spectral radiant exitance of the measured specimen and that of a blackbody at the same temperature of the specimen	one (numeric)	

after image: a visual response that occurs after the stimulus causing it has ceased.

aircraft aeronautical light: any aeronautical light specially provided on an aircraft. See *navigation light system, anti-collision light, ice detection light, fuselage lights, landing light, position lights, taxi light.*

airway beacon: an aeronautical beacon used to indicate a point on the airway.

altitude (in daylighting): the angular distance of a heavenly body measured on the great circle that passes perpendicular to the plane of the horizon through the body and through the zenith. It is measured positively from the horizon to the zenith, from 0 to 90 degrees.

ambient lighting: lighting throughout an area that produces general illumination.

anchor light (aircraft): a light designed for use on a seaplane or amphibian to indicate its position when at anchor or moored.

angle-of-approach lights: aeronautical ground lights arranged so as to indicate a desired angle of descent during an approach to an aerodrome runway. (Also called optical glide path lights.)

angle of collimation: the angle subtended by a light source at a point on an irradiated surface.

angstrom, Å: unit of wavelength equal to 10^{-10} (one ten-billionth) meter.

anti-collision light: a flashing aircraft aeronautical light or system of lights designed to provide a red signal throughout 360 degrees of azimuth for the purpose of giving long-range indication of an aircraft's location to pilots of other aircraft.

aperture color: perceived color of the sky or a patch of color seen through an aperture where it cannot be identified as belonging to a specific object.

apostilb (asb): a lambertian unit of luminance equal to $1/\pi$ candela per square meter. The use of this unit is deprecated.

apparent luminous intensity of an extended

Fig. 1 *Continued*

Quantity†	Symbol†	Defining Equation	Unit	Symbol
Absorptance	α	$\alpha = \Phi_a/\Phi_i\|$	one (numeric)	
Reflectance	ρ	$\rho = \Phi_r/\Phi_i\|$	one (numeric)	
Transmittance	τ	$\tau = \Phi_t/\Phi_i\|$	one (numeric)	
Luminous efficacy	K	$K = \Phi_v/\Phi_e$	lumen per watt‡	lm/W
Luminous efficiency	V	$V = K/K_{maximum}$ ($K_{maximum}$ = maximum value of $K(\lambda)$ function)	one (numeric)	
Luminous energy (quantity of light)	$Q, (Q_v)$	$Q_v = \int_{380}^{770} K(\lambda)Q_{e\lambda}d\lambda$	lumen-hour lumen-second‡ (talbot)	lm·h lm·s
Luminous energy density	$w, (w_v)$	$w = dQ/dV$	lumen-hour per cubic centimeter	lm·h/cm³
Luminous flux	$\Phi, (\Phi_v)$	$\Phi = dQ/dt$	lumen‡	lm
Luminous flux density at a surface Luminous exitance (Luminous emittance§) Illuminance (Illumination§)	$M, (M_v)$ $E, (E_v)$	$M = d\Phi/dA$ $E = d\Phi/dA$	lumen per square foot footcandle (lumen per square foot) lux (lm/m²)‡ phot(lm/cm²)	lm/ft² fc lx ph
Luminous intensity (candlepower)	$I, (I_v)$	$I = d\Phi/d\omega$ (ω = solid angle through which flux from point source is radiated)	candela‡ (lumen per steradian)	cd
Luminance	$L, (L_v)$	$L = d^2\Phi/(d\omega dA \cos\theta)$ $= dI/(dA \cos\theta)$ (θ = angle between line of sight and normal to surface considered)	candela per unit area stilb (cd/cm²) nit (cd/m²‡) footlambert (cd/πft²)§ lambert (cd/πcm²)§ apostilb (cd/πm²)§	cd/in², etc. sb nt, cd/m² fL§ L§ asb§

* The symbols for photometric quantities are the same as those for the corresponding radiometric quantities. When it is necessary to differentiate them the subscripts v and e respectively should be used, e.g., Q_v and Q_e.

† Quantities may be restricted to a narrow wavelength band by adding the word spectral and indicating the wavelength. The corresponding symbols are changed by adding a subscript λ, e.g., Q_λ, for a spectral concentration or a λ in parentheses, e.g., $K(\lambda)$, for a function of wavelength.

‡ International System (SI) unit

§ Use is deprecated

‖ Φ_i = incident flux, Φ_a = absorbed flux, Φ_r = reflected flux, Φ_t = transmitted flux

source at a specific distance: See *equivalent luminous intensity of an extended source.*

approach-light beacon: an aeronautical ground light placed on the extended centerline of the runway at a fixed distance from the runway threshold to provide an early indication of position during an approach to a runway.

NOTE: The runway threshold is the beginning of the runway usable for landing.

approach lights: a configuration of aeronautical ground lights located in extension of a runway or channel before the threshold to provide visual approach and landing guidance to pilots. See *angle-of-approach lights, approach-light beacon, VASIS.*

arc discharge: an electric discharge characterized by high cathode current densities and a low voltage drop at the cathode.

NOTE: The cathode voltage drop is small compared with that in a glow discharge, and secondary emission plays only a small part in electron emission from the cathode.

arc lamp: a discharge lamp in which the light is emitted by an arc discharge or by its electrodes.

NOTE: The electrodes may be either of carbon (operating in air) or of metal.

artificial pupil: a device or arrangement for confining the light passing through the pupil of the eye to an area smaller than the natural pupil.

atmospheric transmissivity: the ratio of the directly transmitted flux incident on a surface after passing through unit thickness of the atmosphere to the flux that would be incident on the same surface if the flux had passed through a vacuum.

average luminance (of a surface): the average luminance of a surface may be expressed in terms of the total luminous flux (lumens) leaving the surface per unit solid angle and unit area.

average luminance (of a luminaire): the luminous intensity at a given angle divided by the projected area of the luminaire at that angle.

azimuth: the angular distance between the vertical plane containing a given line or celestial body and the plane of the meridian.

B

back light: illumination from behind (and usually above) a subject to produce a highlight along its edge and consequent separation between the subject and its background. See *side back light.*

backing lighting: the illumination provided for scenery in off-stage areas visibile to the audience.

backup lamp: a lighting device mounted on the rear of a vehicle for illuminating the region near the rear of the vehicle while moving or about to move in reverse. It normally can be used only while backing up.

bactericidal (germicidal) effectiveness: the capacity of various portions of the ultraviolet spectrum to destroy bacteria, fungi and viruses.

bactericidal (germicidal) efficiency of radiant flux (for a particular wavelength): the ratio of the bactericidal effectiveness at a particular wavelength to that at wavelength 265.0 nanometers, which is rated as unity.

NOTE: Tentative bactericidal efficiency of various wavelengths of radiant flux is given in Fig. 19-12 in 1981 Application Volume.

bactericidal (germicidal) exposure: the product of bactericidal flux density on a surface and time. It usually is measured in bactericidal microwatt-minutes per square centimeter or bactericidal watt-minutes per square foot.

bactericidal (germicidal) flux: radiant flux evaluated according to its capacity to produce bactericidal effects. It usually is measured in microwatts of ultraviolet radiation weighted in accordance with its bactericidal efficiency. Such quantities of bactericidal flux would be in bactericidal microwatts.

NOTE: Ultraviolet radiation of wavelength 253.7 nanometers usually is referred to as "ultraviolet microwatts" or "UV watts." These terms should not be confused with "bactericidal microwatts" because the radiation has not been weighted in accordance with the values given in Fig. 19-12 in 1981 Application Volume.

bactericidal (germicidal) flux density: the bactericidal flux per unit area of the surface being irradiated. It is equal to the quotient of the incident bactericidal flux divided by the area of the surface when the flux is uniformly distributed. It usually is measured in microwatts per square centimeter or watts per square foot of bactericidally weighted ultraviolet radiation (bactericidal microwatts per square centimeter or bactericidal watts per square foot).

bactericidal lamp: an ultraviolet lamp that radiates a significant portion of its radiative power in the UV-C band (100 to 280 nanometers).

baffle: a single opaque or translucent element to shield a source from direct view at certain angles, or to absorb unwanted light.

balcony lights: luminaires mounted on the front edge of the auditorium balcony.

ballast: a device used with an electric-discharge lamp to obtain the necessary circuit conditions (voltage, current and wave form) for starting and operating. See *reference ballast.*

ballast factor: the fractional flux of a lamp(s) operated on a ballast compared to the flux when operated on the reference ballasting specified for rating lamp lumens.

bar (of lights): a group of three or more aeronautical ground lights placed in a line transverse to the axis, or extended axis, of the runway. See *barette.*

bare (exposed) lamp: a light source with no shielding.

barn doors: a set of adjustable flaps, usually two or four (two-way or four-way) that may be attached to the front of a luminaire (usually a Fresnel spotlight) in order to partially control the shape and spread of the light beam.

barrette (in aviation): a short bar in which the lights are closely spaced so that from a distance they appear to be a linear light.

NOTE: Barettes are usually less than 4.6 meters (15 feet) in length.

base light: uniform, diffuse illumination approaching a shadowless condition, which is sufficient for a television picture of technical acceptability, and which may be supplemented by other lighting.

beacon: a light (or mark) used to indicate a geographic location. See *aerodrome beacon, aeronautical beacon, airway beacon, approach-light beacon, hazard or obstruction beacon, identification beacon, landmark beacon.*

beam angle: the included angle between those points on opposite sides of the beam axis at which the luminous intensity from a theatrical luminaire is 50 per cent of maximum. This angle may be determined from a candlepower curve, or may be approximated by use of an incident light meter.

beam axis of a projector: a line midway between two lines that intersect the candlepower distribution curve at points equal to a stated per cent of its maximum (usually 50 per cent).

beam projector: a luminaire with the light source at or near the focus of a paraboloidal reflector producing near parallel rays of light in a beam of small divergence. Some are equipped with spill rings to reduce spill and glare. In most types, the lamp may be moved toward or away from the reflector to vary the beam spread.

beam spread (in any plane): the angle between the two directions in the plane in which the intensity is equal to a stated percentage of the maximum beam intensity. The percentage typically is 10 per cent for floodlights and 50 per cent for photographic lights.

biconical reflectance, $\rho(\omega_i; \omega_r)$**:** the ratio of reflected flux collected through a conical solid angle to the incident flux limited to a conical solid angle.

NOTE: The directions and extent of each cone must be specified; the solid angle is not restricted to a right-circular cone.

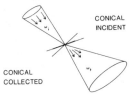

biconical transmittance, $\tau(\omega_i; \omega_t)$: ratio of transmitted flux collected through a conical solid angle to the incident flux limited to a conical solid angle.
NOTE: The directions and extent of each cone must be specified; the solid angle is not restricted to a right-circular cone.

bidirectional reflectance, $\rho(\theta_i, \phi_i; \theta_r, \phi_r)$: ratio of reflected flux collected over an element of solid angle surrounding the given direction to essentially collimated incident flux.
NOTE: The directions of incidence and collections and the size of the solid angle "element" of collection must be specified. In each case of conical incidence or collection, the solid angle is not restricted to a right circular cone, but may be of any cross section, including rectan-

gular, a ring, or a combination of two or more solid angles.
bidirectional reflectance-distribution function (BRDF): the ratio of the differential luminance of a ray $dL_r(\theta_r, \phi_r)$ reflected in a given direction (θ_r, ϕ_r) to the differential luminous flux density $dE_i(\theta_i,$

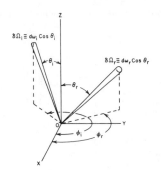

$\phi_i)$ incident from a given direction of incidence, (θ_i, ϕ_i), which produces it.

$$f_r(\theta_i, \phi_i; \theta_r, \phi_r) \equiv dL_r(\theta_r, \phi_r)/dE_i(\theta_i, \phi_i)(sr)^{-1}$$

$$= dL_r(\theta_r, \phi_r)/L_i(\theta_i, \phi_i)d\Omega_i$$

where $d\Omega \equiv d\omega\cos\theta$
NOTE: This distribution function is the basic parameter for describing (geometrically) the reflecting properties of an opaque surface element (negligible internal scattering). It may have any positive value and will approach infinity in the specular direction for ideally specular reflectors. The spectral and polarization aspects must be defined for complete specification, since the BRDF as given above only defines the geometric aspects.

bidirectional transmittance, $\tau(\theta_i, \phi_i; \theta_t, \phi_t)$: ratio of incident flux collected over an element of solid angle surrounding the given direction to essentially collimated incident flux.
NOTE: The direction of incidence, collection and size of the solid angle "element" must be specified.

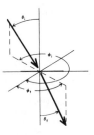

bihemispherical reflectance, $\rho(2\pi; 2\pi)$: ratio of reflected flux collected over the entire hemisphere to the incident flux from the entire hemisphere.

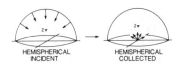

bihemispherical transmittance, $\tau(2\pi; 2\pi)$: ratio of transmitted flux collected over the entire hemisphere to the incident flux from the entire hemisphere.

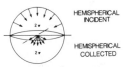

binocular portion of the visual field: that portion of space where the fields of the two eyes overlap.

blackbody: a temperature radiator of uniform temperature whose radiant exitance in all parts of the spectrum is the maximum obtainable from any temperature radiator at the same temperature.

Such a radiator is called a blackbody because it will absorb all the radiant energy that falls upon it. All other temperature radiators may be classed as nonblackbodies. They radiate less in some or all wavelength intervals than a blackbody of the same size and the same temperature.

NOTE: the blackbody is practically realized in the form of a cavity with opaque walls at a uniform temperature and with a small opening for observation purposes. It also is called a full radiator, standard radiator, complete radiator or ideal radiator.

blackbody (Planckian) locus: the locus of points on a chromaticity diagram representing the chromaticities of blackbodies having various (color) temperatures.

"black light:" the popular term for ultraviolet energy near the visible spectrum.

NOTE: For engineering purposes the wavelength range 320–400 nanometers has been found useful for rating lamps and their effectiveness upon fluorescent materials (excluding phosphors used in fluorescent lamps). By confining "black light" applications to this region, germicidal and erythemal effects are, for practical purposes, eliminated.

"black light" flux: radiant flux within the wavelength range 320 to 400 nanometers. It is usually measured in milliwatts. See *fluoren.*

NOTE: Because of the variability of the spectral sensitivity of materials irradiated by "black light" in practice, no attempt is made to evaluate "black light" flux according to its capacity to produce effects.

"black light" flux density: "black light" flux per unit area of the surface being irradiated. It is equal to the incident "black light" flux divided by the area of the surface when the flux is uniformly distributed. It usually is measured in milliwatts per square foot of "black light" flux.

"black light" lamp: An ultraviolet lamp that emits a significant portion of its radiative power in the UV-A band (315–400 nanometers).

blending lighting: general illumination used to provide smooth transitions between the specific lighting areas on a stage.

blinding glare: glare which is so intense that for an appreciable length of time after it has been removed, no object can be seen.

borderlight: a long continuous striplight hanging horizontally above the stage and aimed down to provide general diffuse illumination and/or to light the cyclorama or a drop, usually wired in three or four color circuits.

borderline between comfort and discomfort (BCD): the average luminance of a source in a field of view which produces a sensation between comfort and discomfort.

boundary lights: aeronautical ground lights delimiting the boundary of a land aerodrome without runways. See *range lights.*

bowl: an open top diffusing glass or plastic enclosure used to shield a light source from direct view and to redirect or scatter the light.

bracket (mast arm): an attachment to a lamp post or pole from which a luminaire is suspended.

brightness: See *subjective brightness, luminance, veiling luminance, brightness of a perceived light-source color.*

brightness contrast threshold: when two patches of color are separated by a brightness contrast border as in the case of a bipartite photometric field or in the case of a disk shaped object surrounded by its background, the border between the two patches is a brightness contrast border. The contrast which is just detectable is known as the brightness contrast threshold.

brightness of perceived light-source color: the attribute in accordance with which the source seems to emit more or less luminous flux per unit area.

bulb: a source of electrically powered light. This term is used to distinguish between an assembled unit consisting of a light source in a housing called a *lamp* and the internal source. See *lamp.*

C

candela, cd: the SI unit of luminous intensity. One candela is one lumen per steradian. Formerly, candle. See Fig. 2.

NOTE: The fundamental luminous intensity definition in the SI is the candela in terms of monochromatic radiation at 540×10^{12} hertz (approximately 555 nm). One candela is defined as the luminous intensity, in a given direction, of a source that emits monochromatic radiation of frequency 540×10^{12} hertz and of which the radiant intensity in that direction is 1/683 watt per steradian.

From 1909 until 1948, the unit of luminous intensity in the United States, as well as in France and Great Britain, was the "international candle" which was maintained by a group of carbon-filament vacuum lamps. From 1948 to 1979 the unit of luminous intensity was defined in terms of the luminance of a blackbody at the freezing point of platinum. Since 1948, the internationally accepted name for the unit of luminous intensity is *candela.* The difference between the candela and the old international candle are so small that only measurements of high accuracy are affected.

candlepower, $I = d\Phi/d\omega$**; cp.:** luminous intensity expressed in candelas.

candlepower (intensity) distribution curve: a curve, generally polar, representing the variation of luminous intensity of a lamp or luminaire in a plane through the light center.

NOTE: A vertical candlepower distribution curve is obtained by taking measurements at various angles of elevation in a vertical plane through the light center; unless the plane is specified, the vertical curve is assumed to represent an average such as would be obtained by rotating the lamp or luminaire about its vertical axis. A horizontal candlepower distribution curve represents measurements made at various angles of azimuth in a horizontal plane through the light center.

carbon-arc lamp: an electric-discharge lamp employing an arc discharge between carbon electrodes.

One or more of these electrodes may have cores of special chemicals that contribute importantly to the radiation.

cavity ratio, CR: a number indicating cavity proportions calculated from length, width and height. See *ceiling cavity ratio, floor cavity ratio and room cavity ratio.*

ceiling area lighting: a general lighting system in which the entire ceiling is, in effect, one large luminaire.

> NOTE: Ceiling area lighting includes *luminous ceilings* and *louvered ceilings.*

ceiling cavity: the cavity formed by the ceiling, the plane of the luminaires, and the wall surfaces between these two planes.

ceiling cavity ratio, CCR: a number indicating ceiling cavity proportions calculated from length, width and height.

ceiling projector: a device designed to produce a well-defined illuminated spot on the lower portion of a cloud for the purpose of providing a reference mark for the determination of the height of that part of the cloud.

ceiling ratio: the ratio of the luminous flux reaching the ceiling directly to the upward component from the luminaire.

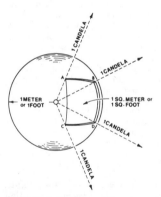

Fig. 2. Relationship between candelas, lumens, lux and footcandles.

A uniform point source (luminous intensity or candlepower = one candela) is shown at the center of a sphere of one meter or one foot radius. It is assumed that the sphere surface has zero reflectance.

The illuminance at any point on the sphere is one lux (one lumen per square meter) when the radius is one meter, or one footcandle (one lumen per square foot) when the radius is one foot.

The solid angle subtended by the area, *A, B, C, D* is one steradian. The flux density is therefore one lumen per steradian, which corresponds to a luminous intensity of one candela, as originally assumed.

The sphere has a total area of 12.57 (4π) square meters or square feet, and there is a luminous flux of one lumen falling on each square meter or square foot. Thus the source provides a total of 12.57 lumens.

central (foveal) vision: the seeing of objects in the central or foveal part of the visual field, approximately two degrees in diameter. It permits seeing much finer detail than does peripheral vision.

central visual field: that region of the visual field corresponding to the foveal portion of the retina.

channel: an enclosure containing the ballast, starter, lamp holders and wiring for a fluorescent lamp, or a similar enclosure on which filament lamps (usually tubular) are mounted.

channel lights: aeronautical ground lights arranged along the sides of a channel of a water aerodrome. See *taxi-channel lights.*

characteristic curve: a curve which expresses the relationship between two variable properties of a light source, such as candlepower and voltage, flux and voltage, etc.

chromatic adaptation: the process by which the chromatic properties of the visual system are modified by the observation of stimuli of various chromaticities and luminances. See *state of chromatic adaptation.*

chromatic contrast thresholds (color contrast thresholds): two patches of color juxtaposed and separated only by a color contrast border cannot be perceived as different in chromaticness or separated by a contrast border if the difference in chromaticity on the two sides of the border is reduced below the threshold of visibility. A contrast border can involve both differences in luminance and chromaticity on the two sides of the border.

chromaticity coordinates (of a color), *x, y, z*: the ratios of each of the tristimulus values of the light to the sum of the three tristimulus values.

chromaticity diagram: a plane diagram formed by plotting one of the three chromaticity coordinates against another.

chromaticity difference thresholds (color difference thresholds): two patches of color of the same luminance may have a threshold difference in chromaticity which makes them perceptibly different in chromaticness. The difference may be a difference in hue or saturation or a combination of the two. The threshold can involve a combination of differences in both luminance and chromaticity.

chromaticity of a color: consists of the dominant or complementary wavelength and purity aspects of the color taken together, or of the aspects specified by the chromaticity coordinates of the color taken together.

chromaticness: the attribute of a visual sensation according to which the (perceived) color of an area appears to be more or less chromatic.

CIE standard chromaticity diagram: a diagram in which the *x* and *y* chromaticity coordinates are plotted in rectangular coordinates.

circling guidance lights: aeronautical ground lights provided to supply additional guidance during a circling approach when the circling guidance furnished by the approach and runway lights is inadequate.

clear sky: a sky that has less than 30 per cent cloud cover.

clearance lamp: a lighting device mounted on a vehicle for the purpose of indicating the overall width and height of the vehicle.

clerestory: that part of a building rising clear of the roofs or other parts and whose walls contain windows for lighting the interior.

cloudy sky: a sky having more than 70 per cent cloud cover.

coefficient of attenuation, μ: the decrement in flux per unit distance in a given direction within a medium and defined by the relation: $\Phi_x = \Phi_o e^{-\mu x}$ where Φ_x is the flux at any distance x from a reference point having flux Φ_o.

coefficient of beam utilization, CBU: the ratio of the luminous flux (lumens) reaching a specified area directly from a floodlight or projector to the total beam luminous flux (lumens).

coefficient of utilization, CU: the ratio of the luminous flux (lumens) from a luminaire calculated as received on the work-plane to the luminous flux emitted by the luminaire's lamps alone.

coffer: a recessed panel or dome in the ceiling.

cold-cathode lamp: an electric-discharge lamp whose mode of operation is that of a glow discharge, and having electrodes so spaced that most of the light comes from the positive column between them.

color: the characteristic of light by which a human observer may distinguish between two structure-free patches of light of the same size and shape. See *light-source color* and *object color.*

color comparison, or color grading (CIE, object color inspection): the judgment of equality, or of the amount and character of difference, of the color of two objects viewed under identical illumination.

color contrast thresholds: See *chromaticity difference thresholds.*

color correction (of a photograph or printed picture): the adjustment of a color reproduction process to improve the perceived-color conformity of the reproduction to the original.

color discrimination: the perception of differences between two or more colors.

color matching: action of making a color appear the same as a given color.

color-matching functions (spectral tristimulus values), $\bar{x}(\lambda) = X_\lambda/\Phi_{e\lambda}; \bar{y}(\lambda) = Y_\lambda/\Phi_{e\lambda}; \bar{z}(\lambda) = Z_\lambda/\Phi_{e\lambda}$**:** the tristimulus values per unit wavelength interval and unit spectral radiant flux.

NOTE: Color-matching functions have been adopted by the International Commission on Illumination. They are tabulated as functions of wavelength throughout the spectrum and are the basis for the evaluation of radiant energy as light and color. The standard values adopted by the CIE in 1931 are given in Fig. 5-1.[*] The $\bar{y}$ values are identical with the values of spectral luminous efficiency for photopic vision.

The $\bar{x}$, $\bar{y}$, and $\bar{z}$ values for the 1931 Standard Observer are based on a two-degree bipartite field, and are recommended for predicting matches for stimuli subtending between one and four degrees. Supplementary data based on a 10-degree bipartite field were adopted in 1964 for use for angular subtenses greater than four degrees.

Tristimulus computational data for CIE standard sources *A* and *C* are given in Fig. 5-3.[*]

color preference index, CPI: See *flattery index.*

color rendering: general expression for the effect of a light source on the color appearance of objects in conscious or subconscious comparison with their color appearance under a reference light source.

color rendering improvement (of a light source): the adjustment of spectral composition to improve color rendering.

color rendering index (of a light source) (CRI): measure of the degree of color shift objects undergo when illuminated by the light source as compared with the color of those same objects when illuminated by a reference source of comparable color temperature.

color temperature of a light source: the absolute temperature of a blackbody radiator having a chromaticity equal to that of the light source. See also *correlated color temperature* and *distribution temperature.*

colorfulness: See *chromaticness.*

colorimetric purity (of a light), p_c: the ratio L_1/L_2 where L_1 is the luminance of the single frequency component that must be mixed with a reference standard to match the color of the light and L_2 is the luminance of the light. See *excitation purity.*

colorimetric shift: the change of chromaticity and luminance factor of an object color due to change of the light source. See *adaptive color shift* and *resultant color shift.*

colorimetry: the measurement of color.

compact-arc lamp: See *short-arc lamp.*

compact source iodide (CSI): an ac arc source utilizing a mercury vapor arc with metal halide additives to produce illumination in the 4000- to 6000-K range. Requires a ballast and ignition system for operation.

comparison lamp: a light source having a constant but not necessarily known, luminous intensity with which standard and test lamps are compared successively.

complementary wavelength (of a light), λ_c: the wavelength of radiant energy of a single frequency that, when combined in suitable proportion in the light, matches the color of a reference standard. See *dominant wavelength.*

complete diffusion: diffusion in which the diffusing medium redirects the flux incident by scattering so that none is in an image-forming state.

cone: a retinal receptor which dominates the retinal response when the luminance level is high and provides the basis for the perception of color.

configuration factor, $C_{2\text{-}1}$: the ratio of illuminance on a surface at point 2 (due to flux directly received from lambertian surface 1) to the exitance of surface 1. It is used in *flux transfer theory.* $C_{2\text{-}1} = (E_2)/(M_1)$

conical-directional reflectance, ρ ($\omega_i; \theta_r, \phi_r$): ratio of reflected flux collected over an element of solid angle surrounding the given direction to the incident flux limited to a conical solid angle.

NOTE: The direction and extent of the cone must be specified and the direction of collection and size of the solid angle "element" must be specified.

CONICAL INCIDENT — DIRECTIONAL COLLECTED

conical-directional transmittance, τ $(\omega_i;\theta_t,\phi_t)$: ratio of transmitted flux collected over an element of solid angle surrounding the direction to the incident flux limited to a conical solid angle.

NOTE: The direction and extent of the cone must be specified and the direction of collection and size of the solid angle "element" must be specified.

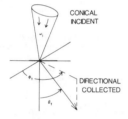

CONICAL INCIDENT

DIRECTIONAL COLLECTED

conical-hemispherical reflectance, ρ $(\omega_i; 2\pi)$: ratio of reflected flux collected over the entire hemisphere to the incident flux limited to a conical solid angle.

NOTE: The direction and extent of the cone must be specified.

CONICAL INCIDENT — HEMISPHERICAL COLLECTED

conical-hemispherical transmittance, τ $(\omega_t; 2\pi)$: ratio of transmitted flux collected over the entire

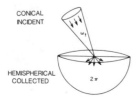

CONICAL INCIDENT

HEMISPHERICAL COLLECTED 2π

hemisphere to the incident flux limited to a conical solid angle.

NOTE: The direction and extent of the cone must be specified.

conspicuity: the capacity of a signal to stand out in relation to its background so as to be readily discovered by the eye.

contrast: See *luminance contrast.*

contrast rendition factor, CRF: the ratio of visual task contrast with a given lighting environment to the contrast with sphere illumination. Also known as *contrast rendering factor.*

contrast sensitivity: the ability to detect the presence of luminance differences. Quantitatively, it is equal to the reciprocal of the contrast threshold.

cornice lighting: lighting comprising sources shielded by a panel parallel to the wall and attached to the ceiling, and distributing light over the wall.

correlated color temperature (of a light source): the absolute temperature of a blackbody whose chromaticity most nearly resembles that of the light source.

cosine law: the law that the illuminance on any surface varies as the cosine of the angle of incidence. The angle of incidence θ is the angle between the normal to the surface and the direction of the incident light. The inverse-square law and the cosine law can be combined as $E = (I \cos \theta)/d^2$. See *cosine-cubed law* and *inverse-square law.*

cosine-cubed law: an extension of the cosine law in which the distance d between the source and surface is replaced by $h/\cos \theta$, where h is the perpendicular distance of the source from the plane in which the point is located. It is expressed by $E = (I \cos^3 \theta)/h^2$.

counter-key light: illumination on a subject from a direction that is opposite to that of the key light.

country beam: See *upper (driving) beams.*

course light: an aeronautical ground light, supplementing an airway beacon, for indicating the direction of the airway and to identify by a coded signal the location of the airway beacon with which it is associated.

cove lighting: lighting comprising sources shielded by a ledge or horizontal recess, and distributing light over the ceiling and upper wall.

critical fusion frequency, cff: See *flicker fusion frequency.*

critical flicker frequency, cff: See *flicker fusion frequency.*

cross light: equal illuminance in front of the subject from two directions at substantially equal and opposite angles with the optical axis of the camera and a horizontal plane.

cucoloris: an opaque cutout panel mounted between a light source (sun, arc, etc.) and a target surface in order to project a shadow pattern (clouds or leaves are typical) upon scenery, cyclorama or acting area.

cut-off angle (of a luminaire): the angle, measured up from nadir, between the vertical axis and the first line of sight at which the bare source is not visible.

D

dark adaptation: the process by which the retina becomes adapted to a luminance less than about 0.034 candela per square meter.

daylight availability: the amount of light from the sun and the sky at a specific location, time, day and sky condition.

daylight factor: a measure of daylight illuminance at a point on a given plane expressed as a ratio of the illuminance on the given plane at that point to the simultaneous exterior illuminance on a horizontal plane from the whole of an unobstructed sky of assumed or known luminance distribution. Direct sunlight is excluded from both interior and exterior values of illuminance.

daylight lamp: a lamp producing a spectral distribution approximating that of a specified daylight.

densitometer: a photometer for measuring the optical density (common logarithm of the reciprocal of the transmittance) of materials.

diffuse reflectance: the ratio of the flux leaving a surface or medium by diffuse reflection to the incident flux.

diffuse reflection: the process by which incident flux is re-directed over a range of angles.

diffuse transmission: the process by which the incident flux passing through a surface or medium is scattered.

diffuse transmittance: the ratio of the diffusely transmitted flux leaving a surface or medium to the incident flux.

diffused lighting: lighting provided on the work-plane or on an object, that is not predominantly incident from any particular direction.

diffuser: a device to redirect or scatter the light from a source, primarily by the process of diffuse transmission.

diffusing panel: a translucent material covering the lamps in a luminaire to reduce the luminance by distributing the flux over an extended area.

diffusing surfaces and media: those that redistribute some of the incident flux by scattering in all directions. See *complete diffusion, incomplete diffusion, perfect diffusion, narrow-angle diffusion, wide-angle diffusion.*

digital display: See *numerical display.*

dimmer: a device used to control the intensity of light emitted by a luminaire by controlling the voltage or current available to it.

direct component: that portion of the light from a luminaire which arrives at the work-plane without being reflected by room surfaces. See *indirect component.*

direct glare: glare resulting from high luminances or insufficiently shielded light sources in the field of view. It usually is associated with bright areas, such as luminaires, ceilings and windows which are outside the visual task or region being viewed.

direct-indirect lighting: a variant of general diffuse lighting in which the luminaires emit little or no light at angles near the horizontal.

direct lighting: lighting by luminaires distributing 90 to 100 per cent of the emitted light in the general direction of the surface to be illuminated. The term usually refers to light emitted in a downward direction.

direct ratio: the ratio of the luminous flux reaching the work-plane directly to the downward component from the luminaire.

directional-conical reflectance, ρ $(\theta_i, \phi_i; \omega_r)$: ratio of reflected flux collected through a conical solid angle to essentially collimated incident flux.
NOTE: The direction of incidence must be specified, and the direction and extent of the cone must be specified.

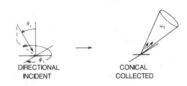

directional-conical transmittance, τ $(\theta_i, \phi_i; \omega_t)$: ratio of transmitted flux collected through a conical solid angle to essentially collimated incident flux.
NOTE: The direction of incidence must be specified, and the direction and extent of the cone must be specified.

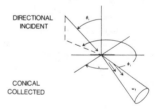

directional-hemispherical reflectance, ρ $(\theta_i, \phi_i; 2\pi)$: ratio of reflected flux collected over the entire hemisphere to essentially collimated incident flux.
NOTE: The direction of incidence must be specified.

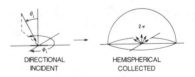

directional-hemispherical transmittance, τ $(\theta_i, \phi_i; 2\pi)$: ratio of transmitted flux collected over the entire hemisphere to essentially collimated incident flux.
NOTE: The direction of incidence must be specified.

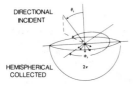

DIRECTIONAL INCIDENT

HEMISPHERICAL COLLECTED

directional lighting: lighting provided on the work-plane or on an object predominantly from a preferred direction. See *accent lighting, key light, cross light.*

disability glare: glare resulting in reduced visual performance and visibility. It often is accompanied by discomfort. See *veiling luminance.*

disability glare factor (DGF): a measure of the visibility of a task in a given lighting installation in comparison with its visibility under reference lighting conditions, expressed in terms of the ratio of luminance contrasts having an equivalent effect upon task visibility. The value of DGF takes account of the equivalent veiling luminance produced in the eye by the pattern of luminances in the task surround.

discomfort glare: glare producing discomfort. It does not necessarily interfere with visual performance or visibility.

discomfort glare factor: the numerical assessment of the capacity of a single source of brightness, such as a luminaire, in a given visual environment for producing discomfort. (This term is obsolete and is retained for reference and literature searches.) See *glare* and *discomfort glare.*

discomfort glare rating (DGR): a numerical assessment of the capacity of a number of sources of luminance, such as luminaires, in a given visual environment for producing discomfort. It is the net effect of the individual values of index of sensation for all luminous areas in the field of view. See *discomfort glare factor.*

distal stimuli: in the physical space in front of the eye one can identify points, lines and surfaces and three dimensional arrays of scattering particles which constitute the distal physical stimuli which form optical images on the retina. Each element of a surface or volume to which an eye is exposed subtends a solid angle at the entrance pupil. Such elements of solid angle make up the field of view and each has a specifiable luminance and chromaticity. Points and lines are specific cases which have to be dealt with in terms of total candlepower and candlepower per unit length.

distribution temperature (of a light source): the absolute temperature of a blackbody whose relative spectral distribution is the same (or nearly so) in the visible region of the spectrum as that of the light source.

dominant wavelength (of a light), λ_d: the wavelength of radiant energy of a single frequency that, when combined in suitable proportion with the ra-

diant energy of a reference standard, matches the color of the light. See *complementary wavelength.*

downlight: a small direct lighting unit which directs the light downward and can be recessed, surface mounted or suspended.

downward component: that portion of the luminous flux from a luminaire emitted at angles below the horizontal. See *upward component.*

driving beam: See *upper beam.*

dual headlighting system: headlighting by means of two double units, one mounted on each side of the front end of a vehicle. Each unit consists of two lamps mounted in a single housing. The upper or outer lamps may have two filaments supplying the lower beam and part of the upper beam, respectively. The lower or inner lamps have one filament providing the primary source of light for the upper beam.

dust-proof luminaire: a luminaire so constructed or protected that dust will not interfere with its successful operation.

dust-tight luminaire: a luminaire so constructed that dust will not enter the enclosing case.

E

effective ceiling cavity reflectance, ρ_{cc}: a number giving the combined reflectance effect of the walls and ceiling of the ceiling cavity. See *ceiling cavity ratio.*

effective floor cavity reflectance, ρ_{fc}: a number giving the combined reflectance effect of the walls and floor of the floor cavity. See *floor cavity ratio.*

efficacy: See *luminous efficacy of a source of light* and *spectral luminous efficacy of radiant flux.*

efficiency: See *luminaire efficiency, luminous efficacy of a source of light* and *spectral luminous efficiency of radiant flux.*

electric discharge: See *arc discharge, gaseous discharge* and *glow discharge.*

electric-discharge lamp: a lamp in which light (or radiant energy near the visible spectrum) is produced by the passage of an electric current through a vapor or a gas. See *fluorescent lamp, cold-cathode lamp, hot-cathode lamp, carbon-arc lamp, glow lamp, fluorescent lamp, high intensity discharge lamp.*

> NOTE: Electric-discharge lamps may be named after the filling gas or vapor that is responsible for the major portion of the radiation; *e.g.* mercury lamps, sodium lamps, neon lamps, argon lamps, etc.
>
> A second method of designating electric-discharge lamps is by psysical dimensions or operating parameters; *e.g.* short-arc lamps, high-pressure lamps, low-pressure lamps, etc.
>
> A third method of designating electric-discharge lamps is by their application; in addition to lamps for illumination there are photochemical lamps, bactericidal lamps, blacklight lamps, sun lamps, etc.

electroluminescence: the emission of light from a phosphor excited by an electromagnetic field.

electromagnetic spectrum: a continuum of electric and magnetic radiation encompassing all wavelengths. See *regions of electromagnetic spectrum.*

elevation: the angle between the axis of a searchlight drum and the horizontal. For angles above the horizontal, elevation is positive, and below the horizontal negative.

ellipsoidal reflector spotlight: a spotlight in which a lamp and an ellipsoidal reflector are mounted in a fixed relationship directing a beam of light into an aperture where it may be shaped by a pattern, iris, shutter system or other insertion. The beam then passes through a single or compound lens system that focuses it as required, producing a sharply defined beam with variable edge definition.

emergency lighting: lighting designed to supply illumination essential to safety of life and property in the event of failure of the normal supply.

emissivity, ϵ: the ratio of the radiance (for directional emissivity) or radiant exitance (for hemispherical emissivity) of an element of surface of a temperature radiator to that of a blackbody at the same temperature.

emittance, ϵ: (1) The ratio of the radiance in a given direction (for directional emittance) or radiant exitance (for hemispherical emittance) of a sample of a thermal radiator to that of a blackbody radiator at the same temperature. (2) See *exitance*. Use of the term with this meaning is deprecated.

enclosed and gasketed: See *vapor-tight*.

equal interval (isophase) light: a rhythmic light in which the light and dark periods are equal.

equipment operating factor: the fractional flux of a high intensity discharge (HID) lamp-ballast-luminaire combination in a given operating position compared to the flux of the lamp-luminaire combination (a) operated in the position for rating lamp lumens and (b) using the reference ballasting specified for rating lamp lumens.

equivalent contrast, $\tilde{C}$: a numerical description of the relative visibility of a task. It is the contrast of the standard visibility reference task giving the same visibility as that of a task whose contrast has been reduced to threshold when the background luminances are the same. See *visual task evaluator*.

equivalent contrast, $\tilde{C}_e$: the actual equivalent contrast in a real luminous environment with nondiffuse illumination. The actual equivalent contrast ($\tilde{C}_e$) may be less than the equivalent contrast due to veiling reflection. $\tilde{C}_e = \tilde{C} \times CRF$. See *contrast rendition factor*.

equivalent luminous intensity (of an extended source at a specified distance): the intensity of a point source which would produce the same illuminance at that distance. Formerly, *apparent luminous intensity of an extended source*.

equivalent sphere illumination (ESI): the level of sphere illumination which would produce task visibility equivalent to that produced by a specific lighting environment.

erythema: the temporary reddening of the skin produced by exposure to ultraviolet energy.
  NOTE: The degree of erythema is used as a guide to dosages applied in ultraviolet therapy.

erythemal effectiveness: the capacity of various portions of the ultraviolet spectrum to produce erythema.

erythemal efficiency of radiant flux (for a particular wavelength): the ratio of the erythemal effectiveness of a particular wavelength to that of wavelength 296.7 nanometers, which is rated as unity.
  NOTE: This term formerly was called "relative erythemal factor."
  The erythemal efficiency of radiant flux of various wavelengths for producing a minimum perceptible erythema (MPE) is given in Fig. 19-12 in the 1981 Application Volume. These values have been accepted for evaluating the erythemal effectiveness of sun lamps.

erythemal exposure: the product of erythemal flux density on a surface and time. It usually is measured in erythemal microwatt-minutes per square centimeter.
  NOTE: For average untanned skin a minimum perceptible erythema requires about 300 microwatt-minutes per square centimeter of radiation at 296.7 nanometers.

erythemal flux: radiant flux evaluated according to its capacity to produce erythema of the untanned human skin. It usually is measured in microwatts of ultraviolet radiation weighted in accordance with its erythemal efficiency. Such quantities of erythemal flux would be in erythemal microwatts. See *erythemal efficiency of radiant flux*.
  NOTE: A commonly used practical unit of erythemal flux is the erythemal unit (EU) or E-viton (erytheme) which is equal to the amount of radiant flux that will produce the same erythemal effect as 10 microwatts of radiant flux at wavelength 296.7 nanometers. See also *erythemal unit or E-viton*.

erythemal flux density: the erythemal flux per unit area of the surface being irradiated. It is equal to the quotient of the incident erythemal flux divided by the area of the surface when the flux is uniformly distributed. It usually is measured in microwatts per square centimeter of erythemally weighted ultraviolet radiation (erythemal microwatts per square centimeter). See *Finsen*.

erythemal threshold: See *minimal perceptible erythema*.

erythemal unit, EU: a unit of erythemal flux that is equal to the amount of radiant flux that will produce the same erythemal effect as 10 microwatts of radiant flux at wavelength 296.7 nanometers. Also called E-viton.

E-viton (erytheme): See *erythemal unit*.

exitance: See *luminous exitance* and *radiant exitance*.

exitance coefficient: a coefficient similar to the coefficient of utilization used to determine wall and ceiling exitances. These coefficients are numerically equal to the previously used luminance coefficients.

excitation purity (of a light), p_e: the ratio of the distance on the CIE (x,y) chromaticity diagram between the reference point and the light point to the distance in the same direction between the reference point and the spectrum locus or the purple boundary. See *colorimetric purity*.

explosion-proof luminaire: a luminaire which is completely enclosed and capable of withstanding an explosion of a specific gas or vapor that may occur within it, and preventing the ignition of a specific gas or vapor surrounding the enclosure by sparks, flashes or explosion of the gas or vapor within. It must operate at such an external temperature that a surrounding flammable atmosphere will not be ignited thereby.

eye light: illumination on a person to provide a specular reflection from the eyes (and teeth) without adding a significant increase in light on the subject.

F

far (long wavelength) infrared: region of the electromagnetic spectrum extending from 5,000 to 1,000,000 nm.

far ultraviolet: region of the electromagnetic spectrum extending from 100 to 200 nm.

fay light: a luminaire that uses incandescent parabolic reflector lamps with a dichroic coating to provide "daylight" illumination.

fenestra method: a procedure for predicting the interior illuminance received from daylight through windows.

fenestration: any opening or arrangement of openings (normally filled with media for control) for the admission of daylight.

field angle: the included angle between those points on opposite sides of the beam axis at which the luminous intensity from a theatrical luminaire is 10 per cent of the maximum value. This angle may be determined from a candlepower curve, or may be approximated by use of an incident light meter.

fill light: supplementary illumination to reduce shadow or contrast range.

film (or aperture) color: perceived color of the sky or a patch of color seen through an aperture.

filter: a device for changing, usually by transmission, the magnitude and/or the spectral composition of the flux incident on it. Filters are called *selective* (or *colored*) or *neutral*, according to whether or not they alter the spectral distribution of the incident flux.

filter factor: the transmittance of "black light" by a filter.

NOTE: The relationship among these terms is illustrated by the following formula for determining the luminance of fluorescent materials exposed to "black light":

candelas per square meter

$$= \frac{I}{\pi^*} \frac{\text{fluorens}}{\text{square meter}} \times \text{glow factor} \times \text{filter factor}$$

When integral-filter "black light" lamps are used, the filter factor is dropped from the formula because it already has been applied in assigning fluoren ratings to these lamps.

Finsen: a suggested practical unit of erythemal flux density equal to one E-viton per square centimeter.

---

* π is omitted when luminance is in footlamberts and the area is in square feet.

fixed light: a light having a constant luminous intensity when observed from a fixed point.

fixture: See *luminaire.*

flashing light: a rhythmic light in which the periods of light are of equal duration and are clearly shorter than the periods of darkness. See *group flashing light, interrupted quick-flashing light, quick-flashing light.*

flashtube: a tube of glass or fused quartz with electrodes at the ends and filled with a gas, usually xenon. It is designed to produce high intensity flashes of light of extremely short duration.

flattery index (of a light source) (R_f): measure appraising a light source for appreciative viewing of colored objects, or for promoting an optimistic viewpoint by flattery (to make more pleasant), or for enhancing the perception of objects in terms of color. Also sometimes called *color preference index (CPI).*

flicker fusion frequency, fff: the frequency of intermittent stimulation of the eye at which flicker disappears. It also is called critical fusion frequency (cff) or critical flicker frequency (cff).

flicker photometer: See *visual photometer.*

floodlight: a projector designed for lighting a scene or object to a luminance considerably greater than its surroundings. It usually is capable of being pointed in any direction and is of weatherproof construction. See *heavy duty floodlight, general purpose floodlight, ground-area open floodlight, ground-area floodlight with reflector insert.*

floodlighting: a system designed for lighting a scene or object to a luminance greater than its surroundings. It may be for utility, advertising or decorative purposes.

floor cavity: the cavity formed by the work-plane, the floor, and the wall surfaces between these two planes.

floor cavity ratio, FCR: a number indicating floor cavity proportions calculated from length, width and height.

floor lamp: a portable luminaire on a high stand suitable for standing on the floor. See *torchere.*

fluoren: a unit of "black light" flux equal to one milliwatt of radiant flux in the wavelength range 320 to 400 nanometers.

fluorescence: the emission of light (luminescence) as the result of, and only during, the absorption of radiation of other (mostly shorter) wavelengths.

fluorescent lamp: a low-pressure mercury electric-discharge lamp in which a fluorescing coating (phosphor) transforms some of the ultraviolet energy generated by the discharge into light. See *instant start fluorescent lamp, preheat (switch start) fluorescent lamp, rapid start fluorescent lamp.*

flush mounted or recessed: a luminaire which is mounted above the ceiling (or behind a wall or other surface) with the opening of the luminaire level with the surface.

flux transfer theory: a method of calculating the illuminance in a room by taking into account the interreflection of the light flux from the room surfaces based on the average flux transfer between surfaces.

fog (adverse-weather) lamps: units which may be used in lieu of headlamps or in connection with the lower beam headlights to provide road illumination under conditions of rain, snow, dust or fog.

follow spot: any instrument operated so as to follow the movement of an actor. Follow spots are usually high intensity controlled beam luminaires.

footcandle, fc: the unit of illuminance when the foot is taken as the unit of length. It is the illuminance on a surface one square foot in area on which there is a uniformly distributed flux of one lumen, or the illuminance produced on a surface all points of which are at a distance of one foot from a directionally uniform point source of one candela. .
See Fig. 2

footcandle meter: See *illuminance meter.*

footlambert, fL: a unit of luminance equal to $1/\pi$ candela per square foot, or to the uniform luminance of a perfectly diffusing surface emitting or reflecting light at the rate of one lumen per square foot, or to the average luminance of any surface emitting or reflecting light at that rate. See *units of luminance.* The use of this unit is deprecated.

> NOTE: The average luminance of any reflecting surface in footlamberts is, therefore, the product of the illumination in footcandles by the luminous reflectance of the surface.

footlights: a set of striplights at the front edge of the stage platform used to soften face shadows cast by overhead luminaires and to add general toning lighting from below.

form factor, f_{1-2}: the ratio of the flux directly received by surface 2 (and due to lambertian surface 1) to the total flux emitted by surface 1. It is used in *flux transfer theory.*

$$f_{1-2} = (\Phi_{1-2})/(\Phi_1)$$

formation light: a navigation light especially provided to facilitate formation flying.

fovea: a small region at the center of the retina, subtending about two degrees, containing only cones but no rods and forming the site of most distinct vision.

foveal vision: See *central vision.*

Fresnel spotlight: a luminaire containing a lamp and a Fresnel lens (stepped "flat" lens with a textured back) which has variable field and beam angles obtained by changing the spacing between lamp and lens (flooding and spotting). The Fresnel produces a smooth, soft edge, defined beam of light.

fuselage lights: aircraft aeronautical lights, mounted on the top and bottom of the fuselage, used to supplement the navigation light.

G

gas-filled lamp: an incandescent lamp in which the filament operates in a bulb filled with one or more inert gases.

gaseous discharge: the emission of light from gas atoms excited by an electric current.

general color rendering index (R_a): measure of the average shift of 8 standardized colors chosen to be of intermediate saturation and spread throughout the range of hues. If the Color Rendering Index is not qualified as to the color samples used, R_a is assumed.

general diffuse lighting: lighting involving luminaires which distribute 40 to 60 per cent of the emitted light downward and the balance upward, sometimes with a strong component at 90 degrees (horizontal). See *direct-indirect lighting.*

general lighting: lighting designed to provide a substantially uniform level of illumination throughout an area, exclusive of any provision for special local requirements. See *direct lighting, semi-direct lighting, general diffuse lighting, direct-indirect lighting, semi-indirect lighting, indirect lighting, ceiling area lighting, localized general lighting.*

general purpose floodlight (GP): a weatherproof unit so constructed that the housing forms the reflecting surface. The assembly is enclosed by a cover glass.

germicidal effectiveness: See *bactericidal effectiveness.*

germicidal efficiency of radiant flux: See *bactericidal efficiency of radiant flux.*

germicidal exposure: See *bactericidal exposure.*

germicidal flux and flux density: See *bactericidal flux and flux density.*

germicidal lamp: a low pressure mercury lamp in which the envelope has high transmittance for 254-nanometer radiation. See *bactericidal lamp.*

glare: the sensation produced by luminance within the visual field that is sufficiently greater than the luminance to which the eyes are adapted to cause annoyance, discomfort, or loss in visual performance and visibility. See *blinding glare, direct glare, disability glare, discomfort glare.*

> NOTE: The magnitude of the sensation of glare depends upon such factors as the size, position and luminance of a source, the number of sources and the luminance to which the eyes are adapted.

globe: a transparent or diffusing enclosure intended to protect a lamp, to diffuse and redirect its light, or to change the color of the light.

glossmeter: an instrument for measuring gloss as a function of the directionally selective reflecting properties of a material in angles near to and including the direction giving specular reflection.

glow discharge: an electric discharge characterized by a low, approximately constant, current density at the cathode, low cathode temperature, and a high, approximately constant, voltage drop.

glow factor: a measure of the visible light response of a fluorescent material to "black light." It is equal to π^* times the luminance in candelas per square meter produced on the material divided by the in-

---

*π is omitted when the luminance is in footlamberts and flux density is in milliwatts per square foot.

cident "black light" flux density in milliwatts per square meter. It may be measured in lumens per milliwatt.

glow lamp: an electric-discharge lamp whose mode of operation is that of a glow discharge, and in which light is generated in the space close to the electrodes.

goniophotometer: a photometer for measuring the directional light distribution characteristics of sources, luminaires, media and surfaces.

graybody: a temperature radiator whose spectral emissivity is less than unity and the same at all wavelengths.

ground-area open floodlight (O): a unit providing a weatherproof enclosure for the lamp socket and housing. No cover glass is required.

ground-area open floodlight with reflector insert (OI): a weatherproof unit so constructed that the housing forms only part of the reflecting surface. An auxiliary reflector is used to modify the distribution of light. No cover glass is required.

ground light: visible radiation from the sun and sky reflected by surfaces below the plane of the horizon.

group flashing light: a flashing light in which the flashes are combined in groups, each including the same number of flashes, and in which the groups are repeated at regular intervals. The duration of each flash is clearly less than the duration of the dark periods between flashes, and the duration of the dark periods between flashes is clearly less than the duration of the dark periods between groups.

H

hard light: light that causes an object to cast a sharply defined shadow.

hazard or obstruction beacon: an aeronautical beacon used to designate a danger to air navigation.

hazardous location: an area where ignitable vapors or dust may cause a fire or explosion created by energy emitted from lighting or other electrical equipment or by electrostatic generation.

headlamp: a major lighting device mounted on a vehicle and used to provide illumination ahead of it. Also called headlight. See *multiple-beam headlamp* and *sealed beam headlamp.*

headlight: an alternate term for *headlamp.*

heavy duty floodlight (HD): a weatherproof unit having a substantially constructed metal housing into which is placed a separate and removable reflector. A weatherproof hinged door with cover glass encloses the assembly but provides an unobstructed light opening at least equal to the effective diameter of the reflector.

hemispherical-conical reflectance, ρ **(2π; ω_r):** ratio of reflected flux collected over a conical solid angle to the incident flux from the entire hemisphere.
NOTE: The direction and extent of the cone must be specified.

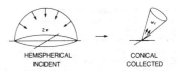

hemispherical-conical transmittance, τ **(2π; ω_t):** ratio of transmitted flux collected over a conical solid angle to the incident flux from the entire hemisphere.
NOTE: The direction and extent of the cone must be specified.

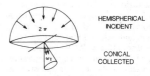

hemispherical-directional reflectance, ρ **(2π; θ_r, ϕ_r):** ratio of reflected flux collected over an element of solid angle surrounding the given direction to the incident flux from the entire hemisphere.
NOTE: The direction of collection and the size of the solid angle "element" must be specified.

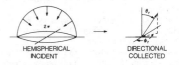

hemispherical-directional transmittance, τ **(2π; θ_t, ϕ_t):** ratio of transmitted flux collected over an element of solid angle surrounding the given direction to the incident flux from the entire hemisphere.
NOTE: The direction of collection and size of the solid angle "element" must be specified.

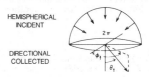

hemispherical reflectance: the ratio of all of the flux leaving a surface or medium by reflection to the incident flux. See *hemispherical transmittance.*
NOTE: If reflectance is not preceded by an adjective descriptive of the angles of view, hemispherical reflectance is implied.

hemispherical transmittance: the ratio of the transmitted flux leaving a surface or medium to the incident flux.

NOTE: If transmittance is not preceded by an adjective, descriptive of the angles of view, hemispherical reflectance is implied.

high intensity discharge (HID) lamp: an electric discharge lamp in which the light producing arc is stabilized by wall temperature, and the arc tube has a bulb wall loading in excess of three watts per square centimeter. HID lamps include groups of lamps known as mercury, metal halide, and high pressure sodium.

high-key lighting: a type of lighting that, applied to a scene, results in a picture having gradations falling primarily between gray and white; dark grays or blacks are present, but in very limited areas. See *low-key lighting.*

high mast lighting: illumination of a large area by means of a group of luminaires which are designed to be mounted in fixed orientation at the top of a high mast, generally 20 meters (65 feet) or higher.

high pressure sodium (HPS) lamp: high intensity discharge (HID) lamp in which light is produced by radiation from sodium vapor operating at a partial pressure of about 1.33×10^4 Pa (100 torr). Includes clear and diffuse-coated lamps.

horizontal plane of a searchlight: the plane which is perpendicular to the vertical plane through the axis of the searchlight drum and in which the train lies.

hot-cathode lamp: an electric-discharge lamp whose mode of operation is that of an arc discharge. The cathodes may be heated by the discharge or by external means.

house lights: the general lighting system installed in the audience area (house) of a theatre, film or television studio or arena.

hue of a perceived color: the attribute that determines whether the color is red, yellow, green, blue, or the like.

hue of a perceived light-source color: the attribute that determines whether the color is red, yellow, green, blue, or the like. See *hue of a perceived color.*

hydrargyrum, medium-arc-length, iodide (HMI): an ac arc light source utilizing mercury vapor and metallic iodide additives for an approximation of daylight (5600 K) illumination. Requires a ballast and ignition system for operation.

I

ice detection light: an inspection light designed to illuminate the leading edge of an aircraft wing to check for ice formation.

ideal radiator: See *blackbody*

identification beacon: an aeronautical beacon emitting a coded signal by means of which a particular point of reference can be identified.

ignitor: a device, either by itself or in association with other components, that generates voltage pulses to start discharge lamps without preheating of electrodes.

illuminance, $E = d\Phi/dA$: the density of the luminous flux incident on a surface; it is the quotient of the luminous flux by the area of the surface when the latter is uniformly illuminated. See Fig. 7 for units and conversion factors.

illuminance (lux or footcandle) meter: an instrument for measuring illuminance on a plane. Instruments which accurately respond to more than one spectral distribution are color corrected, *i.e.*, the spectral response is balanced to $V(\lambda)$ or $V'(\lambda)$. Instruments which accurately respond to more than one spatial distribution of incident flux are cosine corrected, *i.e.*, the response to a source of unit luminous intensity, illuminating the detector from a fixed distance and from different directions decreases as the cosine of the angle between the incident direction and the normal to the detector surface. The instrument is comprised of some form of photodetector, with or without a filter, driving a digital or analog readout through appropriate circuitry.

illumination: the act of illuminating or state of being illuminated. This term has been used for density of luminous flux on a surface (illuminance) and such use is to be deprecated.

incandescence: the self-emission of radiant energy in the visible spectrum due to the thermal excitation of atoms or molecules.

incandescent filament lamp: a lamp in which light is produced by a filament heated to incandescence by an electric current.
NOTE: Normally, the filament is of coiled or coiled-coil (doubly coiled) tungsten wire. However, it may be uncoiled wire, a flat strip, or of material other than tungsten.

incomplete diffusion (partial diffusion): that in which the diffusing medium partially re-directs the incident flux by scattering while the remaining fraction of incident flux is redirected without scattering, *i.e.*, a fraction of the incident flux can remain in an image-forming state.

index of sensation (M) (of a source): a number which expresses the effects of source luminance, *solid angle factor, position index,* and the field luminance on *discomfort glare rating.*

indirect component: the portion of the luminous flux from a luminaire arriving at the work-plane after being reflected by room surfaces. See *direct component.*

indirect lighting: lighting by luminaires distributing 90 to 100 per cent of the emitted light upward.

infrared lamp: a lamp that radiates predominantly in the infrared; the visible radiation is not of principal interest.

infrared radiation: for practical purposes any radiant energy within the wavelength range of 770 to 10^6 nanometers. This radiation is arbitrarily divided as follows:
Near (short wavelength) infrared 770–1400 nm
Intermediate infrared 1400–5000 nm
Far (long wavelength) infrared 5000–1,000,000 nm
NOTE: In general, unlike ultraviolet energy, infrared energy is not evaluated on a wavelength basis but rather in terms of all of such energy incident upon a surface.

Examples of these applications are industrial heating, drying, baking and photoreproduction. However, some applications, such as infrared viewing devices, involve detectors sensitive to a restricted range of wavelengths; in such cases the spectral characteristics of the source and receiver are of importance.

initial luminous exitance: the density of luminous flux leaving a surface within an enclosure before interreflections occur.

NOTE: For light sources this is the luminous exitance as defined in *luminous flux density at a surface*. For non-self-luminous surfaces it is the reflected luminous exitance of the flux received directly from sources within the enclosure or from daylight.

instant start fluorescent lamp: a fluorescent lamp designed for starting by a high voltage without preheating of the electrodes.

NOTE: Also known as a cold-start lamp in some countries.

integrating photometer: a photometer that enables total luminous flux to be determined by a single measurement. The usual type is the Ulbricht sphere with associated photometric equipment for measuring the indirect luminance of the inner surface of the sphere. (The measuring device is shielded from the source under measurement.)

intensity: a shortening of the terms *luminous intensity* and *radiant intensity*. Often misused for level of illumination or illuminance.

interflectance: an alternate term for *room utilization factor*.

interflectance method: a lighting design procedure for predetermining the luminances of walls, ceiling and floor and the average illuminance on the work-plane based on integral equations. It takes into account both direct and reflected flux.

interflected component: the portion of the luminous flux from a luminaire arriving at the workplane after being reflected one or more times from room surfaces, as determined by the *flux transfer theory*.

interflection: the multiple reflection of light by the various room surfaces before it reaches the work-plane or other specified surface of a room.

inter-reflectance: the portion of the luminous flux (lumens) reaching the work-plane that has been reflected one or more times as determined by the flux transfer theory.

interrupted quick-flashing light: a quick flashing light in which the rapid alternations are interrupted by periods of darkness at regular intervals.

inverse-square law: the law stating that the illuminance E at a point on a surface varies directly with the intensity I of a point source, and inversely as the square of the distance d between the source and the point. If the surface at the point is normal to the direction of the incident light, the law is expressed by $E = I/d^2$.

NOTE: For sources of finite size having uniform luminance, this gives results that are accurate within one percent when d is at least five times the maximum dimension of the source as viewed from the point on the surface. Even though practical interior luminaires do not have uniform luminance, this distance, d, is frequently used as the minimum for photometry of such luminaires, when the magnitude of the measurement error is not critical.

iris: an assembly of flat metal leaves arranged to provide an easily adjustable near-circular opening, placed near the focal point of the beam (as in an ellipsoidal reflector spotlight), or in front of the lens to act as a mechanical dimmer as in older types of carbon arc follow spotlights.

irradiance, E: the density of radiant flux incident on a surface.

isocandela line: a line plotted on any appropriate set of coordinates to show directions in space, about a source of light, in which the intensity is the same. A series of such curves, usually for equal increments of intensity, is called an isocandela diagram.

isolux (isofootcandle) line: a line plotted on any appropriate set of coordinates to show all the points on a surface where the illuminance is the same. A series of such lines for various illuminance values is called an isolux (isofootcandle) diagram.

K

key light: the apparent principal source of directional illumination falling upon a subject or area.

kicker: a luminaire used to provide an additional highlight or accent on a subject.

klieg light: a high intensity carbon arc spotlight, typically used in motion picture lighting.

L

laboratory reference standards: the highest ranking order of standards at each laboratory.

lambert, L: a lambertian unit of luminance equal to $1/\pi$ candela per square centimeter. The use of this unit is deprecated.

lambertian surface: a surface that emits or reflects light in accordance with Lambert's cosine law. A lambertian surface has the same luminance regardless of viewing angle.

Lambert's cosine law, $I_\theta = I_0 \cos \theta$: the law stating that the luminous intensity in any direction from an element of a perfectly diffusing surface varies as the cosine of the angle between that direction and the perpendicular to the surface element.

lamp: a generic term for a man-made source of light. By extension, the term is also used to denote sources that radiate in regions of the spectrum adjacent to the visible.

NOTE: A lighting unit consisting of a lamp with shade, reflector, enclosing globe, housing, or other accessories is also called a "lamp." In such cases, in order to distinguish between the assembled unit and the light source within it, the latter is often called a "bulb" or "tube," if it is electrically powered. See also *luminaire*.

lamp burnout factor: the fractional loss of task illuminance due to burned out lamps left in place for long periods.

lamp lumen depreciation factor, LLD: the multiplier to be used in illumination calculations to relate the initial rated output of light sources to the anticipated minimum rated output based on the relamping program to be used.

lamp position factor: The fractional flux of a high intensity discharge (HID) lamp at a given operating position compared to the flux when the lamp is operated in the position at which the lamp lumens are rated.

lamp post: a standard support provided with the necessary internal attachments for wiring and the external attachments for the bracket and luminaire.

lamp shielding angle: the angle between the plane of the baffles or louver grid and the plane most nearly horizontal that is tangent to both the lamps and the louver blades. See Fig. 3.
    NOTE: The lamp shielding angle frequently is larger than the louver shielding angle, but never smaller. See *louver shielding angle.*

landing direction indicator: a device to indicate visually the direction currently designated for landing and take-off.

landing light: an aircraft aeronautical light designed to illuminate a ground area from the aircraft.

landmark beacon: an aeronautical beacon used to indicate the location of a landmark used by pilots as an aid to enroute navigation.

laser: an acronym for *Light Amplification by Stimulated Emission of Radiation.* The laser produces a highly monochromatic and coherent (spatial and temporal) beam of radiation. A steady oscillation of nearly a single electromagnetic mode is maintained in a volume of an active material bounded by highly reflecting surfaces, called a resonator. The frequency of oscillation varies according to the material used and by the methods of initially exciting or pumping the material.

lateral width of a light distribution: (in roadway lighting) the lateral angle between the reference line and the width line, measured in the cone of maximum candlepower. This angular width includes the line of maximum candlepower. See *reference line* and *width line.*

lens: a glass or plastic element used in luminaires to change the direction and control the distribution of light rays.

level of illumination: See *illuminance.*

life performance curve: a curve which represents the variation of a particular characteristic of a light source (luminous flux, intensity, etc.) throughout the life of the source.
    NOTE: Life performance curves sometimes are called maintenance curves as, for example, lumen maintenance curves.

life test of lamps: a test in which lamps are operated under specified conditions for a specified length of time, for the purpose of obtaining information on lamp life. Measurements of photometric and electrical characteristics may be made at specified intervals of time during this test.

light: radiant energy that is capable of exciting the retina and producing a visual sensation. The visible portion of the electromagnetic spectrum extends from about 380 to 770 nm.

NOTE: The subjective impression produced by stimulating the retina is sometimes designated as light. Visual sensations are sometimes arbitrarily defined as sensations of light, and in line with this concept it is sometimes said that light cannot exist until an eye has been stimulated. Electrical stimulation of the retina or the visual cortex is described as producing flashes of light. In illuminating engineering, however, light is a physical entity—radiant energy weighted by the luminous efficiency function. It is a physical stimulus which can be applied to the retina. (See *spectral luminous efficacy of radiant flux, values of spectral luminous efficiency for photopic vision.*)

light adaptation: the process by which the retina becomes adapted to a luminance greater than about 3.4 candelas per square meter. See also *dark adaptation.*

light center (of a lamp): the center of the smallest sphere that would completely contain the light-emitting element of the lamp.

light center length (of a lamp): the distance from the light center to a specified reference point on the lamp.

light loss factor, LLF: a factor used in calculating illuminance after a given period of time and under given conditions. It takes into account temperature and voltage variations, dirt accumulation on luminaire and room surfaces, lamp depreciation, maintenance procedures and atmosphere conditions. Formerly called *maintenance factor.*

light meter: See *illuminance meter.*

light-source color: the color of the light emitted by the source.
    NOTE: The color of a point source may be defined by its luminous intensity and chromaticity coordinates; the color of an extended source may be defined by its luminance and chromaticity coordinates. See *perceived light-source color, color temperature, correlated color temperature.*

lighting effectiveness factor, LEF$_V$: the ratio of equivalent sphere illumination to ordinary measured or calculated illumination.

lightness (of a perceived patch of surface color): the attribute by which it is perceived to transmit or reflect a greater or lesser fraction of the incident light.

light-watt: radiation weighted by the spectral luminous efficiency for photopic vision.

linear light: a luminous signal having a perceptible physical length.

linear polarization: the process by which the transverse vibrations of light waves are oriented in a specific plane. Polarization may be obtained by using either transmitting or reflecting media.

Linnebach projector: a lensless scenic projector, using a concentrated source in a black box and a slide or cutout between the source and the projection surface.

liquid crystal display, (LED): a display made of material whose reflectance or transmittance changes when an electric field is applied.

local lighting: lighting designed to provide illuminance over a relatively small area or confined space without providing any significant general surrounding lighting.

localized general lighting: lighting that utilizes luminaires above the visual task and contributes also to the illumination of the surround.

long-arc lamp: an arc lamp in which the distance between the electrodes is large.

NOTE: This type of lamp (*e.g.*, xenon) is generally of high pressure. The arc fills the discharge tube and is therefore wall stabilized.

louver: a series of baffles used to shield a source from view at certain angles or to absorb unwanted light. The baffles usually are arranged in a geometric pattern.

louver shielding angle, θ: the angle between the horizontal plane of the baffles or louver grid and the plane at which the louver conceals all objects above. See Fig. 4 and *lamp shielding angle*.

NOTE: The planes usually are so chosen that their intersection is parallel with the louvered blade.

louvered ceiling: a ceiling area lighting system comprising a wall-to-wall installation of multicell louvers shielding the light sources mounted above it. See *luminous ceiling*.

lower (passing) beams: one or more beams directed low enough on the left to avoid glare in the eyes of oncoming drivers, and intended for use in congested areas and on highways when meeting other vehicles within a distance of 300 meters (1000 feet). Formerly "traffic beam."

low pressure mercury lamp: a discharge lamp (with or without a phosphor coating) in which the partial pressure of the mercury vapor during operation does not exceed 100 Pa.

low pressure sodium lamp: a discharge lamp in which light is produced by radiation from sodium vapor operating at a partial pressure of 0.1 to 1.5 Pa (approximately 10^{-3} to 10^{-2} torr).

low-key lighting: a type of lighting that, applied to a scene, results in a picture having gradations falling primarily between middle gray and black, with comparatively limited areas of light grays and whites. See *high-key lighting*.

lumen, lm: SI unit of luminous flux. Radiometrically, it is determined from the radiant power. Photometrically, it is the luminous flux emitted within a unit solid angle (one steradian) by a point source having a uniform luminous intensity of one candela.

lumen-hour, lm·h: a unit of quantity of light (luminous energy). It is the quantity of light delivered in one hour by a flux of one lumen.

lumen (or flux) method: a lighting design procedure used for predetermining the relation between the number and types of lamps or luminaires, the room characteristics, and the average illuminance on the work-plane. It takes into account both direct and reflected flux.

lumen-second, lm·s: a unit of quantity of light, the SI unit of luminous energy (also called a talbot). It is the quantity of light delivered in one second by a luminous flux of one lumen.

luminaire: a complete lighting unit consisting of a

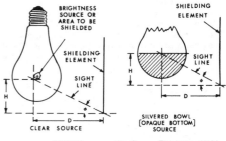

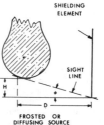

Fig. 3. The lamp shielding angle is formed by a sight line tangent to the lowest part of the brightness area to be shielded. *H* is the vertical distance from the brightness source to the bottom of the shielding element. *D* is the horizontal distance from the brightness source to the shielding element. Lamp shielding angle $\phi = \tan^{-1} H/D$.

lamp or lamps together with the parts designed to distribute the light, to position and protect the lamps and to connect the lamps to the power supply.

luminaire ambient temperature factor: the fractional loss of task illuminance due to improper operating temperature of a gas discharge lamp.

luminaire dirt depreciation factor, LDD: the multiplier to be used in illuminance calculations to relate the initial illuminance provided by clean, new luminaires to the reduced illuminance that they will provide due to dirt collection on the luminaires at the time at which it is anticipated that cleaning procedures will be instituted.

luminaire efficiency: the ratio of luminous flux (lumens) emitted by a luminaire to that emitted by the lamp or lamps used therein.

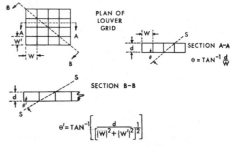

Fig. 4. Louver shielding angles θ and θ'.

luminaire surface depreciation factor: the loss of task illuminance due to permanent deterioration of luminaire surfaces.

luminance, $L = d^2\Phi/(d\omega\, dA\, \cos\theta)$ **(in a direction and at a point of a real or imaginary surface):** the quotient of the luminous flux at an element of

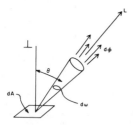

the surface surrounding the point, and propagated in directions defined by an elementary cone containing the given direction, by the product of the solid angle of the cone and the area of the orthogonal projection of the element of the surface on a plane perpendicular to the given direction. The luminous flux may be leaving, passing through, and/or arriving at the surface. Formerly, *photometric brightness.*

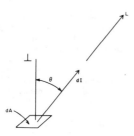

By introducing the concept of luminous intensity, luminance may be expressed as $L = dI/(dA\, \cos\theta)$. Here, luminance at a point of a surface in a direction is interpreted as the quotient of luminous intensity in the given direction produced by an element of the surface surrounding the point by the area of the orthogonal projection of the element of surface on a plane perpendicular to the given direction. (Luminance may be measured at a receiving surface by using $L = dE/(d\omega\, \cos\theta)$. This value may be less than the luminance of the emitting surface due to the attenuation of the transmitting media.)

NOTE: In common usage the term "brightness" usually refers to the strength of sensation which results from viewing surfaces or spaces from which light comes to

the eye. This sensation is determined in part by the definitely measurable luminance defined above and in part by conditions of observation such as the state of adaptation of the eye.

In much of the literature the term brightness, used alone, refers to both luminance and sensation. The context usually indicates which meaning is intended. Previous usage notwithstanding, neither the term brightness nor the term photometric brightness should be used to denote the concept of luminance.

luminance coefficient: a coefficient similar to the coefficient of utilization used to determine wall and ceiling luminances. An obsolete term, see *exitance coefficient.*

luminance contrast: the relationship between the luminances of an object and its immediate background. It is equal to $(L_1 - L_2)/L_1$, $(L_2 - L_1)/L_1$ or $\Delta L/L_1$, where L_1 and L_2 are the luminances of the background and object, respectively. The form of the equation must be specified. The ratio $\Delta L/L_1$ is known as Weber's fraction.

NOTE: See last paragraph of the note under *luminance.* Because of the relationship among luminance, illuminance and reflectance, contrast often is expressed in terms of reflectance when only reflecting surfaces are involved. Thus, contrast is equal to $(\rho_1 - \rho_2)/\rho_1$ or $(\rho_2 - \rho_1)/\rho_1$ where ρ_1 and ρ_2 are the reflectances of the background and object, respectively. This method of computing contrast holds only for perfectly diffusing surfaces; for other surfaces it is only an approximation unless the angles of incidence and view are taken into consideration. (See *reflectance.*)

luminance difference: the difference in luminance between two areas. It usually is applied to contiguous areas, such as the detail of a visual task and its immediate background, in which case it is quantitatively equal to the numerator in the formula for luminance contrast.

NOTE: See last paragraph of the note under *luminance.*

luminance factor, β: the ratio of the luminance of a surface or medium under specified conditions of incidence, observation, and light source, to the luminance of a completely reflecting or transmitting, perfectly diffusing surface or medium under the same conditions.

NOTE: Reflectance or transmittance cannot exceed unity, but luminance factor may have any value from zero to values approaching infinity.

luminance factor of room surfaces: factors by which the average work-plane illuminance is multiplied to obtain the average luminances of walls, ceilings and floors.

luminance ratio: the ratio between the luminances of any two areas in the visual field.

luminance threshold: the minimum perceptible difference in luminance for a given state of adaptation of the eye.

luminescence: any emission of light not ascribable directly to incandescence. See *electroluminescence, fluorescence, phosphorescence.*

luminosity factor: previously used term for *spectral luminous efficacy of radiant flux.*

luminous ceiling: a ceiling area lighting system comprising a continuous surface of transmitting material of a diffusing or light-controlling character with light sources mounted above it. See *louvered ceiling.*

luminous density, $w = dQ/dV$: quantity of light (luminous energy) per unit volume.

luminous efficacy of radiant flux: the quotient of the total luminous flux by the total radiant flux. It is expressed in lumens per watt.

luminous efficacy of a source of light: the quotient of the total luminous flux emitted by the total lamp power input. It is expressed in lumens per watt.

NOTE: The term luminous efficiency has in the past been extensively used for this concept.

luminous efficiency: See *spectral luminous efficiency of radiant flux.*

luminous energy: See *quantity of light.*

luminous exitance, $M = d\Phi/dA$: the density of luminous flux leaving a surface at a point. Formerly luminous emittance.

NOTE: This is the total luminous flux emitted, reflected, and transmitted from the surface into a complete hemisphere.

luminous flux, Φ: the time rate of flow of light.

luminous flux density at a surface, $d\Phi/dA$: the luminous flux per unit area at a point on a surface.

NOTE: This need not be a physical surface; it may equally well be a mathematical plane. Also see *illuminance* and *luminous exitance.*

luminous intensity, $I = d\Phi/d\omega$ **(of a point source of light in a given direction):** the luminous flux per unit solid angle in the direction in question. Hence, it is the luminous flux on a small surface centered on and normal to that direction divided by the solid angle (in steradians) which the surface subtends at the source. Luminous intensity may be expressed in candelas or in lumens per steradian (lm/sr).

NOTE: Mathematically a solid angle must have a point as its apex; the definition of luminous intensity, therefore, applies strictly only to a point source. In practice, however, light emanating from a source whose dimensions are negligible in comparison with the distance from which it is observed may be considered as coming from a point. Specifically, this implies that with change of distance (1) the variation in solid angle subtended by the source at the receiving point approaches 1/(distance)² and that (2) the average luminance of the projected source area as seen from the receiving point does not vary appreciably.

luminous intensity distribution curve: See *candlepower distribution curve.*

luminous reflectance: any reflectance (regardless of beam geometry) in which both the incident and reflected flux are weighted by the spectral luminous efficiency of radiant flux, $V(\lambda)$; *i.e.,* where they are expressed as luminous flux. Thus, luminous reflectance is not a unique property of a reflecting surface but depends also on the spectral distribution of the incident radiant flux.

NOTE: Unless otherwise qualified, the term luminous reflectance is meant by the term reflectance.

luminous transmittance: any transmittance (regardless of beam geometry) in which both the incident and transmitted flux are weighted by the spectral luminous efficiency of radiant flux, $V(\lambda)$; *i.e.,* where they are expressed as luminous flux. Thus, luminous transmittance is not a unique property of a transmitting element but depends also on the spectral distribution of the incident radiant flux.

NOTE: Unless otherwise qualified, the term luminous transmittance is meant by the term transmittance.

lux, lx: the SI unit of illuminance. One lux is one lumen per square meter (lm/m²). (See Fig. 7 for conversion values.)

lux meter: See *illuminance meter.*

M

maintenance factor, MF: a factor formerly used to denote the ratio of the illuminance on a given area after a period of time to the initial illuminance on the same area. See *light loss factor.*

matte surface: one from which the reflection is predominantly diffuse, with or without a negligible specular component. See *diffuse reflection.*

mean horizontal intensity (candlepower): the average intensity (candelas) of a lamp in a plane perpendicular to the axis of the lamp and which passes through the luminous center of the lamp.

mean spherical luminous intensity: average value of the luminous intensity in all directions for a source. Also, the quotient of the total emitted luminous flux of the source divided by 4π.

mean zonal candlepower: the average intensity (candelas) of a symmetrical luminaire or lamp at an angle to the luminaire or lamp axis which is in the middle of the zone under consideration.

mechanical equivalent of light: See *spectral luminous efficacy of radiant flux,* $K(\lambda) = \Phi_{v\lambda}/\Phi_{e\lambda}$.

mercury lamp: a high intensity discharge (HID) lamp in which the major portion of the light is produced by radiation from mercury operating at a partial pressure in excess of 10^5 Pa (approximately 1 atmosphere). Includes clear, phosphor-coated (mercury-fluorescent), and self-ballasted lamps.

mercury-fluorescent lamp: an electric-discharge lamp having a high-pressure mercury arc in an arc tube, and an outer envelope coated with a fluorescing substance (phosphor) which transforms some of the ultraviolet energy generated by the arc into light.

mesopic vision: vision with fully adapted eyes at luminance conditions between those of photopic and scotopic vision, that is, between about 3.4 and 0.034 candelas per square meter.

metal halide lamp: a high intensity discharge (HID) lamp in which the major portion of the light is produced by radiation of metal halides and their products of dissociation-possibly in combination with metallic vapors such as mercury. Includes clear and phosphor coated lamps.

metamers: lights of the same color but of different spectral energy distribution.

middle ultraviolet: a portion of the electromagnetic spectrum in the range of 200–300 nanometers.

minimal perceptible erythema, MPE: the erythemal threshold.

mired: a unit of reciprocal color temperature; microreciprocal degrees; $1/T_k$ times 10^6.

NOTE: The unit of thermodynamic temperature is now denoted by "kelvin" and not "degree-kelvin". Consequently the acronym "mirek" for microreciprocal-kelvin occasionally has been used in the literature for "mired."

mirek: see *mired.*

modeling light: that illumination which reveals the depth, shape and texture of a subject. Key light, cross light, counter-key light, side light, back light, and eye light are types of modeling light.

modulation threshold: in the case of a square wave or sine wave grating, manipulation of luminance differences can be specified in terms of modulation and the threshold may be called the modulation threshold.

$$\text{modulation} = \frac{L_{max} - L_{min}}{L_{max} + L_{min}}$$

Periodic patterns that are not sine wave can be specified in terms of the modulation of the fundamental sine wave component. The number of periods or cycles per degree of visual angle represents the spatial frequency.

monocular visual field: the visual field of a single eye. See *binocular visual field.*

mounting height (roadway): the vertical distance between the roadway surface and the center of the apparent light source of the luminaire.

mounting height above the floor, MH_f: the distance from the floor to the light center of the luminaire or to the plane of the ceiling for recessed equipment.

mounting height above the work-plane, MH_{wp}: the distance from the work-plane to the light center of the luminaire or to the plane of the ceiling for recessed equipment.

multiple-beam headlamp: a headlamp so designed to permit the driver of a vehicle to use any one of two or more distributions of light on the road.

Munsell chroma, C: an index of perceived chroma of the object color defined in terms of the luminance factor (Y) and chromaticity coordinates (x, y) for CIE Standard Illuminant C and the CIE 1931 Standard Observer.

Munsell color system: a system of surface-color specification based on perceptually uniform color scales for the three variables: Munsell hue, Munsell value, and Munsell chroma. For an observer of normal color vision, adapted to daylight, and viewing a specimen when illuminated by daylight and surrounded with a middle gray to white background, the Munsell hue, value and chroma of the color correlate well with the hue, lightness and perceived chroma.

Munsell hue, H: an index of the hue of the perceived object color defined in terms of the luminance factor (Y) and chromaticity coordinates (x, y) for CIE Standard Illuminant C and the CIE 1931 Standard Observer.

Munsell Value, V: an index of the lightness of the perceived object color defined in terms of the luminance factor (Y) for CIE Standard Illuminant C and the CIE 1931 Standard Observer.

NOTE: The exact definition gives Y as a 5th power function of V so that tabular or iterative methods are needed to find V as a function of Y. However, V can be estimated within ± 0.1 by $V = 11.6 \ (Y/100)^{1/3} -1.6$ or within ± 0.6 by $V = Y^{1/2}$ where Y is the luminance factor expressed in per cent.

N

nanometer, nm: unit of wavelength equal to 10^{-9} meter. See Fig. 11.

narrow-angle diffusion: that in which flux is scattered at angles near the direction that the flux would take by regular reflection or transmission. See *wide-angle-diffusion.*

narrow angle luminaire: a luminaire that concentrates the light within a cone of comparatively small solid angle. See *wide angle luminaire.*

national standard of light: a primary standard of light which has been adopted as a national standard. See *primary standard of light.*

navigation lights: an alternate term for *position lights.*

navigation light system: a set of aircraft aeronautical lights provided to indicate the position and direction of motion of an aircraft to pilots of other aircraft or to ground observers.

near infrared: the region of the electromagnetic spectrum from 770 to 1400 nanometers.

near ultraviolet: the region of the electromagnetic spectrum from 300 to 380 nanometers.

night: the hours between the end of evening civil twilight and the beginning of morning civil twilight.

NOTE: Civil twilight ends in the evening when the center of the sun's disk is six degrees below the horizon and begins in the morning when the center of the sun's disk is six degrees below the horizon.

nit, nt: a unit of luminance equal to one candela per square meter.

NOTE: Candela per square meter is the International Standard (SI) unit of luminance.

numerical display (digital display): an electrically operated display of digits. Tungsten filaments, gas discharges, light emitting diodes, liquid crystals, projected numerals, illuminated numbers and other principles of operation may be used.

O

object color: the color of the light reflected or transmitted by the object when illuminated by a standard light source, such as CIE source A, B, C or D_{65}. See *standard source* and *perceived object color.*

obstruction beacon: See *hazard beacon*

obstruction lights: aeronautical ground lights provided to indicate obstructions.

occulting light: a rhythmic light in which the periods of light are clearly longer than the periods of darkness.

orientation: the relation of a building with respect to compass directions.

Ostwald color system: a system of describing colors in terms of color content, white content and black content.

overcast sky: one that has 100 per cent cloud cover; the sun is not visible.

overhang: the distance between a vertical line passing through the luminaire and the curb or edge of the roadway.

ozone-producing radiation: ultraviolet energy shorter than about 220 nanometers that decomposes oxygen O_2 thereby producing ozone O_3. Some ultraviolet sources generate energy at 184.9 nanometers that is particularly effective in producing ozone.

P

PAR lamp: See *pressed reflector lamp.*

parking lamp: a lighting device placed on a vehicle to indicate its presence when parked.

partial diffusion: See *incomplete diffusion.*

partly cloudy sky: one that has 30 to 70 per cent cloud cover.

passing beams: See *lower beams.*

pendent luminaire: See *suspended luminaire.*

perceived light-source color: the color perceived to belong to a light source.

perceived object color: the color perceived to belong to an object, resulting from characteristics of the object, of the incident light, and of the surround, the viewing direction and observer adaptation. See *object color.*

perfect diffusion: that in which flux is uniformly scattered such that the luminance (radiance) is the same in all directions.

perimeter lights: aeronautical ground lights provided to indicate the perimeter of a landing pad for helicopters.

peripheral vision: the seeing of objects displaced from the primary line of sight and outside the central visual field.

peripheral visual field: that portion of the visual field that falls outside the region corresponding to the foveal portion of the retina.

phosphor mercury lamp: See *mercury-fluorescent lamp.*

phosphorescence: the emission of light (luminescence) as the result of the absorption of radiation, and continuing for a noticeable length of time after excitation.

phot, ph: the unit of illuminance when the centimeter is taken as the unit of length; it is equal to one lumen per square centimeter.

photochemical radiation: energy in the ultraviolet, visible and infrared regions capable of producing chemical changes in materials.

NOTE: Examples of photochemical processes are accelerated fading tests, photography, photoreproduction and chemical manufacturing. In many such applications a specific spectral region is of importance.

photoelectric receiver: a device that reacts electrically in a measurable manner in response to incident radiant energy.

photoflash lamp: a lamp in which combustible metal or other solid material is burned in an oxidizing atmosphere to produce light of high intensity and short duration for photographic purposes.

photoflood lamp: an incandescent filament lamp of high color temperature for lighting objects for photography or videography.

photometer: an instrument for measuring photometric quantities such as luminance, luminous intensity, luminous flux and illuminance. See *densitometer, goniophotometer, illuminance meter, integrating photometer, reflectometer, spectrophotometer, transmissometer.*

photometry: the measurement of quantities associated with light.

NOTE: Photometry may be visual in which the eye is used to make a comparison, or physical in which measurements are made by means of physical receptors.

photometric brightness: a term formerly used for *luminance.*

photopic vision: vision mediated essentially or exclusively by the cones. It is generally associated with adaptation to a luminance of at least 3.4 candelas per square meter. See *scotopic vision.*

physical photometer: an instrument containing a physical receptor (photoemissive cell, barrier-layer cell, thermopile, etc.) and associated filters, that is calibrated so as to read photometric quantities directly. See *visual photometer.*

pilot house control: a mechanical means for controlling the elevation and train of a searchlight from a position on the other side of the bulkhead or deck on which it is mounted.

Planck radiation law: an expression representing the spectral radiance of a blackbody as a function of the wavelength and temperature. This law commonly is expressed by the formula

$$L_\lambda = dI_\lambda/dA' = c_{1L}\lambda^{-5}[e^{(c_2/\lambda T)} - 1]^{-1}$$

in which L_λ is the spectral radiance, dI_λ is the spectral radiant intensity, dA' is the projected area ($dA \cos \theta$) of the aperture of the blackbody, e is the base of natural logarithms (2.71828), T is absolute temperature, c_{1L} and c_2 are constants designated as the first and second radiation constants.

NOTE: The designation c_{1L} is used to indicate that the equation in the form given here refers to the radiance L, or to the intensity I per unit projected area A', of the source. Numerical values are commonly given not for c_{1L} but for c_1. which applies to the total flux radiated from a blackbody aperture, that is, in a hemisphere (2π steradians), so that, with the Lambert cosine law taken into account, $c_1 = \pi c_{1L}$. The currently recommended value of c_1 is 3.741832×10^{-16} W·m^2 or 3.741832×10^{-12} W·cm^2. Then c_{1L} is $1.1910621 \times 10^{-16}$ W·m^2·sr^{-1} or $1.1910621 \times 10^{-12}$ W·cm^2·sr^{-1}. If, as is more convenient, wavelengths are expressed in micrometers and area in square centimeters. $c_{1L} = 1.1910621 \times 10^4$ W μm^4·cm^{-2}·sr^{-1}, L_λ being given in W cm^{-2}·sr^{-1}·μm^{-1}. The currently recommended value of c_2 is 1.438786×10^{-2} m·K. The Planck law in the following form gives the energy radiated from the blackbody in a given wavelength interval $(\lambda_1 - \lambda_2)$:

$$Q = \int_{\lambda_1}^{\lambda_2} Q_\lambda \, d\lambda = Atc_1 \int_{\lambda_1}^{\lambda_2} \lambda^{-5}(e^{(c_2/\lambda T)} - 1)^{-1} \, d\lambda$$

If A is the area of the radiation aperture or surface in square centimeters, t is time in seconds, λ is wavelength in micrometers, and $c_1 = 3.7418 \times 10^4$ $W \cdot \mu m^4 \cdot cm^{-2}$, then Q is the total energy in watt seconds emitted from this area (that is, in the solid angle 2π), in time t, within the wavelength interval $(\lambda_1 - \lambda_2)$.

planckian locus: See *blackbody locus.*

point of fixation: a point or object in the visual field at which the eyes look and upon which they are focused.

point of observation: the midpoint of the base line connecting the centers of the entrance pupils of the two eyes. For practical purposes, the center of the pupil of the eye often is taken as the point of observation.

point by point method: a method of lighting calculation now called *point method.*

point method: a lighting design procedure for predetermining the illuminance at various locations in lighting installations, by use of luminaire photometric data.

> NOTE: The direct component of illuminance due to the luminaires and the interreflected component of illuminance due to the room surfaces are calculated separately. The sum is the total illuminance at a point.

point source: a source of radiation the dimensions of which are small enough, compared with the distance between the source and the irradiated surface, for them to be neglected in calculations and measurements.

polarization: the process by which the transverse vibrations of light waves are oriented in a specific plane. Polarization may be obtained by using either transmitting or reflecting media.

pole (roadway lighting): a standard support generally used where overhead lighting distribution circuits are employed.

portable lighting: lighting involving lighting equipment designed for manual portability.

portable luminaire: a lighting unit that is not permanently fixed in place. See *table lamp* and *floor lamp.*

portable traffic control light: a signalling light producing a controllable distinctive signal for purposes of directing aircraft operations in the vicinity of an aerodrome.

position index, P: a factor which represents the relative average luminance for a sensation at the borderline between comfort and discomfort (BCD), for a source located anywhere within the visual field.

position lights: aircraft aeronautical lights forming the basic or internationally recognized navigation light system.

> NOTE: The system is composed of a red light showing from dead ahead to 110 degrees to the left, a green light showing from dead ahead to 110 degrees to the right, and a white light showing to the rear through 140

degrees. Position lights are also called navigation lights.

prefocus lamp: a lamp in which, during manufacture, the luminous element is accurately adjusted to a specified position with respect to the physical locating element (usually the base).

preheat (switch start) fluorescent lamp: a fluorescent lamp designed for operation in a circuit requiring a manual or automatic starting switch to preheat the electrodes in order to start the arc.

pressed reflector lamp: an incandescent filament or electric discharge lamp of which the outer bulb is formed of two pressed parts that are fused together; namely, a reflectorized bowl and a cover which may be clear or patterned for optical control.

> NOTE: Often called a projector or PAR lamp.

primary (light): any one of three lights in terms of which a color is specified by giving the amount of each required to match it by additive combination.

primary line of sight: the line connecting the point of observation and the point of fixation.

primary standard of light: a light source by which the unit of light is established and from which the values of other standards are derived. See *national standard of light.*

> NOTE: A satisfactory primary (national) standard must be reproducible from specifications (see *candela*). Primary (national) standards usually are found in national physical laboratories such as the National Bureau of Standards.

projection lamp: a lamp with physical and luminous characteristics suited for projection systems (*e.g.,* motion picture projectors, microfilm viewers, etc.).

projector: a lighting unit that by means of mirrors and lenses, concentrates the light to a limited solid angle so as to obtain a high value of luminous intensity. See *floodlight, searchlight, signalling light.*

protective lighting: a system intended to facilitate the nighttime policing of industrial and other properties.

proximal stimuli: the distribution of illuminance on the retina constitutes the proximal stimulus.

pupil (pupillary aperture): the opening in the iris that admits light into the eye. See *artificial pupil.*

Purkinje phenomenon: the reduction in subjective brightness of a red light relative to that of a blue light when the luminances are reduced in the same proportion without changing the respective spectral distributions. In passing from photopic to scotopic vision, the curve of spectral luminous efficiency changes, the wavelength of maximum efficiency being displaced toward the shorter wavelengths.

purple boundary: the straight line drawn between the ends of the spectrum locus on a chromaticity diagram.

Q

quality of lighting: pertains to the distribution of luminance in a visual environment. The term is used in a positive sense and implies that all luminances contribute favorably to visual performance, visual comfort, ease of seeing, safety, and esthetics for the specific visual tasks involved.

quantity of light (luminous energy, $Q = \int \Phi dt$): the product of the luminous flux by the time it is maintained. It is the time integral of luminous flux (compare *light* and *luminous flux*.)

quartz-iodine lamp: an obsolete term for *tungsten-halogen lamp*.

quick-flashing light: a rhythmic light exhibiting very rapid regular alternations of light and darkness. There is no restriction on the ratio of the durations of the light to the dark periods.

R

radiance, $L = d^2\Phi/(d\omega dA \cos \theta) = dI/(dA \cos \theta)$ (in a direction, at a point of the surface of a source, of a receiver, or of any other real or virtual surface): the quotient of the radiant flux leaving, passing through, or arriving at an element of the surface surrounding the point, and propagated in directions defined by an elementary cone containing the given direction, by the product of the solid angle of the cone and the area of the orthogonal projection of the element of the surface on a plane perpendicular to the given direction.

NOTE: In the defining equation θ is the angle between the normal to the element of the source and the given direction.

radiant energy density, $w = dQ/dV$: radiant energy per unit volume; *e.g.*, joules per cubic meter.

radiant energy, Q: energy traveling in the form of electromagnetic waves. It is measured in units of energy such as joules, ergs, or kilowatt-hours. See *spectral radiant energy*.

radiant exitance, M: the density of radiant flux leaving a surface. It is expressed in watts per unit area of the surface.

radiant flux, $\Phi = dQ/dt$: the time rate of flow of radiant energy. It is expressed preferably in watts, or in joules per second. See *spectral radiant flux*.

radiant flux density at a surface: the quotient of radiant flux of an element of surface to the area of that element; *e.g.*, watts per square meter. When referring to radiant flux emitted from a surface, this has been called *radiant emittance* (deprecated). The preferred term for radiant flux leaving a surface is *radiant exitance*. The radiant exitance per unit wavelength interval is called *spectral radiant exitance*. When referring to radiant flux incident on a surface, it is called *irradiance* (E).

radiant intensity, $I = d\Phi/d\omega$: the radiant flux proceeding from the source per unit solid angle in the direction considered; *e.g.*, watts per steradian. See *spectral radiant intensity*.

radiator: an emitter of radiant energy.

radiometry: the measurement of quantities associated with radiant energy.

range lights: groups of color-coded boundary lights provided to indicate the direction and limits of a preferred landing path normally on an aerodrome without runways but exceptionally on an aerodrome with runways.

rapid start fluorescent lamp: a fluorescent lamp designed for operation with a ballast that provides a low-voltage winding for preheating the electrodes and initiating the arc without a starting switch or the application of high voltage.

reaction time: the interval between the beginning of a stimulus and the beginning of the response of an observer.

recessed luminaire: See *flush mounted luminaire*.

reciprocal color temperature: color temperature T_k expressed on a reciprocal scale ($1/T_k$). An important use stems from the fact that a given small increment in reciprocal color temperature is approximately equally perceptible regardless of color temperature. Also, color temperature conversion filters for sources approximating graybody sources change the reciprocal color temperature by nearly the same amount anywhere on the color temperature scale. See *mired*.

recoverable light loss factors: factors which give the fractional light loss that can be recovered by cleaning or lamp replacement.

redirecting surfaces and media: those surfaces and media that change the direction of the flux without scattering the redirected flux.

reference ballast: a ballast specially constructed to have certain prescribed characteristics for use in testing electric discharge lamps and other ballasts.

reference line (roadway lighting): either of two radial lines where the surface of the cone of maximum candlepower is intersected by a vertical plane parallel to the curb line and passing through the light-center of the luminaire.

reference standard: an alternate term for *secondary standard*.

reflectance of a surface or medium, $\rho = \Phi_r/\Phi_i$: the ratio of the reflected flux to the incident flux. See *biconical reflectance, bidirectional reflectance, bihemispherical reflectance, conical-directional reflectance, conical-hemispherical reflectance, diffuse reflectance, directional hemispherical reflectance, effective ceiling cavity reflectance, effective floor cavity reflectance, reflectance factor, hemispherical reflectance, hemispherical-conical reflectance, hemispherical directional reflectance, luminous reflectance, spectral reflectance, specular reflectance*.

NOTE: Measured values of reflectance depend upon the angles of incidence and view and on the spectral character of the incident flux. Because of this dependence, the angles of incidence and view and the spectral characteristics of the source should be specified. See *reflection*.

reflectance factor, R: ratio of the radiant (or luminous) flux reflected in directions delimited to that reflected in the same directions by a perfect reflecting diffuser identically irradiated (or illuminated).

reflected glare: glare resulting from specular reflections of high luminances in polished or glossy surfaces in the field of view. It usually is associated with reflections from within a visual task or areas in close proximity to the region being viewed. See *veiling reflection*.

reflection: a general term for the process by which

the incident flux leaves a surface or medium from the incident side, without change in frequency.

NOTE: Reflection is usually a combination of regular and diffuse reflection. See *regular (specular) reflection, diffuse reflection and veiling reflection.*

reflectivity: reflectance of a layer of a material of such a thickness that there is no change of reflectance with increase in thickness.

reflectometer: a photometer for measuring reflectance.

NOTE: Reflectometers may be visual or physical instruments.

reflector: a device used to redirect the luminous flux from a source by the process of reflection. See *retroreflector.*

reflector lamp: an incandescent filament or electric discharge lamp in which the outer blown glass bulb is coated with a reflecting material so as to direct the light (*e.g.*, R- or ER-type lamps). The light transmitting region may be clear, frosted, patterned or phosphor coated.

reflex reflector: See *retro-reflector.*

refraction: the process by which the direction of a ray of light changes as it passes obliquely from one medium to another in which its speed is different.

refractor: a device used to redirect the luminous flux from a source, primarily by the process of refraction

regions of electromagnetic spectrum: for convenience of reference the electromagnetic spectrum is arbitrarily divided as follows:

Vacuum ultraviolet

Extreme ultraviolet	10–100 nm
Far ultraviolet	100–200 nm
Middle ultraviolet	200–300 nm
Near ultraviolet	300–380 nm
Visible	380–770 nm
Near (short wavelength) infrared	770–1400 nm
Intermediate infrared	1400–5000 nm
Far (long wavelength) infrared	5000–1,000,000 nm

NOTE: The spectral limits indicated above have been chosen as a matter of practical convenience. There is a gradual transition from region to region without sharp delineation. Also, the division of the spectrum is not unique. In various fields of science the classifications may differ due to the phenomena of interest.

regressed luminaire: a luminaire mounted above the ceiling with the opening of the luminaire above the ceiling line. See *flush-mounted, surface-mounted, suspended* and *troffer.*

regular (specular) reflectance: the ratio of the flux leaving a surface or medium by regular (specular) reflection to the incident flux. See *regular (specular) reflection.*

regular (specular) reflection: that process by which incident flux is re-directed at the specular angle. See *specular angle.*

regular transmission: that process by which incident flux passes through a surface or medium without scattering. See *regular transmittance.*

regular transmittance: the ratio of the regularly transmitted flux leaving a surface or medium to the incident flux.

relative contrast sensitivity RCS: the relation between the reciprocal of the luminous contrast of a task at visibility threshold and the background luminance expressed as a percentage of the value obtained under a very high level of diffuse task illumination.

relative erythemal factor: See *erythemal efficiency of radiant flux.*

relative luminosity: previously used term for *spectral luminous efficiency of radiant flux.*

relative luminosity factor: previously used term for *spectral luminous efficiency of radiant flux.*

resolving power: the ability of the eye to perceive the individual elements of a grating or any other periodic pattern with parallel elements. It is measured by the number of cycles per degree that can be resolved. The resolution threshold is the period of the pattern that can be just resolved. Visual acuity, in such a case, is the reciprocal of one-half of the period expressed in minutes. The resolution threshold for a pair of points or lines is the distance between them when they can just be distinguished as two, not one, expressed in minutes of arc.

resultant color shift: the difference between the perceived color of an object illuminated by a test source and that of the same object illuminated by the reference source, taking account of the state of chromatic adaptation in each case; *i.e.*, the resultant of colorimetric shift and adaptive color shift.

retina: a membrane lining the more posterior part of the inside of the eye. It comprises photoreceptors (cones and rods) that are sensitive to light and nerve cells that transmit to the optic nerve the responses of the receptor elements.

retro-reflector (reflex reflector): a device designed to reflect light in a direction close to that at which it is incident, whatever the angle of incidence.

rhythmic light: a light, when observed from a fixed point, having a luminous intensity that changes periodically. See *equal interval light, flashing light, group flashing light, interrupted quick-flashing light, quick flashing light, occulting light.*

ribbon filament lamp: an incandescent lamp in which the luminous element is a tungsten ribbon.

NOTE: This type of lamp is often used as a standard in pyrometry and radiometry.

rods: retinal receptors which respond at low levels of luminance even down below the threshold for cones. At these levels there is no basis for perceiving differences in hue and saturation. No rods are found in the center of the fovea.

room cavity: the cavity formed by the plane of the luminaires, the work-plane, and the wall surfaces between these two planes.

room cavity ratio, RCR: a number indicating room cavity proportions calculated from length, width and height.

room index: a letter designation for a range of room ratios.

room ratio: a number indicating room proportions,

calculated from the length, width and ceiling height (or luminaire mounting height) above the work-plane.

room utilization factor (utilance): the ratio of the luminous flux (lumens) received on the work-plane to that emitted by the luminaire.

> NOTE: This ratio sometimes is called interflectance. Room utilization factor is based on the flux emitted by a complete luminaire, whereas coefficient of utilization is based on the flux generated by the bare lamps in a luminaire.

room surface dirt depreciation (RSDD): the fractional loss of task illuminance due to dirt on the room surface.

runway alignment indicator: a group of aeronautical ground lights arranged and located to provide early direction and roll guidance on the approach to a runway.

runway centerline lights: lights installed in the surface of the runway along the centerline indicating the location and direction of the runway centerline; of particular value in conditions of very poor visibility.

runway-edge lights: lights installed along the edges of a runway marking its lateral limits and indicating its direction.

runway-end identification light: a pair of flashing aeronautical ground lights symmetrically disposed on each side of the runway at the threshold to provide additional threshold conspicuity.

runway exit lights: lights placed on the surface of a runway to indicate a path to the taxiway centerline.

runway lights: aeronautical ground lights arranged along or on a runway. See *runway centerline lights, runway-edge lights, runway-end identification light, runway-exit lights.*

runway threshold: the beginning of the runway usable for landing.

runway visibility: the meteorological visibility along an identified runway. Where a transmissometer is used for measurement, the instrument is calibrated in terms of a human observer; *e.g.,* the sighting of dark objects against the horizon sky during daylight and the sighting of moderately intense unfocused lights of the order of 25 candelas at night. See *visibility (meteorological).*

runway visual range, RVR: in the United States an instrumentally derived value, based on standard calibrations, representing the horizontal distance a pilot will see down the runway from the approach end; it is based on the sighting of either high intensity runway lights or on the visual contrast of other targets—whichever yields the greater visual range.

S

saturation of a perceived color: the attribute according to which it appears to exhibit more or less chromatic color judged in proportion to its brightness. In a given set of viewing conditions, and at luminance levels that result in photopic vision, a stimulus of a given chromaticity exhibits approximately constant saturation for all luminances.

scoop: a floodlight consisting of a lamp in an ellipsoidal or paraboloidal matte reflector, usually in a fixed relationship, though some types permit adjustment of the beam shape.

scotopic vision: vision mediated essentially or exclusively by the rods. It is generally associated with adaptation to a luminance below about 0.034 candela per square meter. *See photopic vision.*

sealed beam lamp. A pressed glass reflector lamp (PAR) that provides a closely controlled beam of light.

> NOTE: This term is generally applied in transportation lighting (automotive headlamps, aircraft landing lights, ect.) to distinguish from similar devices in which the light source is replaceable within the reflector-lens unit.

sealed beam headlamp: an integral optical assembly designed for headlighting purposes, identified by the name "Sealed Beam" branded on the lens.

searchlight: a projector designed to produce an approximately parallel beam of light, and having an optical system with an aperture of 20 centimeters (8 inches) or more.

secondary standard source: a constant and reproducible light source calibrated directly or indirectly by comparison with a primary standard. This order of standard also is designated as a reference standard.

> NOTE: National secondary (reference) standards are maintained at national physical laboratories; laboratory secondary (reference) standards are maintained at other photometric laboratories.

self-ballasted lamp: any arc discharge lamp of which the current-limiting device is an integral part.

> NOTE: Known as blended lamp in some countries.

semi-direct lighting: lighting by luminaires distributing 60 to 90 per cent of their emitted light downward and the balance upward.

semi-indirect lighting: lighting by luminaires distributing 60 to 90 per cent of their emitted light upward and the balance downward.

service period: the number of hours per day for which daylighting provides a specified illuminance. It often is stated as a monthly average.

set light: the separate illumination of the background or set, other than that provided for principal subjects or areas.

shade: a screen made of opaque or diffusing material designed to prevent a light source from being directly visible at normal angles of view.

shielding angle (of a luminaire): the angle between a horizontal line through the light center and the line of sight at which the bare source first becomes visible. See *cut-off angle (of a luminaire).*

short-arc lamp: an arc lamp in which the distance between the electrodes is small (on the order of 1 to 10 millimeters).

> NOTE: This type of lamp (*e.g.,* xenon or mercury) is generally of very high pressure.

side back light: illumination from behind the subject in a direction not parallel to a vertical plane through the optical axis of the camera. See *back light*.

side light: lighting from the side to enhance subject modeling and place the subject in depth; apparently separated from background.

side marker lamp: lights indicating the presence and overall length of a vehicle seen from the side.

signal shutter: a device that modulates a beam of light by mechanical means for the purpose of transmitting intelligence.

signalling light: a projector used for directing light signals toward a designated target zone.

size threshold: the minimum perceptible size of an object. It also is defined as the size that can be detected some specific fraction of the times it is presented to an observer, usually 50 per cent. It usually is measured in minutes of arc. See *visual acuity*.

sky factor: the ratio of the illuminance on a horizontal plane at a given point inside a building due to the light received directly from the sky, to the illuminance due to an unobstructed hemisphere of sky of uniform luminance equal to that of the visible sky.

sky light: visible radiation from the sun redirected by the atmosphere.

sky luminance distribution function: for a specified sky condition, the luminance of each point in the sky relative to the zenith luminance.

soft light: diffuse illumination that produces soft edged poorly defined shadows on the background when an object is placed in its path. Also, a luminaire designed to produce such illumination.

solar efficacy: the ratio of the solar illuminance constant to the solar irradiance constant. The current accepted value is 94.2 lumens per watt.

solar illuminance constant: the solar illuminance at normal incidence on a surface in free space at the earth's mean distance from the sun. The current accepted value is 127.5 klx (11,850 footcandles).

solar irradiance constant: the solar irradiance at normal incidence on a surface in free space at the earth's mean distance from the sun. The current accepted value is 1353 watts per square meter (125.7 watts per square foot).

solar radiation simulator: a device designed to produce a beam of collimated radiation having a spectrum, flux density, and geometric characteristics similar to those of the sun outside the earth's atmosphere.

solid angle, ω: a measure of that portion of space about a point bounded by a conic surface whose vertex is at the point. It can be measured by the ratio of intercepted surface area of a sphere centered on that point to the square of the sphere's radius. It is expressed in steradians.

NOTE: Solid angle is a convenient way of expressing the area of light sources and luminaires for computations of discomfort glare factors. It combines into a single number the projected area A_p of the luminaire and the distance D between the luminaire and the eye. It usually is computed by means of the approximate formula

$$\omega = \frac{A_p}{D^2}$$

in which A_p and D^2 are expressed in the same units. This formula is satisfactory when the distance D is greater than about three times the maximum linear dimension of the projected area of the source. Larger projected areas should be sub-divided into several elements.

solid angle factor, Q: a function of the solid angle (ω) subtended by a source at a viewing location and is given by: $Q = 20.4 \omega + 1.52 \omega^{0.2} - 0.075$. See *index of sensation*.

spacing: for roadway lighting the distance between successive lighting units, measured along the center line of the street.

spacing-to-mounting height ratio, S/MH_{up}: the ratio of the distance between luminaire centers to the mounting height above the work-plane.

special color rendering index (R_i): measure of the color shift of various standardized special colors including saturated colors, typical foliage, and Caucasian skin. It also can be defined for other color samples when the spectral reflectance distributions are known.

spectral directional emissivity, ϵ $(2\pi, \lambda, T)$ (of an element of surface of a temperature radiator at any wavelength and in a given direction): the ratio of its spectral radiance at that wavelength and in the given direction to that of a blackbody at the same temperature and wavelength.

$$\epsilon(2\pi, \lambda, T) \equiv L_\lambda (\lambda, \theta, \phi, T)/L_{\lambda, \text{blackbody}}(\lambda, T)$$

spectral hemispherical emissivity, $\epsilon(\lambda, T)$ (of an element of surface of a temperature radiator): the ratio of its spectral radiant exitance to that of a blackbody at the same temperature.

$$\epsilon(\lambda, T) = \int \epsilon(\theta, \phi, \lambda, T) \cos\theta \cdot d\omega$$
$$\equiv M_\lambda(\lambda, T)/M_{\lambda, \text{blackbody}}(\lambda, T)$$

NOTE: Hemispherical emissivity is frequently called "total" emissivity. "Total" by itself is ambiguous, and should be avoided since it may also refer to "spectral-total" (all wavelengths) as well as directional-total (all directions).

spectral (spectroscopic) lamp: a discharge lamp that radiates a significant portion of its radiative power in a line spectrum and which, in combination with filters, may be used to obtain monochromatic radiation.

spectral luminous efficacy of radiant flux, $K(\lambda)$ $= \Phi_{v\lambda}/\Phi_{e\lambda}$: the quotient of the luminous flux at a given wavelength by the radiant flux at that wavelength. It is expressed in lumens per watt.

NOTE: This term formerly was called "luminosity factor." The reciprocal of the maximum luminous efficacy of radiant flux is sometimes called "mechanical equivalent of light;" that is, the ratio between radiant and luminous flux at the wavelength of maximum luminous efficacy. The most probable value is 0.00146 watt per lumen, corresponding to 683 lumens per watt as the maximum possible luminous efficacy. For scotopic vision values (Fig. 5) the maximum luminous efficacy is 1754 "scotopic" lumens per watt.

spectral luminous efficiency for photopic vision, $V(\lambda)$: See *values of spectral luminous efficiency for photopic vision.*

spectral luminous efficiency for scotopic vision, $V'(\lambda)$: See *values of spectral luminous efficiency for scotopic vision.*

spectral luminous efficiency of radiant flux: the ratio of the luminous efficacy for a given wavelength to the value for the wavelength of maximum luminous efficacy. It is dimensionless.

> NOTE: The term *spectral luminous efficiency* replaces the previously used terms *relative luminosity* and *relative luminosity factor.*

spectral radiant energy, $Q_\lambda = dQ/d\lambda$: radiant energy per unit wavelength interval at wavelength λ; *e.g.*, joules/nanometer.

spectral radiant exitance: See *radiant flux density.*

spectral radiant flux, $\Phi_\lambda = d\Phi/d\lambda$: radiant flux per unit wavelength interval at wavelengh λ; *e.g.*, watts per nanometer.

spectral radiant intensity, $I_\lambda = dI/d\lambda$: radiant intensity per unit wavelength interval; *e.g.*, watts per (steradian-nanometer).

spectral reflectance of a surface or medium, $\rho(\lambda)$ $= \Phi_{r\lambda}/\Phi_{i\lambda}$: the ratio of the reflected flux to the incident flux at a particular wavelength, λ, or within a small band of wavelengths, $\Delta\lambda$, about λ.

> NOTE: The terms *hemispherical, regular,* or *diffuse reflectance* may each be considered restricted to a specific region of the spectrum and may be so designated by the addition of the adjective "spectral."

Fig. 5. Scotopic Spectral Luminous Efficiency Values, $V'(\lambda)$

(Unity at Wavelength of Maximum Luminous Efficacy)

Wavelength λ (nanometers)	Relative Value	Wavelength λ (nanometers)	Relative Value
380	0.000589	590	0.0655
390	.002209	600	.03315
400	.00929	610	.01593
410	.03484	620	.00737
420	.0966	630	.003335
430	.1998	640	.001497
440	.3281	650	.000677
450	.455	660	.0003129
460	.567	670	.0001480
470	.676	680	.0000715
480	.793	690	.00003533
490	.904	700	.00001780
500	.982	710	.00000914
510	.997	720	.00000478
520	.935	730	.000002546
530	.811	740	.000001379
540	.650	750	.000000760
550	.481	760	.000000425
560	.3288	770	.0000002413
570	.2076	780	.0000001390
580	.1212		

spectral-total directional emissivity, ϵ (θ,ϕ,T) (of an element of surface of a temperature radiator in a given direction): the ratio of its radiance to that of a blackbody at the same temperature.

$$\epsilon(\theta,\phi,T) \equiv L\ (\theta,\phi,T)/L_{\text{blackbody}}(T)$$

where θ and ϕ are directional angles and T is temperature.

spectral-total hemispherical emissivity, ϵ (of an element of surface of a temperature radiator): the ratio of its radiant exitance to that of a blackbody at the same temperature.

$$\epsilon = \int \epsilon(\theta,\phi)\cdot\cos\theta\cdot d\omega$$
$$\int\int \epsilon(\lambda,\theta,\phi)\cdot\cos\theta\cdot d\omega\cdot d\lambda \equiv M(T)/M_{\text{blackbody}}\ (T)$$

spectral transmittance of a medium $\tau(\lambda) = \Phi_{t\lambda}/ \Phi_{i\lambda}$: the ratio of the transmitted flux to the incident flux at a particular wavelength, λ or within a small band of wavelengths, $\Delta\lambda$, about λ.

> NOTE: The terms *hemispherical, regular,* or *diffuse transmittance* may each be considered restricted to a specific region of the spectrum and may be so designated by the addition of the adjective "spectral."

spectral tristimulus values: See *color matching functions.*

spectrophotometer: an instrument for measuring the transmittance and reflectance of surfaces and media as a function of wavelength.

spectroradiometer: an instrument for measuring radiant flux as a function of wavelength.

spectrum locus: the locus of points representing the colors of the visible spectrum in a chromaticity diagram.

specular angle: the angle between the perpendicular to the surface and the reflected ray that is numerically equal to the angle of incidence and that lies in the same plane as the incident ray and the perpendicular but on the opposite side of the perpendicular to the surface.

specular reflectance: see *regular reflectance.*

specular reflection: see *regular reflection.*

specular surface: one from which the reflection is predominantly regular. See *regular (specular) reflection.*

speed of light: the speed of all radiant energy, including light, is 2.997925×10^8 meters per second in vacuum (approximately 186,000 miles per second). In all material media the speed is less and varies with the material's index of refraction, which itself varies with wavelength.

speed of vision: the reciprocal of the duration of the exposure time required for something to be seen.

sphere illumination: the illumination on a task from a source providing equal luminance in all directions about that task, such as an illuminated sphere with the task located at the center.

spherical reduction factor: the ratio of the *mean spherical luminous intensity* to the *mean horizontal intensity.*

spotlight: any of several different types of luminaires with relatively narrow beam angle designed to illuminate a specifically defined area. In motion pictures, generic for Fresnel lens luminaires.

standard illuminant: a hypothetical light source of specified relative spectral power distribution.
NOTE: The CIE has specified spectral power distributions for standard illuminants A, B, and C and several D-illuminants.

standard source: in colorimetry, a source that has a specified spectral distribution and is used as a standard.

standard source A: a tungsten filament lamp operated at a color temperature of 2856 K, and approximating a blackbody operating at that temperature.

standard source B: an approximation of noon sunlight having a correlated color temperature of approximately 4874 K. It is obtained by a combination of Source A and a special filter consisting of a layer one centimeter thick of each of the following solutions, contained in a double cell constructed of nonselective optical glass:

No. 1

Copper sulphate ($CuSO_4 \cdot 5H_2O$)	2.452 g
Mannite ($C_6H_8(OH)_6$)	2.452 g
Pyridine (C_5H_5N)	30.0 cm³
Distilled water to make	1000.0 cm³

No. 2

Cobalt-ammonium sulphate	
($CoSO_4 \cdot (NH_4)_2SO_4 \cdot 6H_2O$)	21.71 g
Copper sulphate ($CuSO_4 \cdot 5H_2O$)	16.11 g
Sulphuric acid (density 1.835)	10.0 cm³
Distilled water to make	1000.0 cm³

standard source C: an approximation of daylight provided by a combination of direct sunlight and clear sky having a correlated color temperature of approximately 6774 K. It is obtained by a combination of Source A plus a cell identical with that used for Source B, except that the solutions are:

No. 1

Copper sulphate ($CuSO_4 \cdot 5H_2O$)	3.412 g
Mannite ($C_6H_8(OH)_6$)	3.412 g
Pyridine (C_5H_5N)	30.0 cm³
Distilled water to make	1000.0 cm³

No. 2

Cobalt-ammonium sulphate	
($CoSO_4 \cdot (NH_4)_2SO_4 \cdot 6H_2O$)	30.580 g
Copper sulphate ($CuSO_4 \cdot 5H_2O$)	22.520 g
Sulphuric acid (density 1.835)	10.0 cm³
Distilled water to make	1000.0 cm³

starter: a device used in conjunction with a ballast for the purpose of starting an electric-discharge lamp.

state of chromatic adaptation: the condition of the eye in equilibrium with the average color of the visual field.

Stefan-Boltzmann law: the statement that the radiant exitance of a blackbody is proportional to the fourth power of its absolute temperature; that is,

$$M = \sigma T^4$$

NOTE: The currently recommended value of the Stefan-Boltzmann constant σ is 5.67032×10^{-8} $W \cdot m^{-2}K^{-4}$ or 5.67032×10^{-12} $W \cdot cm^{-2}K^{-4}$.

steradian, sr (unit solid angle): a solid angle subtending an area on the surface of a sphere equal to the square of the sphere radius.

stilb: a CGS unit of luminance. One stilb equals one candela per square centimeter. The use of this term is deprecated.

Stiles-Crawford effect: the reduced luminous efficiency of rays entering the peripheral portion of the pupil of the eye.

stop lamp: a device giving a steady warning light to the rear of a vehicle or train of vehicles, to indicate the intention of the operator to diminish speed or to stop.

stray light (in the eye): light from a source that is scattered onto parts of the retina lying outside the retinal image of the source.

street lighting luminaire: a complete lighting device consisting of a light source and ballast, where appropriate, together with its direct appurtenances such as globe, reflector, refractor housing, and such support as is integral with the housing. The pole, post or bracket is not considered part of the luminaire.

street lighting unit: the assembly of a pole or lamp post with a bracket and a luminaire.

striplight (theatrical): a compartmentalized luminaire, with each compartment containing a lamp, reflector and color frame holder, wired in rotation in three- or four-circuit and used as borderlights, footlights or cyclorama lighting from above or below.

stroboscopic lamp (strobe light): a flash tube designed for repetitive flashing.

subjective brightness: the subjective attribute of any light sensation giving rise to the percept of luminous magnitude, including the whole scale of qualities of being bright, light, brilliant, dim or dark. See *saturation of a perceived color*.
NOTE: The term brightness often is used when referring to the measurable luminance. While the context usually makes it clear as to which meaning is intended, the preferable term for the photometric quantity is *luminance*, thus reserving *brightness* for the subjective sensation.

sun bearing: the angle measured in the plane of the horizon between a vertical plane at a right angle to the window wall and the position of this plane after it has been rotated to contain the sun.

sun light: direct visible radiation from the sun.

sunlamp: an ultraviolet lamp that radiates a significant portion of its radiative power in the UV-B band (280 to 315 nanometers).

supplementary lighting: lighting used to provide an additional quantity and quality of illumination that cannot readily be obtained by a general lighting system and that supplements the general lighting level, usually for specific work requirements.

supplementary standard illuminant D₅₅: a representation of a phase of daylight at a correlated color temperature of approximately 5500 K.

supplementary standard illuminant D₇₅: a representation of a phase of daylight at a correlated color temperature of approximately 7500 K.

surface mounted luminaire: a luminaire mounted

directly on the ceiling.

suspended (pendant) luminaire: a luminaire hung from a ceiling by supports.

switch start fluorescent lamps: see *preheat fluorescent lamp.*

T

table lamp: a portable luminaire with a short stand suitable for standing on furniture.

tail lamp: a lamp used to designate the rear of a vehicle by a warning light.

talbot, T.: a unit of light; equal to one lumen-second.

tanning lamp: an ultraviolet lamp that radiates a significant portion of its radiative power in the UV-A and/or B band.

task lighting: lighting directed to a specific surface or area that provides illumination for visual tasks.

task-ambient lighting: a combination of task lighting and ambient lighting within an area such that the general level of ambient lighting is lower than and complementary to the task lighting.

taxi-channel lights: aeronautical ground lights arranged along a taxi-channel of a water aerodrome to indicate the route to be followed by taxiing aircraft.

taxi light: an aircraft aeronautical light designed to provide necessary illumination for taxiing.

taxiway lights: aeronautical ground lights provided to indicate the route to be followed by taxiing aircraft. See *taxiway-centerline lights, taxiway-edge lights, taxiway holding-post light.*

taxiway-centerline lights: taxiway lights placed along the centerline of a taxiway except that on curves or corners having fillets, these lights are placed a distance equal to half the normal width of the taxiway from the outside edge of the curve or corner.

taxiway-edge lights: taxiway lights placed along or near the edges of a taxiway.

taxiway holding-post light: a light or group of lights installed at the edge of a taxiway near an entrance to a runway, or to another taxiway, to indicate the position at which the aircraft should stop and obtain clearance to proceed.

temperature radiator: a radiator whose radiant flux density (radiant exitance) is determined by its temperature and the material and character of its surface, and is independent of its previous history. See *blackbody* and *graybody.*

thermopile: a thermal radiation detector consisting of a number of thermocouples interconnected in order to increase the sensitivity to incident radiant flux.

threshold: the value of a variable of a physical stimulus (such as size, luminance, contrast or time) that permits the stimulus to be seen a specific percentage of the time or at a specific accuracy level. In many psychophysical experiments, thresholds are presented in terms of 50 per cent accuracy or accurately 50 per cent of the time. However, the threshold also is expressed as the value of the physical variable that permits the object to be just barely seen. The threshold may be determined by merely detecting the presence of an object or it may be determined by discriminating certain details of the object. See *absolute luminance threshold, brightness contrast threshold, luminance threshold, modulation size threshold.*

threshold lights: runway lights placed to indicate the longitudinal limits of that portion of a runway, channel or landing path usable for landing.

top light: illumination of a subject directly from above employed to outline the upper margin or edge of the subject.

torchere: an indirect floor lamp sending all or nearly all of its light upward.

tormentor light: luminaire mounted directly behind the sides of the stage arch.

total emissivity: See *spectral-total directional emissivity* and *spectral-total hemispherical emissivity.*

touchdown zone lights: barettes of runway lights installed in the surface of the runway between the runway edge lights and the runway centerline lights to provide additional guidance during the touchdown phase of a landing in conditions of very poor visibility.

traffic beam: See *lower (passing) beams.*

train: the angle between the vertical plane through the axis of the searchlight drum and the plane in which this plane lies when the search light is in a position designated as having zero train.

transient adaptation factor, TAF: a factor which reduces the *equivalent contrast* due to readaptation from one luminous background to another.

transmission: a general term for the process by which incident flux leaves a surface or medium on a side other than the incident side, without change in frequency.

> NOTE: Transmission through a medium is often a combination of regular and diffuse transmission. See *regular transmission, diffuse transmission,* and *transmittance.*

transmissometer: a photometer for measuring transmittance.

> NOTE: Transmissometers may be visual or physical instruments.

transmittance, $\tau = \Phi_t/\Phi_i$: the ratio of the transmitted flux to the incident flux.

> NOTE: Measured values of transmittance depend upon the angle of incidence, the method of measurement of the transmitted flux, and the spectral character of the incident flux. Because of this dependence complete information on the technique and conditions of measurement should be specified.
>
> It should be noted that transmittance refers to the ratio of flux emerging to flux incident; therefore, reflections at the surface as well as absorption within the material operate to reduce the transmittance.

tristimulus values of a light, *X, Y, Z*: the amounts of each of three primaries required to match the color of the light.

troffer: a recessed lighting unit, usually long and installed with the opening flush with the ceiling. The term is derived from "trough" and "coffer."

troland: a unit of retinal illuminance which is based upon the fact that retinal illuminance is proportional to the product of the luminance of the distal stimulus and the area of entrance pupil. One troland is the retinal illuminance produced when the luminance of the distal stimulus is one candela per square meter and the area of the pupil is one square millimeter.

NOTE: The troland makes no allowance for interocular attenuation or for the *Stiles-Crawford effect.*

tube: See *lamp.*

tungsten-halogen lamp: a gas filled tungsten incandescent lamp containing a certain proportion of halogens in an inert gas whose pressure exceeds three atmospheres.

NOTE: The tungsten-iodine lamp (UK) and quartz-iodine lamp (USA) belong to this category.

turn signal operating unit: that part of a signal system by which the operator of a vehicle indicates the direction a turn will be made, usually by a flashing light.

U

ultraviolet lamp: a lamp which radiates a significant portion of its radiative power in the ultraviolet (UV) part of the spectrum; the visible radiation is not of principal interest.

ultraviolet radiation: for practical purposes any radiant energy within the wavelength range 10 to 380 nanometers. See *regions of electromagnetic spectrum.*

NOTE: On the basis of practical applications and the effect obtained, the ultraviolet region often is divided into the following bands:

Ozone-producing	180–220 nanometers
Bactericidal (germicidal)	220–300 nanometers
Erythemal	280–320 nanometers
"Black light"	320–400 nanometers

There are no sharp demarcations between these bands, the indicated effects usually being produced to a lesser extent by longer and shorter wavelengths. For engineering purposes, the "black light" region extends slightly into the visible portion of the spectrum. Another division of the ultraviolet spectrum often used by photobiologists is given by the CIE:

UV-A	315-400 nanometers
UV-B	280-315 nanometers
UV-C	100–280 nanometers

units of luminance: the luminance of a surface in a specified direction may be expressed in luminous intensity per unit of projected area of surface or in luminous flux per unit of solid angle and per unit of projected surface area.

NOTE: Typical units are the candela per square meter (lumen per steradian and per square meter) and the candela per square foot (lumen per steradian and per square foot).

The luminance of a surface in a specified direction is also expressed (incorrectly) in lambertian units as the number of lumens per unit area that would leave the surface *if the luminance in all directions within the hemisphere on the side of the surface being considered were the same as the luminance in the specified direction.*

NOTE: A typical unit in this system is the footlambert, equal to one lumen per square foot.

This method of specifying luminance is equivalent to stating the number of lumens that would leave the surface *if the surface were replaced by a perfectly diffusing surface with a luminance in all directions within the hemisphere equal to the luminance of the actual surface in the direction specified.* In practice no surface follows exactly the cosine formula of emission or reflection; hence the luminance is not uniform but varies with the angle from which it is viewed. For this reason, this practice is denigrated.

unrecoverable light loss factors: factors which give the fractional light loss that cannot be recovered by cleaning or lamp replacement.

upper (driving) beams: one or more beams intended for distant illumination and for use on the open highway when not meeting other vehicles. Formerly "country beam." See *lower (passing) beams.*

upward component: that portion of the luminous flux from a luminaire emitted at angles above the horizontal. See *downward component.*

utilance: See *room utilization factor.*

V

vacuum lamp: an incandescent lamp in which the filament operates in an evacuated bulb.

valance: a longitudinal shielding member mounted across the top of a window or along a wall and usually parallel to the wall, to conceal light sources giving both upward and downward distributions.

valance lighting: lighting comprising light sources shielded by a panel parallel to the wall at the top of a window.

values of spectral luminous efficiency for photopic vision, $V(\lambda)$: values for spectral luminous efficiency at 10-nanometer intervals (see Fig. 6) were provisionally adopted by the CIE in 1924 and were adopted in 1933 by the International Committee on Weights and Measures as a basis for the establishment of photometric standards of types of sources differing from the primary standard in spectral distribution of radiant flux. These values are given in the second column of Fig. 6; the intermediate values given in the other columns have been interpolated.

NOTE: These standard values of spectral luminous efficiency were determined by observations with a two-degree photometric field having a moderately high luminance, and photometric evaluations based upon them consequently do not apply exactly to other conditions of observation. Power in watts weighted in accord with these standard values are often referred to as *light-watts.*

values of spectral luminous efficiency for scotopic vision $V'(\lambda)$: values of spectral luminous efficiency at 10-nanometer intervals (see Fig. 5) were provisionally adopted by the CIE in 1951.

Fig. 6. Photopic Spectral Luminous Efficiency, $V(\lambda)$
(Unity at Wavelength of Maximum Luminous Efficacy)

Wavelength λ (nanometers)	Standard Values	Values Interpolated at Intervals of One Nanometer								
		1	2	3	4	5	6	7	8	9
380	0.00004	.000045	.000049	.000054	.000058	.000064	.000071	.000080	.000090	.000104
390	.00012	.000138	.000155	.000173	.000193	.000215	.000241	.000272	.000308	.000350
400	.0004	.00045	.00049	.00054	.00059	.00064	.00071	.00080	.00090	.00104
410	.0012	.00138	.00156	.00174	.00195	.00218	.00244	.00274	.00310	.00352
420	.0040	.00455	.00515	.00581	.00651	.00726	.00806	.00889	.00976	.01066
430	.0116	.01257	.01358	.01463	.01571	.01684	.01800	.01920	.02043	.02170
440	.023	.0243	.0257	.0270	.0284	.0298	.0313	.0329	.0345	.0362
450	.038	.0399	.0418	.0438	.0459	.0480	.0502	.0525	.0549	.0574
460	.060	.0627	.0654	.0681	.0709	.0739	.0769	.0802	.0836	.0872
470	.091	.0950	.0992	.1035	.1080	.1126	.1175	.1225	.1278	.1333
480	.139	.1448	.1507	.1567	.1629	.1693	.1761	.1833	.1909	.1991
490	.208	.2173	.2270	.2371	.2476	.2586	.2701	.2823	.2951	.3087
500	.323	.3382	.3544	.3714	.3890	.4073	.4259	.4450	.4642	.4836
510	.503	.5229	.5436	.5648	.5865	.6082	.6299	.6511	.6717	.6914
520	.710	.7277	.7449	.7615	.7776	.7932	.8082	.8225	.8363	.8495
530	.862	.8739	.8851	.8956	.9056	.9149	.9238	.9320	.9398	.9471
540	.954	.9604	.9661	.9713	.9760	.9803	.9840	.9873	.9902	.9928
550	.995	.9969	.9983	.9994	1.0000	1.0002	1.0001	.9995	.9984	.9969
560	.995	.9926	.9898	.9865	.9828	.9786	.9741	.9691	.9638	.9581
570	.952	.9455	.9386	.9312	.9235	.9154	.9069	.8981	.8890	.8796
580	.870	.8600	.8496	.8388	.8277	.8163	.8046	.7928	.7809	.7690
590	.757	.7449	.7327	.7202	.7076	.6949	.6822	.6694	.6565	.6437
600	.631	.6182	.6054	.5926	.5797	.5668	.5539	.5410	.5282	.5156
610	.503	.4905	.4781	.4658	.4535	.4412	.4291	.4170	.4049	.3929
620	.381	.3690	.3570	.3449	.3329	.3210	.3092	.2977	.2864	.2755
630	.265	.2548	.2450	.2354	.2261	.2170	.2082	.1996	.1912	.1830
640	.175	.1672	.1596	.1523	.1452	.1382	.1316	.1251	.1188	.1128
650	.107	.1014	.0961	.0910	.0862	.0816	.0771	.0729	.0688	.0648
660	.061	.0574	.0539	.0506	.0475	.0446	.0418	.0391	.0366	.0343
670	.032	.0299	.0280	.0263	.0247	.0232	.0219	.0206	.0194	.0182
680	.017	.01585	.01477	.01376	.01281	.01192	.01108	.01030	.00956	.00886
690	.0082	.00759	.00705	.00656	.00612	.00572	.00536	.00503	.00471	.00440
700	.0041	.00381	.00355	.00332	.00310	.00291	.00273	.00256	.00241	.00225
710	.0021	.001954	.001821	.001699	.001587	.001483	.001387	.001297	.001212	.001130
720	.00105	.000975	.000907	.000845	.000788	.000736	.000688	.000644	.000601	.000560
730	.00052	.000482	.000447	.000415	.000387	.000360	.000335	.000313	.000291	.000270
740	.00025	.000231	.000214	.000198	.000185	.000172	.000160	.000149	.000139	.000130
750	.00012	.000111	.000103	.000096	.000090	.000084	.000078	.000074	.000069	.000064
760	.00006	.000056	.000052	.000048	.000045	.000042	.000039	.000037	.000035	.000032

NOTE: These values of spectral luminous efficiency were determined by observation by young dark-adapted observers using extra-foveal vision at near-threshold luminance.

vapor-tight luminaire: a luminaire designed and approved for installation in damp or wet locations. It also is described as "enclosed and gasketed."

VASIS (Visual Approach Slope Indicator System): the system of angle-of-approach lights accepted as a standard by the International Civil Aviation Organization, comprising two bars of lights located at each side of the runway near the threshold and showing red or white or a combination of both (pink) to the approaching pilot depending upon his position with respect to the glide path.

veiling luminance: a luminance superimposed on the retinal image which reduces its contrast. It is this veiling effect produced by bright sources or areas in the visual field that results in decreased visual performance and visibility.

veiling reflection: regular reflections superimposed upon diffuse reflections from an object that partially or total obscure the details to be seen by reducing the contrast. This sometimes is called reflected glare.

vertical plane of a searchlight: the plane through the axis of the searchlight drum which contains the elevation angle. See *horizontal plane of a searchlight*

visibility: the quality or state of being perceivable by the eye. In many outdoor applications, visibility is defined in terms of the distance at which an object can be just perceived by the eye. In indoor applications it usually is defined in terms of the contrast or size of a standard test object, observed under standardized view-conditions, having the same threshold as the given object. See *visibility (meteorological)*.

visibility (meteorological): a term that denotes the greatest distance, expressed in kilometers or miles, that selected objects (visibility markers) or lights of moderate intensity (25 candelas) can be seen and identified under specified conditions of observation.

visibility level, VL: a contrast multiplier to be

applied to the visibility reference function to provide the luminance contrast required at different levels of task background luminance to achieve visibility for specified conditions relating to the task and observer.

visibility performance criteria function, VL8: a function representing the luminance contrast required to achieve 99 percent visual certainty for the same task used for the visibility reference function, including the effects of dynamic presentation and uncertainty in task location.

visibility reference function, VL1: a function representing the luminance contrast required at different levels of task background luminance to achieve visibility threshold for the visibility reference task consisting of a 4 minute disk exposed for 1/5 second.

vision: See *central vision, foveal vision, mesopic vision, peripheral vision, photopic vision* and *scotopic vision.*

visual acuity: a measure of the ability to distinguish fine details. Quantitatively, it is the reciprocal of the minimum angular separation in minutes of two lines of width subtending one minute of arc when the lines are just resolvable as separate. See *size threshold.*

visual angle: the angle subtended by an object or detail at the point of observation. It usually is measured in minutes of arc.

visual approach slope indicator system: see *VASIS*

visual comfort probability, VCP: the rating of a lighting system expressed as a per cent of people who, when viewing from a specified location and in a specified direction, will be expected to find it acceptable in terms of discomfort glare. *Visual comfort probability* is related to *discomfort glare rating (DGR).*

visual field: the locus of objects or points in space that can be perceived when the head and eyes are kept fixed. The field may be monocular or binocular. See *monocular visual field, binocular visual field, central visual field, peripheral visual field.*

visual perception: the interpretation of impressions transmitted from the retina to the brain in terms of information about a physical world displayed before the eye.

NOTE: Visual perception involves any one or more of the following: recognition of the presence of something (object, aperture or medium); identifying it; locating it in space; noting its relation to other things; identifying its movement, color, brightness or form.

visual performance: the quantitative assessment of the performance of a visual task, taking into consideration speed and accuracy.

visual photometer: a photometer in which the equality of brightness of two surfaces is established visually. See *physical photometer.*

NOTE: The two surfaces usually are viewed simultaneously side by side. This method is used in portable visual luminance meters. This is satisfactory when the color difference between the test source and comparison source is small. However, when there is a color differ-

ence, a flicker photometer provides more precise measurements. In this type of photometer the two surfaces are viewed alternately at such a rate that the color sensations either nearly or completely blend and the flicker due to brightness difference is balanced by adjusting the comparison source.

visual range (of a light or object): the maximum distance at which a particular light (or object) can be seen and identified.

visual surround: includes all portions of the visual field except the visual task.

visual task: conventionally designates those details and objects that must be seen for the performance of a given activity, and includes the immediate background of the details or objects.

NOTE: The term visual task as used is a misnomer because it refers to the visual display itself and not the task of extracting information from it. The task of extracting information also has to be differentiated from the overall task performed by the observer.

visual task evaluator, VTE: a contrast reducing instrument which permits obtaining a value of luminance contrast, called the *equivalent contrast* $\check{C}$ of a standard visibility reference task giving the same visibility as that of a task whose contrast has been reduced to threshold when the background luminances are the same for the task and the reference task. See *equivalent contrast.*

voltage to luminaire factor: the fractional loss of task illuminance due to improper voltage at the luminaire.

W

wavelengths: wavelength is the distance between two successive points of a periodic wave in the direction of propagation, in which the oscillation has the same phase. The three commonly used units are listed in the following table:

Name	Symbol	Value
Micrometer	μm	$1 \ \mu m = 10^{-6}$ m
Nanometer	nm	$1 \ nm = 10^{-9}$ m
Angstrom*	Å	$1 \ \text{Å} = 10^{-10}$ m

Weber's fraction: See *luminous luminance contrast*

wide-angle diffusion: diffusion in which flux is scattered at angles far from the direction that the flux would take by regular reflection or transmission. See *narrow-angle diffusion.*

wide-angle luminaire: a luminaire distributing the light through a comparatively large solid angle. See also *narrow-angle luminaire.*

width line: the radial line (the one that makes the larger angle with the reference line) that passes through the point of one-half maximum candlepower on the lateral candlepower distribution curve plotted on the surface of the cone of maximum candlepower.

---

* The use of this unit is deprecated.

Wien displacement law: an expression representing, in a functional form, the spectral radiance of a blackbody as a function of the wavelength and the temperature.

$$L_\lambda = dI_\lambda/dA' = C_1\lambda^{-5}f(\lambda T)$$

The two principal corollaries of this law are:

$$\lambda_m T = b$$
$$L_m/T^5 = b'$$

which show how the maximum spectral radiance L_m and the wavelength λ_m at which it occurs are related to the absolute temperature T. See *Wien radiation law*.

NOTE: The currently recommended value of b is 2.898 × 10^{-3} m·K or 2.898 × 10^{-1} cm·K. From the Planck radiation law, and with the use of the value of b, c_1, and c_2 as given above, b' is found to be 4.10 × 10^{-12} W· cm^{-3}·K^{-5}·sr^{-1}.

Wien radiation law: an expression representing approximately the spectral radiance of a blackbody as a function of its wavelength and temperature. It commonly is expressed by the formula

$$L_\lambda = dI_\lambda/dA' = c_{1L}\lambda^{-5}e^{-(c_2/\lambda T)}$$

This formula is accurate to one per cent or better for values of λT less than 3000 micrometer-kelvin.

wing clearance lights: aircraft lights provided at the wing tips to indicate the extent of the wing span when the navigation lights are located an appreciable distance inboard of the wings tips.

work-plane: the plane at which work usually is done, and on which the illuminance is specified and measured. Unless otherwise indicated, this is assumed to be a horizontal plane 0.76 meters (30 inches) above the floor.

working standard: a standardized light source for regular use in photometry.

Z

zonal-cavity inter-reflectance method: a procedure for calculating coefficients of utilization, wall luminance coefficients, and ceiling cavity luminance coefficients taking into consideration the luminaire intensity distribution, room size and shape (cavity ratio concepts), and room reflectances. It is based on *flux transfer theory*.

zonal constant: a factor by which the mean candlepower emitted by a source of light in a given angular zone is multiplied to obtain the lumens in the zone.

zonal-factor method: a procedure for predetermining, from typical luminaire photometric data in discrete angular zones, the proportion of luminaire output which would be incident initially (without interreflections) on the work-plane, ceiling, walls and floor of a room.

zonal factor interflection method: a former procedure for calculating coefficients of utilization.

Fig. 7. Illuminance Conversion Factors.

1 lumen = 1/683 light-watt	1 watt-second = 10^7 ergs
1 lumen-hour = 60 lumen-minutes	1 phot = 1 lumen/square centimeter
1 footcandle = 1 lumen/square foot	1 lux = 1 lumen/square meter = 1 metercandle

Multiply Number of → To Obtain By Number of ↓	Foot-candles	Lux	Phot	Milliphot
Footcandles	1	0.0929	929	0.929
Lux	10.76	1	10,000	10
Phot	0.00108	0.0001	1	0.001
Milliphot	1.076	0.1	1,000	1

Fig. 8. Greek Alphabet (Capital and Lower Case).

Capital	Lower Case	Greek Name
A	α	Alpha
B	β	Beta
Γ	γ	Gamma
Δ	δ	Delta
E	ϵ	Epsilon
Z	ζ	Zeta
H	η	Eta
Θ	θ	Theta
I	ι	Iota
K	κ	Kappa
Λ	λ	Lambda
M	μ	Mu
N	ν	Nu
Ξ	ξ	Xi
O	o	Omicron
Π	π	Pi
P	ρ	Rho
Σ	σ, ς	Sigma
T	τ	Tau
Υ	υ	Upsilon
Φ	φ, ϕ	Phi
X	χ	Chi
Ψ	ψ	Psi
Ω	ω	Omega

Fig. 9. Unit Prefixes.

Prefix	Symbol	Factor by Which the Unit is Multiplied
exa	E	1,000,000,000,000,000,000 = 10^{18}
peta	P	1,000,000,000,000,000 = 10^{15}
tera	T	1,000,000,000,000 = 10^{12}
giga	G	1,000,000,000 = 10^9
mega	M	1,000,000 = 10^6
kilo	k	1,000 = 10^3
hecto	h	100 = 10^2
deka	da	10 = 10^1
deci	d	0.1 = 10^{-1}
centi	c	0.01 = 10^{-2}
milli	m	0.001 = 10^{-3}
micro	μ	0.000,001 = 10^{-6}
nano	n	0.000,000,001 = 10^{-9}
pico	p	0.000,000,000,001 = 10^{-12}
femto	f	0.000,000,000,000,001 = 10^{-15}
atto	a	0.000,000,000,000,000,001 = 10^{-18}

Fig. 10. Partial list of abbreviations and acronyms

A

A	ampere
Å	Angstrom unit
ac	alternating current
AIA	American Institute of Architects
ANSI	American National Standards Institute
ASID	American Society of Interior Designers
ASTM	American Society for Testing and Materials
ASHRAE	American Society of Heating, Refrigerating and Air-conditioning Engineers
atm	atmosphere

B

BCD	borderline between comfort and discomfort
BCP	beam candlepower
BL	blacklight
BRDF	bidirectional reflectance-distribution function
Btu	British thermal unit

C

°C	degree Celsius
cal	calorie
CBM	Certified Ballast Manufacturers
CBU	coefficient of beam utilization
CCR	ceiling cavity ratio
cgs	centimeter-gram-second (system)
CIE	Commission Internationale de l'Eclairage (International Commission on Illumination)
cm	centimeter
cos	cosine
cp	candlepower
CPI	color preference index
CRF	contrast rendition factor
CRI	color rendering index
CRT	cathode ray tube
CSA	Canadian Standards Association
CSI	compact source iodide
CU	coefficient of utilization
CW	cool white
CWX	cool white deluxe

D

dB	decibel
dc	direct current
DGF	disability glare factor
DGR	disability glare rating
DIC	direct illumination component

E

emf	electromotive force
ESI	equivalent sphere illumination
EU	erythemal unit

F

°F	degree Fahrenheit
fc	footcandle
FCR	floor cavity ratio
fff	flicker fusion frequency
ft	foot
ft^2	square foot

H

h	hour
HID	high intensity discharge
HMI	hydrargyrum, medium-arc-length, iodide

hp	horse power
HPS	high pressure sodium
Hz	hertz

I

IALD	International Association of Lighting Designers
IDSA	Industrial Designers Society of America
IEEE	Institute of Electrical and Electronics Engineers
IERI	Illuminating Engineering Research Institute
in	inch
in^2	square inch
IR	infrared
ISO	International Organization for Standardization

J

J	joule

K

K	kelvin
kcal	kilocalorie
kg	kilogram
kHz	kilohertz
km	kilometer
km^2	square kilometer
km/s	kilometer per second
kV	kilovolt
kVA	kilovolt-ampere
kVAr	reactive kilovolt-ampere
kW	kilowatt
kWh	kilowatt-hour

L

LBO	lamp burnout
LCD	liquid crystal display
LDD	luminaire dirt depreciation
LED	light emitting diode
LEF$_v$	lighting effectiveness factor
LLD	lamp lumen depreciation
LLF	light loss factor
lm	lumen
ln	logarithm (natural)
LPS	low pressure sodium
lx	lux

M

m	meter
m^2	square meter
mA	milliampere
max	maximum
MF	maintenance factor
MH	mounting height
MHz	megahertz
min	minimum
min	minute (time)
mm	millimeter
mm^2	square millimeter
mol wt	molecular weight
MPE	minimal perceptible erythema
mph	mile per hour

N

NBS	National Bureau of Standards
NEC	National Electrical Code
NEMA	National Electrical Manufacturers Association
nm	nanometer

Fig. 10. *Continued*

O		U	
OSA	Optical Society of America	UV	ultraviolet
		UL	Underwriters Laboratories
P			
PAR	pressed reflector lamp	**V**	
pf	power factor	V	volt
R		VA	volt-ampere
		VAr	reactive volt-ampere
R	reflectance factor	VASIS	visual approach slope indicator system
rad	radian	VCP	visual comfort probability
RCR	room cavity ratio	VDU	visual display unit
RCS	relative contrast sensitivity	VHO	very high output (lamp)
rms	root mean square	VI	visibility index
RSDD	room surface dirt depreciation	VL	visibility level
RTP	relative task performance	VTE	visual task evaluator
RVP	relative visual performance	VTP	visual task photometer
RVR	runway visual range		
S		**W**	
s	second (time)	W	watt
sin	sine	WW	warm white
SPD	spectral power distribution	WWX	warm white deluxe
sq	square	μA	microampere
sr	steradian	μV	microvolt
T		μW	microwatt
		ρ_{CC}	effective ceiling cavity reflectance
TAF	transient adaptation factor	ρ_{FC}	effective floor cavity reflectance
tan	tangent	'	minute (angular measure)
temp	temperature	"	second (angular measure)
		°	degree

Fig. 11. Conversion Factors for Units of Length

Multiply Number of → To Obtain Number of ↓ By	Angstroms	Nanometers	Micrometers (Microns)	Millimeters	Centimeters	Meters	Kilometers	Mils	Inches	Feet	Miles
Angstroms	1	10	10^4	10^7	10^8	10^{10}	10^{13}	2.540×10^5	2.540×10^8	3.048×10^9	1.609×10^{13}
Nanometers	10^{-1}	1	10^3	10^6	10^7	10^9	10^{12}	2.540×10^4	2.540×10^7	3.048×10^8	1.609×10^{12}
Micrometers (Microns)	10^{-4}	10^{-3}	1	10^3	10^4	10^6	10^9	2.540×10	2.540×10^4	3.048×10^5	1.609×10^9
Millimeters	10^{-7}	10^{-6}	10^{-3}	1	10	10^3	10^6	2.540×10^{-2}	2.540×10	3.048×10^2	1.609×10^6
Centimeters	10^{-8}	10^{-7}	10^{-4}	10^{-1}	1	10^2	10^5	2.540×10^{-3}	2.540	3.048×10	1.609×10^5
Meters	10^{-10}	10^{-9}	10^{-6}	10^{-3}	10^{-2}	1	10^3	2.540×10^{-5}	2.540×10^{-2}	3.048×10^{-1}	1.609×10^3
Kilometers	10^{-13}	10^{-12}	10^{-9}	10^{-6}	10^{-5}	10^{-3}	1	2.540×10^{-8}	3.048×10^{-5}	3.048×10^{-4}	1.609
Mils	3.937×10^{-6}	3.937×10^{-5}	3.937×10^{-2}	3.937×10	3.937×10^2	3.937×10^4	3.937×10^7	1	10^3	1.2×10^4	6.336×10^7
Inches	3.937×10^{-9}	3.937×10^{-8}	3.937×10^{-5}	3.937×10^{-2}	3.937×10^{-1}	3.937×10	3.937×10^4	10^{-3}	1	12	6.336×10^4
Feet	3.281×10^{-10}	3.281×10^{-9}	3.281×10^{-6}	3.281×10^{-3}	3.281×10^{-2}	3.281	3.281×10^3	8.333×10^3	8.333×10^{-2}	1	5.280×10^3
Miles	6.214×10^{-14}	6.214×10^{-13}	6.214×10^{-10}	6.214×10^{-7}	6.214×10^{-6}	6.214×10^{-4}	6.214×10^{-1}	1.578×10^{-8}	1.578×10^{-5}	1.894×10^{-4}	1

Fig. 12. Conversion from Values in SI Units.

m → ... → in
cm → ...
kcd/m² → ... cd/in²
cd/m² → ... fL
lx* → fc ... ft

	fc	fL	cd/in²	in	ft		fc	fL	cd/in²	in	ft
1	.09	.29	.65	.39	3.3	500	46.5	146.0	322.5	196.9	1641
2	.19	.58	1.29	.79	6.6	510	47.4	148.9	329.0	200.8	1673
3	.28	.88	1.94	1.18	9.8	520	48.3	151.8	335.4	204.7	1706
4	.37	1.17	2.58	1.57	13.1	530	49.2	154.7	341.9	208.7	1739
5	.47	1.46	3.23	1.97	16.4	540	50.2	157.6	348.3	212.6	1772
6	.56	1.75	3.87	2.36	19.7	550	51.1	160.5	354.8	216.5	1805
7	.65	2.04	4.52	2.76	23.0	560	52.0	163.5	361.2	220.5	1837
8	.74	2.34	5.16	3.15	26.2	570	53.0	166.4	367.7	224.4	1870
9	.84	2.63	5.81	3.54	29.5	580	53.9	169.3	374.1	228.3	1903
						590	54.8	172.2	380.6	232.3	1936
100	9.3	29.2	64.5	39.4	328	600	55.7	175.1	387.0	236.2	1969
110	10.2	32.1	71.0	43.3	361	610	56.7	178.1	393.5	240.2	2001
120	11.1	35.0	77.4	47.2	394	620	57.6	181.0	399.9	244.1	2034
130	12.1	37.9	83.9	51.2	427	630	58.5	183.9	406.4	248.0	2067
140	13.0	40.9	90.3	55.1	459	640	59.5	186.8	412.8	252.0	2100
150	13.9	43.8	96.8	59.1	492	650	60.4	189.7	419.3	255.9	2133
160	14.9	46.7	103.2	63.0	525	660	61.3	192.7	425.7	259.8	2165
170	15.8	49.6	109.7	66.9	558	670	62.2	195.6	432.2	263.8	2198
180	16.7	52.5	116.1	70.9	591	680	63.2	198.5	438.6	267.7	2231
190	17.7	55.5	122.6	74.8	623	690	64.1	201.4	445.1	271.7	2264
200	18.6	58.4	129.0	78.7	656	700	65.0	204.3	451.5	275.6	2297
210	19.5	61.3	135.5	82.7	689	710	66.0	207.2	458.0	279.5	2330
220	20.4	64.2	141.9	86.6	722	720	66.9	210.2	464.4	283.5	2362
230	21.4	67.1	148.4	90.6	755	730	67.8	213.1	470.9	287.4	2395
240	22.3	70.1	154.8	94.5	787	740	68.7	216.0	477.3	291.3	2428
250	23.2	73.0	161.3	98.4	820	750	69.7	218.9	483.8	295.3	2461
260	24.2	75.9	167.7	102.4	853	760	70.6	221.8	490.2	299.2	2494
270	25.1	78.8	174.2	106.3	886	770	71.5	224.8	496.7	303.1	2526
280	26.0	81.7	180.6	110.2	919	780	72.5	227.7	503.1	307.1	2559
290	26.9	84.7	187.1	114.2	951	790	73.4	230.6	509.6	311.0	2592
300	27.9	87.6	193.5	118.1	984	800	74.3	233.5	516.0	315.0	2625
310	28.8	90.5	200.0	122.0	1017	810	75.2	236.4	522.5	318.9	2658
320	29.7	93.4	206.4	126.0	1050	820	76.2	239.4	528.9	322.8	2690
330	30.7	96.3	212.9	130.0	1083	830	77.1	242.3	535.4	326.8	2723
340	31.6	99.2	219.3	133.9	1116	840	78.0	245.2	541.8	330.7	2756
350	32.5	102.2	225.8	137.8	1148	850	79.0	248.1	548.3	334.6	2789
360	33.4	105.8	232.2	141.7	1181	860	79.9	251.0	554.7	338.6	2822
370	34.4	108.0	238.7	145.7	1214	870	80.8	254.0	561.2	342.5	2854
380	35.3	110.9	245.1	149.6	1247	880	81.8	256.9	567.6	346.5	2887
390	36.2	113.8	251.6	153.5	1280	890	82.7	259.8	574.1	350.4	2920
400	37.2	116.8	258.0	157.5	1312	900	83.6	262.7	580.5	354.3	2953
410	38.1	119.7	264.5	161.4	1345	910	84.5	265.6	587.0	358.3	2986
420	39.0	122.6	270.9	165.4	1378	920	85.5	268.5	593.4	362.2	3019
430	39.9	125.5	277.4	169.3	1411	930	86.4	271.5	600.0	366.1	3051
440	40.9	128.4	283.8	173.2	1444	940	87.3	274.4	606.3	370.1	3084
450	41.8	131.4	290.3	177.2	1476	950	88.3	277.3	612.8	374.0	3117
460	42.7	134.3	296.7	181.1	1509	960	89.2	280.2	619.2	378.0	3150
470	43.7	137.2	303.2	185.0	1542	970	90.1	283.1	625.7	381.9	3183
480	44.6	140.1	309.6	189.0	1575	980	91.0	286.1	632.1	385.8	3215
490	45.5	143.0	316.1	192.9	1608	990	92.0	289.0	638.6	389.8	3248

* Also useful for converting from ft² to m².

Fig. 13. Conversion to Values in SI Units

ft						ft				
in						in				
cd/in²						cd/in²				
fL						fL				
fc*						fc*				
	lx	cd/m²	kcd/m²	cm	m ×		lx	cd/m²	kcd/m²	cm	m
1	10.76	3.4	1.55	2.54	.30	500	5380	1713	775.0	1270	152.4
2	21.5	6.9	3.00	5.08	.61	510	5488	1747	790.5	1295	155.4
3	32.3	10.3	4.65	7.62	.91	520	5595	1782	806.0	1321	158.5
4	43.0	13.7	6.20	10.16	1.22	530	5703	1816	821.6	1346	161.5
5	53.8	17.1	7.75	12.70	1.52	540	5810	1850	837.0	1372	164.6
6	64.6	20.6	9.30	15.24	1.83	550	5918	1884	852.5	1397	167.6
7	75.3	24.0	10.85	17.78	2.13	560	6026	1919	868.0	1422	170.7
8	86.1	27.4	12.40	20.32	2.44	570	6133	1953	883.5	1448	173.7
9	96.8	30.8	13.95	22.86	2.74	580	6241	1987	899.0	1473	176.8
						590	6348	2021	914.5	1499	179.8
100	1076	343	155.0	254	30.5	600	6456	2056	930.0	1524	182.9
110	1184	377	170.5 *	279	33.5	610	6564	2090	945.5	1549	185.9
120	1291	411	186.0	305	36.6	620	6671	2124	961.0	1575	189.0
130	1399	445	201.5	330	39.6	630	6779	2158	976.5	1600	192.0
140	1506	480	217.0	356	42.7	640	6886	2193	992.0	1626	195.1
150	1614	514	232.5	381	45.7	650	6994	2227	1007.5	1651	198.1
160	1722	548	248.0	406	48.8	660	7102	2261	1023.0	1676	201.2
170	1829	582	263.5	432	51.8	670	7209	2295	1038.5	1702	204.2
180	1937	617	279.0	457	54.9	680	7317	2330	1054.0	1727	207.3
190	2044	651	294.5	483	57.9	690	7424	2364	1069.5	1753	210.3
200	2152	685	310.0	508	61.0	700	7532	2398	1085.0	1778	213.4
210	2260	719	325.5	533	64.0	710	7640	2432	1100.5	1803	216.4
220	2367	754	341.0	559	67.1	720	7747	2467	1116.0	1829	219.5
230	2475	788	356.5	584	70.1	730	7855	2501	1131.5	1854	222.5
240	2582	822	372.0	610	73.2	740	7962	2535	1147.0	1880	225.6
250	2690	857	387.5	635	76.2	750	8070	2570	1162.5	1905	228.6
260	2798	891	403.0	660	79.2	760	8178	2604	1178.0	1930	231.6
270	2905	925	418.5	686	82.3	770	8285	2638	1193.5	1956	234.7
280	3013	959	434.0	711	85.3	780	8393	2672	1209.0	1981	237.7
290	3120	994	449.5	737	88.4	790	8500	2702	1224.5	2007	240.8
300	3228	1028	465.0	762	91.4	800	8608	2741	1240.0	2032	243.8
310	3336	1062	480.5	787	94.5	810	8716	2775	1255.5	2057	246.9
320	3443	1096	496.0	813	97.5	820	8823	2809	1271.0	2083	249.9
330	3551	1131	511.5	838	100.6	830	8931	2844	1286.5	2108	253.0
340	3658	1165	527.0	864	103.6	840	9038	2878	1302.0	2134	256.0
350	3766	1199	542.5	889	106.7	850	9146	2912	1317.5	2159	259.1
360	3874	1233	558.0	914	109.7	860	9254	2946	1333.0	2184	262.1
370	3981	1268	573.5	940	112.8	870	9361	2981	1348.5	2210	265.2
380	4089	1302	589.0	965	115.8	880	9469	3015	1364.0	2235	268.2
390	4196	1336	604.5	991	118.9	890	9576	3049	1379.5	2261	271.3
400	4304	1370	620.0	1016	121.9	900	9684	3083	1395.0	2286	274.3
410	4412	1405	635.5	1041	125.0	910	9792	3118	1410.5	2311	277.4
420	4519	1439	651.0	1067	128.0	920	9899	3152	1426.0	2337	280.4
430	4627	1473	666.5	1092	131.1	930	10010	3186	1441.5	2362	283.5
440	4734	1507	682.0	1118	134.1	940	10110	3220	1457.0	2388	286.5
450	4842	1542	697.5	1143	137.2	950	10220	3255	1472.5	2413	289.6
460	4950	1576	713.0	1168	140.2	960	10330	3289	1488.0	2438	292.6
470	5057	1610	728.5	1194	143.3	970	10440	3323	1503.5	2464	295.7
480	5165	1644	744.0	1219	146.3	980	10540	3357	1519.0	2489	298.7
490	5272	1679	759.5	1245	149.4	990	10650	3392	1534.5	2515	301.8

* Also useful for converting from m² to ft²

Fig. 14. Luminance Conversion Factors.

1 nit = 1 candela/square meter
1 stilb = 1 candela/square centimeter
1 apostilb (international) = 0.1 millilambert = 1 blondel
1 apostilb (German Hefner) = 0.09 millilambert
1 lambert = 1000 millilamberts

Multiply Number of → To Obtain Number of ↓	By ↘	Footlambert*	Candela/ square meter	Millilambert*	Candela/ square inch	Candela/ square foot	Stilb
Footlambert*		1	0.2919	0.929	452	3.142	2,919
Candela/square meter		3.426	1	3.183	1,550	10.76	10,000
Millilambert*		1.076	0.3142	1	487	3.382	3,142
Candela/square inch		0.00221	0.000645	0.00205	1	0.00694	6.45
Candela/square foot		0.3183	0.0929	0.2957	144	1	929
Stilb		0.00034	0.0001	0.00032	0.155	0.00108	1

* Deprecated unit of luminance.

Fig. 15.   Angular Measure, Temperature, Power and Pressure Conversion Equations.

Angle
1 radian = 57.29578 degrees
Temperature
(F to C) C = 5/9 (F − 32)
(C to F) F = 9/5 C + 32
(C to K) K = C + 273
Power
1 kilowatt = 1.341 horsepower
= 56.89 Btu per minute
Pressure
1 atmosphere = 760 millimeters of mercury at 0°C
= 29.92 inches of mercury at 0°C
= 14.7 pounds per square inch
= 101.3 kilopascals

REFERENCES

1. *American National Standard Nomenclature and Definitions for Illuminating Engineering,* ANSI/IES-RP-16, 1980, American National Standard Institute, New York, 1980. Sponsored by the Illuminating Engineering Society.
2. *American National Standard Practice for the Use of Metric (SI) Units in Building Design and Construction,* ANSI/ASTM E621-78, American National Standards Institute, New York, 1978. Sponsored by American Society for Testing and Materials.
3. Barbrow, L. E.: "The Metric System in Illuminating Engineering," *Illum. Eng.,* Vol. 62, p. 638, November, 1967.
4. *Colorimetry, CIE Publication No. 15,* Bureau Central de la Commission Internationale de L'Eclairage, Paris, France, 1971.
5. Commission Internationale de l'Eclairage, *International Lighting Vocabulary,* 3rd Edition, Paris, France, 1970.
6. Committee of Testing Procedures of the IES: "IES General Guide to Photometry," *Illum. Eng.,* Vol. L., p. 201, April, 1955.
7. Committee on Colorimetry of the Optical Society of America: *The Science of Color,* Edward Brothers Inc., New York, 1973.
8. Theatre, Television and Film Lighting Committee of the IES, "A Glossary of Commonly Used Terms in Theatre, Television and Film Lighting," *Light. Des. Appl.,* Vol. 13, p. 43, November 1983.
9. Hollander, A., ed: *Radiation Biology, Volume II, Ultraviolet and Related Radiations,* McGraw-Hill Book Company, New York, 1955.
10. Kaufman, J. E.: "Introducing SI Units," *Illum. Eng.* Vol. 63, p. 537, October, 1968.
11. Koller, L. R.: *Ultraviolet Radiation,* John Wiley & Sons Inc., New York, 1952.
12. Levin, R. E.: "Luminance—A Tutorial Paper," *J. Soc. Motion Pict. Telev. Engineers,* Vol. 77, p. 1005, October, 1968.
13. Nicodemus, F. E., Richmond, J. C., Hsia, J. J., Ginsberg, I. W. and Limperis, T.: *Geometrical Considerations and Nomenclature for Reflectance, NBS Monograph No. 160,* National Bureau of Standards, U.S. Department of Commerce, Washington, D.C., October, 1977.
14. "Nomenclature and Definitions Applicable to Radiometric and Photometric Characteristics of Matter," *ASTM Spec. Tech. Publ. 475,* American Society for Testing and Materials, Philadelphia, 1970.
15. Schapero, M., Cline, D. and Hofstetter, A. W.: *Dictionary of Visual Science,* Chilton Company, Philadelphia and New York, 1960.
16. *Supplement No. 2 to CIE Publication No. 15, Uniform Color Spaces, Color Difference Equations, Psychometric Color Terms,* Bureau Central de la Commission Internationale de l'Eclairage, Paris, France, 1978.
17. Wyszecki, G. and Stiles, W. S.: *Color Science: Concepts and Methods, Quantitative Data and Formulae,* John Wiley & Sons Inc., New York, 1982.

Light Source Data

(From IES Lighting Handbook—*1984 Reference Volume* and *1987 Application Volume*)

Fig. 16 Approximate Luminance of Various Light Sources

Light Source		Approximate Average Luminance (cd/m^2)
Natural Light Sources:		
Sun (at its surface)		2.3×10^9
Sun (as observed from earth's surface)	At meridian	1.6×10^9
Sun (as observed from earth's surface)	Near horizon	6×10^6
Moon (as observed from earth's surface)	Bright spot	2.5×10^3
Clear sky	Average brightness	8×10^3
Overcast sky		2×10^3
Lightning flash		8×10^{10}
Combustion Sources:		
Candle flame (sperm)	Bright spot	1×10^4
Kerosene flame (flat wick)	Bright spot	1.2×10^4
Illuminating gas flame	Fish-tail burner	4×10^3
Welsbach mantle	Bright spot	6.2×10^4
Acetylene flame	Mees burner	1.1×10^5
Photoflash lamps		1.6×10^8 to 4×10^8 Peak
Nuclear Sources:		
Atomic fission bomb	0.1 millisecond after firing—30-m dia. ball	2×10^{12}
Self-luminous paints		0.2 to 0.3
Incandescent Lamps:		
Carbon filament	3.15 lumens per watt	5.2×10^5
Tantalum filament	6.3 lumens per watt	7×10^5
Tungsten filament	Vacuum lamp 10 lumens per watt	2×10^6
Tungsten filament	Gas filled lamp 20 lumens per watt	1.2×10^7
Tungsten filament	750-watt projection lamp 26 lumens per watt	2.4×10^7
Tungsten filament	1200-watt projection lamp 31.5 lumens per watt	3.3×10^7
RF (radio frequency) lamp	24-millimeter diameter disk	6.2×10^7
Blackbody at 6500 K		3×10^9
Blackbody at 4000 K		2.5×10^8
Blackbody at 2042 K		6×10^5
60-watt inside frosted bulb		1.2×10^5
25-watt inside frosted bulb		5×10^4
15-watt inside frosted bulb		3×10^4
10-watt inside frosted bulb		2×10^4
60-watt "white" bulb		3×10^4
Fluorescent Lamps:		
T-17 bulb cool white	420 mA low loading	4.3×10^3
T-12 bulb cool white	430 mA medium loading	8.2×10^3
T-12 bulb cool white	800 mA high loading	1.1×10^4
T-17 grooved bulb cool white	1500 mA extra high loading	1.5×10^4
T-12 bulb cool white	1500 mA extra high loading	1.7×10^4
Electroluminescent Lamps:		
Green color at 120 volts 60 hertz		27
Green color at 600 volts 400 hertz		68
Electric Arcs:		
Plain carbon arc	Positive crater	1.3×10^8 to 1.6×10^8
High intensity carbon arc	13.6 mm rotating positive carbon	7×10^8 to 1.5×10^9
Electric Arc Lamps:		
High pressure mercury arc	Type H33, 2.5 atmospheres	1.5×10^6
High pressure mercury arc	Type H38, 10 atmospheres	1.8×10^6
High intensity mercury short arc	30 atmospheres	2.4×10^8 (4.3×10^9 peak)
Xenon short arc	900 W dc	1.8×10^8
Electronic flash tubes		1×10^9 to 3×10^9
Clear glass neon tube	15 mm 60 mA	1.6×10^3
Clear glass blue tube	15 mm 60 mA	8×10^2
Clear glass fluorescent		
daylight and white	15 mm 60 mA	5×10^3
green	15 mm 60 mA	9.5×10^3
blue and gold	15 mm 60 mA	3×10^3
pink and coral	15 mm 60 mA	2×10^3

Fig. 17 General Service Lamps for 115-, 120-, and 125-Volt Circuits (Will Operate in Any Position but Lumen Maintenance is Best for 40 to 1500 Watts When Burned Vertically Base-Up)

Watts	Bulb and Other Description	Base (see Fig. 36)	Filament (see Fig. 34)	Rated Average Life (hours)	Maximum Over-All Length		Average Light Center Length		Approximate Initial Filament Temperature (K)	Maximum Bare Bulb Temperature*		Base Temperature†		Approximate Initial Lumens	Rated Initial Lumens Per Watt‡	Lamp Lumen Depreciation** (per cent)
					(millimeters)	(inches)	(millimeters)	(inches)		(°C)	(°F)	(°C)	(°F)			
10	S-14 inside frosted or clear	Med.	C-9	1500	89	3½	64	2½	2420	41	106	41	106	80	8.0	89
15	A-15 inside frosted	Med.	C-9	2500	89	3½	60	2 3/8	—	—	—	—	—	126	8.4	83
25	A-19 inside frosted	Med.	C-9	2500	98	3 7/8	64	2½	2550	43	110	42	108	230	9.2	79
34[a,b]	A-19 inside frosted	Med.	CC-6 or CC-8	1500	113	4 7/16	76	3	—	—	—	—	—	410	12.1	
40	A-19 inside frosted and white¶	Med.	C-9	1500	108	4¼	75	2 15/16	2650	127	260	105	221	455	11.4	87.5
50	A-19 inside frosted	Med.	CC-6	1000	113	4 7/16	79	3 1/8	—	—	—	—	—	680	13.6	—
52[a,c]	A-19 clear or inside frosted	Med.	CC-8	1000	113	4 7/16	79	3 1/8	—	—	—	—	—	800	15.4	
55[a]	A-19 clear or white	Med.	CC-8	1000	113	4 7/16	79	3 1/8	—	—	—	—	—	870	15.8	
60	A-19 inside frosted and white¶	Med.	CC-6	1000	113	4 7/16	79	3 1/8	2790	124	25	93	200	860	14.3	93
67[a,d]	A-19 clear or inside frosted	Med.	CC-8	750	113	4 7/16	79	3 1/8	—	—	—	—	—	1130	16.9	
70[a]	A-19 clear or white	Med.	CC-8	750	113	4 7/16	79	3 1/8	—	—	—	—	—	1190	17.0	
75	A-19 inside frosted and white¶	Med.	CC-6	750	113	4 7/16	79	3 1/8	2840	135	275	96	205	1180	15.7	92
90[a]	A-19 clear or inside frosted	Med.	CC-8	750	113	4 7/16	79	3 1/8	—	—	—	—	—	1620	18.0	
95[a]	A-19 clear or white	Med.	CC-8	750	113	4 7/16	79	3 1/8	—	—	—	—	—	1710	18.0	
100	A-19 inside frosted and white¶	Med.	CC-8	750	113	4 7/16	79	3 1/8	2905	149	300	98	208	1740	17.4	90.5
100§	A-19 inside frosted and white	Med.	CC-8	1000	113	4 7/16	79	3 1/8	—	—	—	—	—	1680	16.8	—
100	A-21 inside frosted	Med.	CC-6	750	133	5¼	98	3 7/8	2880	127	260	90	194	1690	16.9	90
135[a]	A-21 clear or inside frosted	Med.	CC-8	750	139	5½	101	4	—	—	—	—	—	2580	19.1	
150	A-21 inside frosted	Med.	CC-8	750	139	5½	101	4	2960	—	—	—	—	2880	19.2	89
150	A-21 white	Med.	CC-8	750	139	5½	101	4	2930	—	—	—	—	2790	18.6	89
150	A-23 inside frosted or clear or white	Med.	CC-6	750	157	6 3/16	117	4 5/8	2925	138	280	99	210	2780	18.5	89
150	PS-25 clear or inside frosted	Med.	C-9	750	176	6 15/16	133	5¼	2910	143	290	99	210	2660	17.7	87.5
200	A-23 inside frosted white or clear	Med.	CC-8	750	160	6 5/16	117	4 5/8	2980	174	345	107	225	4000	20.0	89.5
200	PS-25 clear or inside frosted	Med.	CC-6	750	176	6 15/16	133	5¼	—	—	—	—	—	3800	19.0	—

Watts	Lamp	Base	Bulb													
200	PS-30 clear or inside frosted	Med.	C-9	750	204	8 1/16	152	6	2925	152	305	99	210	3700	18.5	85
300	PS-25 clear or inside frosted	Med.	CC-8	750	176	6 15/16	131	5 3/16	3015	205	401	112	234	6360	21.2	87.5
300	PS-30 clear or inside frosted	Med.	C-9	750	204	8 1/16	152	6	3000	135	275	79	175	6100	20.3	82.5
300	PS-30 clear or inside frosted	Mog.	CC-8	1000	219	8 5/8	177	7	—	—	—	—	—	5960	19.8	—
300	PS-35 clear or inside frosted	Mog.	C-9	1000	238	9 5/8	177	7	2980	166	330	102	215	5860	19.6	86
500	PS-35 clear or inside frosted	Mog.	CC-8	1000	238	9 5/8	177	7	3050	213	415	79	175	10600	21.2	89
500	PS-40 clear or inside frosted	Mog.	C-9	1000	247	9 3/4	177	7	2945	199	390	102	215	10140	20.3	—
750	PS-52 clear or inside frosted	Mog.	C-7A	1000	331	13 1/16	241	9 1/2	2990	—	—	—	—	15660	20.9	—
750	PS-52 clear or inside frosted	Mog.	CC-8 or 2CC-8	1000	331	13 1/16	241	9 1/2	3090	—	—	—	—	17000	22.6	89
1000	PS-52 clear or inside frosted	Mog.	C-7A	1000	331	13 1/16	241	9 1/2	2995	249	480	113	235	21800	21.8	—
1000	PS-52 clear or inside frosted	Mog.	CC-8 or 2CC-8	1000	331	13 1/16	241	9 1/2	3110	—	—	—	—	23600	23.6	89
1500	PS-52 clear or inside frosted	Mog.	C-7A	1000	331	13 1/16	241	9 1/2	3095	266	510	129	265	34000	22.6	78

* Lamp burning base up in ambient temperature of 25°C (77°F).
† At junction of base and bulb.
‡ For 120-volt lamps.
§ Used mainly in Canada.
¶ Lumen and lumen per watt value of white lamps are generally lower than for inside frosted.
** Per cent initial light output at 70 per cent of rated life.

ª Reduced wattage, high efficacy, 120 and 130 volt.
ᵇ Substitute for 40-watt lamp.
ᶜ Substitute for 60-watt lamp.
ᵈ Substitute for 75-watt lamp.

Fig. 18 Tungsten-Halogen Lamps for General Lighting (Burning Position—Any, Except as Noted.)

Watts	Volts	Bulb and Finish	Base*	Filament	Maximum Overall Length (millimeters)	(inches)	Average Light Center Length (millimeters)	(inches)	Rated Life (hours)	Approximate Initial Lumens	Approximate Initial Lumens Per Watt	Lamp Lumen Depreciation† (per cent)	Approximate Color Temperature (K)
							Double-Ended Types						
200	120	T-3 clear	RSC	CC-8	79	3⅛	—	—	1500	3460	17.3	96	2900
300	120	T-3 clear	RSC	C-8	119	4¹¹/₁₆	—	—	2000	5950	19.9	96	2950
300	120	T-3 frosted	RSC	C-8	119	4¹¹/₁₆	—	—	2000	5900	19.3	96	2950
300	120	T-3 clear	RSC	CC-8	79	3⅛	—	—	2000	5650	18.5	—	2900
300	120	T-4 clear	RSC	CC-8	79	3⅛	—	—	2000	5650	18.9	—	3000
300	130	T-4 clear	RSC	CC-8	79	3⅛	—	—	2000	5650	18.9	—	3000
300	120	T-4 frosted	RSC	CC-8	79	3⅛	—	—	2000	5300	17.7	—	3000
350	**	T-3 IR refl.	RSC	C-8	119	4¹¹/₁₆	—	—	2000	10000	28.6	95	—
400	**	T-3 clear	RSC	CC-8	119	4¹¹/₁₆	—	—	2000	9000	22.5	—	—
400	120	T-4 clear	RSC	CC-8	79	3⅛	—	—	2000	7750	19.4	96	2950
400	125	T-4 clear	RSC	CC-8	79	3⅛	—	—	2000	7500	18.7	—	3000
400	130	T-4 clear	RSC	CC-8	79	3⅛	—	—	2000	7800	19.5	—	3000
500	120	T-3 clear	RSC	C-8§	119	4¹¹/₁₆	—	—	1500	10500	20.8	96	—
500	120	T-3 clear	RSC	C-8	119	4¹¹/₁₆	—	—	2000	10950	21.9	96	3000
500	120	T-3 frosted	RSC	C-8	119	4¹¹/₁₆	—	—	2000	10700	21.4	96	3000
500	125	T-3 clear	RSC	C-8	119	4¹¹/₁₆	—	—	2000	10500	21.0	—	3000
500	130	T-3 clear	RSC	C-8	119	4¹¹/₁₆	—	—	2000	10750	21.5	96	3000
900	***	T-3 IR refl.	RSC	C-8	256	10¹/₁₆	—	—	2000	32000	35.6	95	—
900	277	T-3 IR refl.	RSC	C-8	256	10¹/₁₆	—	—	2000	31000	34.4	95	—
1000	120	T-6 clear	RSC or RSC (Rect)	CC-8	143	5⅝	—	—	2000	23400	23.4	96	3000
1000	120	T-6 clear	RSC or RSC (Rect)	CC-8	143	5⅝	—	—	4000	19800	19.8	93	2950
1000	120	T-6 frosted	RSC or RSC (Rect)	CC-8	143	5⅝	—	—	2000	22700	22.7	96	3000
1000	120	T-6 clear	RSC or RSC (Rect)	CC-8	143	5⅝	—	—	500	24500	24.5	—	3100
1000	220	T-3 clear	RSC	C-8	256	10¹/₁₆	—	—	2000	21400	21.4	96	3000
1000	240	T-3 clear	RSC	C-8	256	10¹/₁₆	—	—	2000	21400	21.4	96	3000
1200	***	T-3 clear	RSC	C-8	256	10¹/₁₆	—	—	2000	29000	24.2	—	—
1250	208	T-3 clear	RSC	C-8	256	10¹/₁₆	—	—	2000	28000	22.4	96	3050
1500	208	T-3 clear	RSC	C-8	256	10¹/₁₆	—	—	2000	35800	23.0	96	3050
1500	220	T-3 clear	RSC	C-8	256	10¹/₁₆	—	—	2000	35800	23.2	96	3050
1500	240	T-3 clear	RSC	C-8	256	10¹/₁₆	—	—	2000	35800	23.2	96	3050
1500	240	T-3 frosted	RSC	C-8	256	10¹/₁₆	—	—	2000	32000	21.3	—	3050
1500	277	T-3 clear	RSC	C-8	256	10¹/₁₆	—	—	2000	33700	22.5	96	3050
1500	277	T-3 frosted	RSC	C-8	256	10¹/₁₆	—	—	2000	31600	21.0	—	3050
							Single-Ended Types						
5	12	T-3 clear	G4	C-6	31	1⁷/₃₂	—	—	2000	60	12	—	3000
10	6	T-3 clear	G4	C-6	31	1⁷/₃₂	—	—	2000	120	12	—	3000
10	12	T-3 clear	G4	C-6	31	1⁷/₃₂	—	—	2000	—	—	—	3000
20	††	T-3 clear	G4	C-6	31	1⁷/₃₂	—	—	2000	350	17.5	—	3000
50	12	T-4 clear	GY 6.35	C-6	44	1²³/₃₂	—	—	2000	950	19	—	3000
50	24	T-4 clear	GY 6.35	C-6	44	1²³/₃₂	—	—	2000	900	18	—	3000
75	28	T-3 clear	Min.	CC-6	60	2⅜	30	1³/₁₆	2000	1400	21.4	—	3000
100	12	T-3½ clear	Minican.	—	57	2¼	—	—	1000	2500	25.0	—	—
100	12	T-4 clear	GY 6.35	C-6	44	1²³/₃₂	—	—	2000	2500	25	—	3000
100	24	T-4 clear	GY 6.35	C-6	44	1²³/₃₂	—	—	2000	2000	20	—	3000
100	120	T-4 clear	Minican.	—	70	2¾	—	—	1000	1800	18.0	—	—
100	120	T-4 frosted	Minican.	—	70	2¾	—	—	1000	1750	17.5	—	—
100	120	T-4 clear	D.C. Bay.	—	62	2⁷/₁₆	—	—	1000	1800	18.0	—	—
100	120	T-4 frosted	D.C. Bay.	—	62	2⁷/₁₆	—	—	1000	1750	17.5	—	—
150	120	T-4 clear	Minican.	CC-2V	70	2¾	—	—	1500	2900	19.3	—	3000
150	120	T-4 clear	D.C. Bay	CC-2V	62	2⁷/₁₆	—	—	1500	2900	19.3	—	3000
150	120	T-4 frosted	D.C. Bay	—	62	2¹/₁₆	—	—	1000	2700	18.0	—	—
150	120	T-10 clear or frosted	Med.	C-6	105	4⅛	75	2¹⁵/₁₆	2000	2500	16.7	—	3000

See footnotes on page 49.

Continued on next page.

Fig. 18 *Continued*

Watts	Volts	Bulb and Finish	Base*	Filament	Maximum Overall Length		Average Light Center Length		Rated Life (hours)	Approximate Initial Lumens	Approximate Initial Lumens Per Watt	Lamp Lumen Depreciation† (per cent)	Approximate Color Temperature (K)
					(millimeters)	(inches)	(millimeters)	(inches)					
150	120	T-5 clear or frosted	D.C. Bay.	C-6	95	3¾	—	—	2000	2000	16.7	—	3000
250	120	T-10 clear or frosted	Med.	C-6	105	4⅛	75	2¹⁵/₁₆	2000	4200	16.8	—	3000
250	120	T-5 clear or frosted	D.C. Bay.	C-6	95	3¾	—	—	2000	4200	16.8	—	3000
250	120	T-4 clear	Minican.	CC-8	79	3⅛	41	1⅝	2000	4850	19.4	96	2950
250	120	T-4 frosted	Minican.	CC-8	79	3⅛	41	1⅝	2000	4700	18.8	96	2950
250	120	T-4 clear	D.C. Bay.	CC-8	71	2¹³/₁₆	38	1½	2000	4850	19.4	—	2950
250	120	T-4 clear	D.C. Bay.	CC-8	71	2¹³/₁₆	41	1⅝	2000	4850	19.4	96	2950
250	120	T-4 frosted	D.C. Bay.	CC-8	71	2¹³/₁₆	41	1⅝	2000	4700	18.8	96	2950
250	130	T-4 clear	Minican.	CC-8	79	3⅛	41	1⅝	2000	4850	19.4	—	2950
400	120	T-4 frosted	Minican.	CC-8	92	3⅝	51	2	2000	8550	21.4	96	3050
400	120	T-4 clear	Minican.	CC-8	92	3⅝	51	2	2000	8800	22.0	96	3050
500	120	T-4 clear	Minican.	CC-8	95	3¾	51	2	2000	11500	23.0	—	3000
500	120	T-4 frosted	Minican.	CC-8	95	3¾	51	2	2000	10000	20.0	—	3000
500	120	T-4 frosted	D.C. Bay.	CC-8	87	3⁷/₁₆	54	2⅛	2000	10100	20.2	96	3000
500	120	T-4 clear	D.C. Bay.	CC-8	87	3⁷/₁₆	54	2⅛	2000	10450	20.9	96	3000
500	130	T-4 clear	Minican.	CC-8	95	3¾	51	2	2000	11500	23.0	—	3000
1000	120	T-24 clear	Med. Bipost	CC-8	233	9½	102	4	3000	22400	22.4	93	3050

* RSC = recessed single contact, RSC (Rect) = rectangular recessed single contact.
† Per cent initial light output at 70 per cent of rated life.
§ Lamp provides maximum filament straightness under severe operating conditions.

** 120, 130 volts
*** 208, 220, 240 volts
†† 12 and 24 volts

Fig. 19 Showcase Lamps for 120- and 130-Volt Circuits

Watts	Bulb and Other Description	Base (see Fig. 36)	Filament (see Fig. 34)	Rated Average Life (hours)	Maximum Over-All Length		Approximate Initial Lumens	Lamp Lumen Depreciation‡ (per cent)
					(millimeters)	(inches)		
25*	T-6½ clear or inside frosted	Intermed.	C-8	1000	140	5½	250	76
25*	T-10 clear or inside frosted	Med.	C-8	1000	143	5⅝	255	79
25*	T-10 light inside frosted and side reflectorized	Med.	CC-8	1000	143**	5⅝**	230	80
40*	T-8 clear or inside frosted	Med.	C-23 or C-8	1000	302	11⅞	420	77
40*	T-10 clear or inside frosted	Med.	C-8	1000	143	5⅝	445	77
40*	T-10 light inside frosted and side reflectorized	Med.	CC-8	1000	143**	5⅝**	430	80
60†	T-10 clear	Med.	C-8	1000	143	5⅝	745	—
75*	T-10 clear	Med.	C-23 or C-8	1000	302	11⅞	800	—

* May be burned in any position.
** Exclusive of spring contact.
† Must be burned from base down to horizontal.
‡ Per cent initial light output at 70 per cent of rated life.

Fig. 20   Reflectorized R, ER and PAR Lamps for 120-Volt Circuits

Watts	Bulb	Description	Base (see Fig. 36)	Maximum Over-All Length (millimeters)	(inches)	Approximate Beam Spread (degrees)[c]	Approximate Beam Lumens	Approximate Total Lumens	Approximate Average Candle-power in Central 10 Degree Cone[f]	Rated Averaged Life (hours)
R Lamps for Spotlighting and Floodlighting (Parabolic Reflector)										
30	R-20	Flood	Med.	100	3¹⁵⁄₁₆	85	150	205	290	2000
50	R-20	Flood	Med.	100	3¹⁵⁄₁₆	90	430	435	550	2000
75	R-30	Spot	Med.	136	5⅜	50	400	850	1730	2000
75	R-30	Flood	Med.	136	5⅜	130	610	850	430	2000
150	R-40	Spot	Med.	165	6½	37	835	1825	7000	2000
150	R-40	Flood	Med.	165	6½	110	1550	1825	1200	2000
300	R-40	Spot	Med.	165	6½	35	1800	3600	13500	2000
300	R-40	Flood	Med.	165	6½	115	3000	3600	2500	2000
300	R-40[a]	Spot	Med.	173	6⅞	35	1800	3600	13500	2000
300	R-40[a]	Flood	Med.	173	6⅞	115	3000	3600	2500	2000
300	R-40[a]	Spot	Mog.	184	7¼	35	1800	3600	14000	2000
300	R-40[a]	Flood	Mog.	184	7¼	115	3000	3600	2500	2000
500	R-40[a]	Spot	Mog.	184	7¼	60	3300	6500	22000	2000
500	R-40[a]	Flood	Mog.	184	7¼	120	5700	6500	4750	2000
1000	R-60[a, g]	Spot	Mog.	257	10⅛	32	11500	18300	135000	3000
1000	R-60[a, g]	Flood	Mog.	257	10⅛	110	15500	18300	15500	3000
ER Lamps for Spotlighting and Floodlighting (Ellipsoidal Reflector)										
44	ER-30[b]	Light Inside Frost	Med.	162	6⅜	—	—	420	440	5000
50	ER-30	Light Inside Frost	Med.	162	6⅜	—	—	525	—	2000
67	ER-30[b]	Light Inside Frost	Med.	162	6⅜	—	—	700	1030	5000
75	ER-40	Light Inside Frost	Med.	162	6⅜	—	—	850	—	2000
90	ER-30[b]	Light Inside Frost	Med.	162	6⅜	—	—	950	1450	5000
90	ER-40	—	Med.	187	7⅜	—	—	—	1200	4000
120	ER-40	Light Inside Frost	Med.	187	7⅜	—	—	1475	—	2000
135	ER-40	—	Med.	187	7⅜	—	—	2050	—	4000
R Lamps for General Lighting[d] (Parabolic Reflector)										
500	R-52	Wide Beam	Mog.	300	11¾	90	—	7750	—	2000
750	R-52	Wide Beam	Mog.	300	11¾	110	—	13000	—	2000
1000	R-52[a]	Wide Beam	Mog.	300	11¾	110	—	16300	—	2000
1000	RB-52	Wide Beam	Mog.	322	12¹¹⁄₁₆	130	—	18900	—	2000
PAR Lamps for Spotlighting and Floodlighting[e] (Parabolic Reflector)										
55*	PAR-38	Spot	Med. Skt.	135	5⁵⁄₁₆	30 × 30	—	520	—	2000
55*	PAR-38	Flood	Med. Skt.	135	5⁵⁄₁₆	60 × 60	—	520	—	2000
65*	PAR-38	Spot	Med. Skt.	135	5⁵⁄₁₆	30 × 30	—	—	5100	2000
65*	PAR-38	Flood	Med. Skt.	135	5⁵⁄₁₆	60 × 60	—	—	1850	2000
65*	PAR-38	Spot	Med. Side Prong	110	4⁵⁄₁₆	30 × 30	—	—	—	2000
65*	PAR-38	Flood	Med. Side Prong	110	4⁵⁄₁₆	60 × 60	—	—	—	2000
75	PAR-38	Spot	Med. Skt.	135	5⁵⁄₁₆	30 × 30	465	750	3800	2000
75	PAR-38	Flood	Med. Skt.	135	5⁵⁄₁₆	60 × 60	570	750	1500	2000
80*	PAR-38	Spot	Med. Skt.	135	5⁵⁄₁₆	—	—	—	—	2000
80*	PAR-38	Flood	Med. Skt.	135	5⁵⁄₁₆	—	—	—	—	2000
85*	PAR-38	Spot	Med. Skt.	135	5⁵⁄₁₆	30 × 30	—	—	—	2000
85*	PAR-38	Flood	Med. Skt.	135	5⁵⁄₁₆	60 × 60	—	—	—	2000
90*	PAR-38	Spot	Med. Skt.	135	5⁵⁄₁₆	—	—	—	—	2000
90*	PAR-38	Flood	Med. Skt.	135	5⁵⁄₁₆	—	—	—	—	2000
90[g]	PAR-38	Flood	Med. Skt.	135	5⁵⁄₁₆	30 × 30	735	1450	3580	2000
100	PAR-38	Spot	Med. Skt.	135	5⁵⁄₁₆	30 × 30	—	1250	—	2000
100	PAR-38	Flood	Med. Skt.	135	5⁵⁄₁₆	60 × 60	—	1250	—	2000
120*	PAR-38	Spot	Med. Skt.	135	5⁵⁄₁₆	30 × 30	—	1420	12000	2000

See footnotes on page 51.   Continued on next page.

Fig. 20 *Continued*

Watts	Bulb	Description	Base (see Fig. 36)	Maximum Over-All Length (millimeters)	(inches)	Approximate Beam Spread (degrees)[c]	Approximate Beam Lumens	Approximate Total Lumens	Approximate Average Candle-power in Central 10 Degree Cone[f]	Rated Average Life (hours)
						PAR Lamps for Spotlighting and Floodlighting[a] (Parabolic Reflector)				
120[*]	PAR-38	Flood	Med. Skt.	135	5⁵⁄₁₆	60 × 60	—	1420	4250	2000
120[*]	PAR-38	Spot	Med. Side Prong	110	4⁵⁄₁₆	30 × 30	—	1420	—	2000
120[*]	PAR-38	Flood	Med. Side Prong	110	4⁵⁄₁₆	60 × 60	—	1420	—	2000
150	PAR-38[i]	Spot	Med. Skt.	135	5⁵⁄₁₆	30 × 30	1100	1735	11000	2000
150	PAR-38[i]	Flood	Med. Skt.	135	5⁵⁄₁₆	60 × 60	1350	1735	3700	2000
150	PAR-38	Spot	Med. Side Prong	110	4⁵⁄₁₆	30 × 30	1100	1735	11000	2000
150	PAR-38	Flood	Med. Side Prong	110	4⁵⁄₁₆	60 × 60	1350	1735	3700	2000
200	PAR-46	Narrow Spot	Med. Side Prong	102	4	17 × 23	1200	2325	32500[h]	2000
200	PAR-46	Med. Flood	Med. Side Prong	102	4	20 × 40	1300	2325	11200[h]	2000
250	PAR-38[g]	Spot	Med. Skt.	135	5⁵⁄₁₆	26 × 26	1600	3180	25000	4000
250	PAR-38[g]	Flood	Med. Skt.	135	5⁵⁄₁₆	60 × 60	2400	3180	6500	4000
300	PAR-56[i]	Narrow Spot	Mog. End Prong	127	5	15 × 20	1800	3750	70000[h]	2000
300	PAR-56[i]	Med. Flood	Mog. End Prong	127	5	20 × 35	2000	3750	24000[h]	2000
300	PAR-56[i]	Wide Flood	Mog. End Prong	127	5	30 × 60	2100	3750	10000[h]	2000
500	PAR-64	Narrow Spot	Extended Mog. End Prong	153	6	13 × 20	3000	6000	110000[h]	2000
500	PAR-64	Med. Flood		153	6	20 × 35	3400	6000	35000[h]	2000
500	PAR-64	Wide Flood		153	6	35 × 65	3500	6000	12000[h]	2000
500	PAR-56[g]	Narrow Spot	Mog. End. Prong	127	5	15 × 32	4900	7650	96000	4000
500	PAR-56[g]	Med. Flood	Mog. End. Prong	127	5	20 × 42	5700	7650	43000	4000
500	PAR-56[g]	Wide Flood	Mog. End. Prong	127	5	34 × 66	5725	7650	19000	4000
1000	PAR-64[g]	Narrow Spot	Extended Mog. End Prong	153	6	14 × 31	8500	19400	180000	4000
1000	PAR-64[g]	Med. Flood		153	6	22 × 45	10000	19400	80000	4000
1000	PAR-64[g]	Wide Flood		153	6	45 × 72	13500	19400	33000	4000

[*] High efficacy, reduced wattage.
[a] Heat-Resistant glass bulb.
[b] Krypton filled.
[c] To 10 per cent of maximum candlepower.
[d] Some of these types are also available for 230 to 260-volt circuits.
[e] All PAR lamps have bulbs of molded heat-resistant glass.
[f] Central cone defined as 5-degree cone for all spots and 10-degree cone for all floods.
[g] Halogen cycle lamps.
[h] Horizontal operation. May be slightly lower for vertical operation.
[i] Also available with an interference filter reflector.

Fig. 21 Three-Lite Lamps for 115-, 120-, and 125-Volt Circuits (For Base-Down Burning Only)

Watts	Bulb and Other Description	Base	Filament (see Fig. 34)	Rated Average Life (hours)	Maximum Over-all Length (millimeters)	(inches)	Light Center Length (millimeters)	(inches)	Approximate Initial Lumens	Approximate Initial Lumens Per Watt	Lamp Lumen Depreciation* (per cent)
15 135 150	A-21	3 contact Med.	C-2R/CC-8	1500 1200 1200	135	5⁵⁄₁₆	98	3⅞	85 2300 2385	5.7 17.0 15.9	
30 70 100	A-21 or T-19 white	3 contact Med.	C-8 or 2CC-6 or 2CC-8 or C-2R/CC-8	1200	135	5⁵⁄₁₆	95	3¾	290 1035 1315	9.7 14.8 13.2	86
50 100 150	A-23 or A-21 white	3 contact Med.	2CC-6 or 2CC-8 or C-2R/CC-8	1200	151	5¹⁵⁄₁₆	98	3⅞	600 1600 2200	12.0 16.0 14.7	85
50 100 150	PS-25 or T-19 white	3 contact Med.	2C-2R or 2CC-6 or 2CC-8	1200	151	5¹⁵⁄₁₆	98	3⅞	575 1450 2025	11.5 14.5 13.5	—
50 100 150	PS-25 inside frosted	3 contact Mog.	2C-2R or 2CC-8	1200	173	6¹³⁄₁₆	127	5	640 1630 2270	12.8 16.3 15.1	—
50 135 185	A-21	3 contact Med.	C-2R/CC-8	1500 1200 1200	112	4⁷⁄₁₆	79	3⅛	580 2330 2910	11.6 17.3 15.7	
100 200 300	PS-25 white	3 contact Mog.	2CC-6 or 2CC-8 or C-2R/CC-8	1200	173	6¹³⁄₁₆	105	4⅛	1360 3400 4760	13.6 17.0 15.9	72
50 200 250	PS-25, A-23 or T-21 white	3 contact Med.	2C-2R or 2CC-6 or 2CC-8 or C-2R/CC-8	1200	151	5¹⁵⁄₁₆	98	3⅞	600 3660 4210	12.0 18.3 16.8	72

* Per cent initial light output at 70 per cent of rated life.

Fig. 22 Energy Saving Fluorescent Lamps*

	Rapid Start						Preheat Start		Instant Start		
Nominal length											
Millimeters	900	1200	2400	2400	2400	1200	1200	1500	1200	2400	2400
Inches	36	48	96	96	96	48	48	60	48	96	96
Bulb	T-12 38-mm	T-12 38-mm	T-12 38-mm	T-12 38-mm	PG-17 54-mm	PG-17 54-mm	T-12 38-mm	T-17 54-mm	T-12 38-mm	T-8 25-mm	T-12 38-mm
Base	Med. Bipin	Med. Bipin	Recess D.C.	Recess D.C.	Recess D.C.	Recess D.C.	Med. Bipin	Mog. Bipin	Single Pin	Single Pin	Single Pin
Lamp amperes	0.453	0.45	0.81	1.58	1.57	1.53	0.45				0.44
Approx. lamp volts	64	84	134	137	144	64	84				153
Approx. lamp watts[a]	25	34–35[d]	95	185–195	185	95	34	82–84	30–32	40–41	60
Lamp watts replaced[a]	30	40	110	215	215	110	40	90	40	50	75
Rated life (hours)[b]	18000	20000	12000	10000–12000	12000	12000	15000	9000	9000	7500	12000
Initial lumens[c]											
Cool White	2000	2770	8300	13000	14000	6550	2800	6175	2550	3725	5500
Deluxe Cool White		1925	6000								4000
Warm White	2050	2820	8500		13000		2900				5700
Deluxe Warm White		1925	5720								3870
White		2820	8300					6400		3795	5600
Daylight		2300						5400			4800
Supermarket White		2050									4250
Royal White		2900	8800								5700
Lite-White		2925	8800	13800	14900					2700	5850
Lite White Deluxe		2925		8800						2700	5850
Warm Lite Deluxe		2925		8800						2700	5850
Regal White		2450									5100
Super Saver II		3050								2750	6000
3 K		2900	8800							2700	5800
Chroma 50											4100
Optima 50		2025									
Super White		2630									5220
Optima 32		2260									4200
Vita-Lite		2010									4015
Ultralume 83		2980									5860
Ultralume 84		2980									5860
Ultralume 85		2980									5860
Colortone 50											4100
Design 50		2250									
SP 30		2900									5950
SP 35		2900	8550								5800
SP 41		2850	8600								5800

* The life and light output ratings of fluorescent lamps are based on their use with ballasts that provide proper operating characteristics. Ballasts that do not provide proper electrical values may substantially reduce either lamp life or light output, or both.

[a] For rapid start lamps, includes watts for cathode heat.

[b] Rated life under specified test conditions with three hours per start. At longer burning intervals per start, longer life can be expected.

[c] At 100 hours. Where color is made by more than one manufacturer, lumens represent average of manufacturers.

[d] Also in 32-watt cathode-cutout, but with reduced life.

Note: All electrical and lumen values apply only under standard photometric conditions.

Fig. 23 Typical Hot-Cathode Fluorescent Lamps (Rapid Starting)[a]

	Circline				U-Shaped		Lightly Loaded Lamps						
Nominal length (millimeters)	165 dia.	210 dia.	300 dia.	400 dia.	600	600	900	1200	1200	1200	900	1200	1500
(inches)	6½ dia.	8¼ dia.	12 dia.	16 dia.	24	24	36	48	48	48	36	48	60
Bulb	T-9	T-9	T-10	T-10	T-12	T-12	T-12	T-12	T-12	T-10	T-8	T-8	T-8
Base	4-Pin	4-Pin	4-Pin	4-Pin	Med. Bipin	Med. Bipin	Med. Bipin	Med. Bipin	Med. Bipin	Med. Bipin	Med. Bipin	Med. Bipin	Med. Bipin
Leg spacing					92 mm (3⅝ in)	152 mm (6 in)							
Approx. lamp current (amperes)	0.38	0.37	0.425	0.415	0.42	0.43	0.43	0.43	0.43	0.42	0.265	0.265	0.265
Approx. lamp volts	48	61	81	108	103	100	81	101	101	104	100	135	172
Approx. lamp watts[f]	20	22.5	33	41.5	41	40.5	32.4	41	41	41	25	32	40
Rated life (hours)[b]	12000	12000	12000	12000	12000	12000	18000	20000	15000	24000	20000	20000	20000
Lamp lumen depreciation (LLD)[c]		72	82	77	84	84	81	84					
Initial lumens[d]													
Cool White	800	1065	1870	2580	2900	2935	2210	3150	3250	3200	*	*	*
Deluxe Cool White		875	1425	2000	2020	2065	1555	2200		2270			
Warm White	825	1065	1835	2550	2850	2965	2235	3175	3250	3250			
Deluxe Warm White	630	800	1375	1950	1980	2040	1505	2165					
White		1100	1870	2650	2850	2965	2255	3185	3250	3250			
Daylight		906	1550	2165			1900	2615	2650	2700			
Factors for Calculating Luminance[e]													
Candelas per Square Meter	11.5	8.4	4.6	3.3	2.65	2.65	3.58	2.65	2.65	3.18	5.42	4.03	3.18
Candelas per Square Foot	1.07	0.78	0.43	0.31	0.25	0.25	0.33	0.25	0.25	0.30	0.50	0.37	0.30

	Medium Loaded Lamps									
Nominal length (millimeters)	600	750	900	1050	1200	1500	1600	1800	2100	2400
(inches)	24	30	36	42	48	60	64	72	84	96
Bulb	T-12	T-12	T-12	T-12	T-12	T-12	T-12	T-12	T-12	T-12
Base	Recess D.C.	Recess D.C.	Recess D.C.	Recess D.C.	Recess D.C.	Recess D.C.	Recess D.C.	Recess D.C.	Recess D.C.	Recess D.C.
Approx. lamp current (amperes)	0.8	0.8	0.8	0.8	0.8	0.8	0.8	0.78	0.8	0.79
Approx. lamp volts	41		59		78	98		117	135	153
Approx. lamp watts[f]	37		50	42	63	75.5		87	100	113
Rated life (hours)[b]	9000	9000	9000	9000	12000	12000	12000	12000	12000	12000
Lamp lumen depreciation (LLD)[c]	77		77	77	82	82	82	82	82	82
Initial lumens[d]										
Cool White	1700	2290	2885	3516	4300	5400	5800	6650	7800	9150
Deluxe Cool White					3050			4550		6533
Warm White	1700				4300	5500		6500		9200
Deluxe Warm White										6475
White	1400		2476	3100	4300	4650	4900	6475	6867	9200
Daylight								5600		7800
Factors for Calculating Luminance[e]										
Candelas per Square Meter	5.54	4.35	3.58	3.04	2.65	2.1	1.95	1.7	1.44	1.24
Candelas per Square Foot	0.51	0.40	0.33	0.28	0.25	0.20	0.18	0.16	0.13	0.12

Nominal length (millimeters) / (inches) / Bulb	1200 48 T-10[a]	1800 72 T-10[a]	2400 96 T-10[a]	1200 48 T-12[i]	1500 60 T-12	1800 72 T-12[i]	2400 96 T-12[i]	Highly Loaded Lamps 1200 48 PG-17	1800 72 PG-17	2400 96 PG-17
Base	Recess D.C.	Recess D.C.	Recess D.C.	Recess D.C.	Recess D.C.	Recess D.C.	Recess D.C.	Recess D.C.	Recess D.C.	Recess D.C.
Approx. lamp current (amperes)	1.5	1.5	1.5	1.5	1.5	1.52	1.5	1.5	1.5	1.5
Approx. lamp volts	80	120	160[h]	84		125	163	84	125	163
Approx. lamp watts[f]	105[h]	150[h]	205[h]	116		168	215	116	168	215
Rated life (hours)[b]	9000	9000	9000	9000	9000	9000	9000	9000	9000	9000
Lamp lumen depreciation (LLD)[c]	66	66	66	69		72	72	69	69	69
Initial lumens[d]										
Cool White	6700[h]	10000[h]	14000[h]	6900	8950	10640	15250	7450	11500	16000
Deluxe Cool White	4690[h]	7000[h]	9800[h]	4900		7400	10750	5200		11200
Warm White				6700		10500	14650	7000		15000
White						10500	15000		9300	
Daylight				5700		9300	12650	6000		13300
Factors for Calculating Luminance[e]										
Candelas per Square Meter	3.18	2.04	1.49	2.65	2.1	1.7	1.24	1.91[j]	1.2[j]	0.89[j]
Candelas per Square Foot	0.30	0.19	0.14	0.25	0.20	0.16	0.12	0.18[j]	0.11[j]	0.08[j]

[a] The life and light output ratings of fluorescent lamps are based on their use with ballasts that provide proper operating characteristics. Ballasts that do not provide proper electrical values may substantially reduce either lamp life or light output, or both.

[b] Rated life under specified test conditions at three hours per start. At longer burning intervals per start, longer life can be expected.

[c] Per cent of initial light output at 70 per cent rated life at three hours per start. Average for cool white lamps. Approximate values.

[d] At 100 hours. Where lamp is made by more than one manufacturer, light output is the average of manufacturers. For the lumen output of other colors of fluorescent lamps, multiply the cool white lumens by the relative light output value from the table below.

[e] To calculate approximate lamp luminance, multiply the lamp lumens of the lamp color desired by the appropriate factor. Factors derived using method by E. A. Linsday, "Brightness of Cylindrical Fluorescent Sources", Illuminating Engineering, Vol. XXXIX, January, 1944, p. 23.

[f] Includes watts for cathode heat.

[g] A jacketed T-10 design is also available for use in applications where lamps are directly exposed to cold temperatures.

[h] Peak value. At 25 °C (77 °F) lumen and wattage values are lower.

[i] These lamps available in several variations (outdoor, low temperature, jacketed) with the same, or slightly different, ratings.

[j] Average luminance for center section of lamp. Parts of surface of lamp will have higher luminance.

Note: All electrical and lumen values apply only under standard photometric conditions.

[k] Initial Lumens for 3100 K and 4100 K color lamps (GTE Sylvania) are: 25-W, 2150; 32-W, 2900; and 40-W, 3650.

Approximate Light Output of Various Fluorescent Lamp Colors Compared to Cool White

Blue	0.40		Incandescent/Fluorescent Modern White Deluxe	0.55	(GTE Sylvania)
Green	1.38	A		0.77	(North American Philips Lamp)
Pink	0.37		Sign White	0.76	(GE)
Gold	0.72		Design White	0.78	(GTE Sylvania)
Red	0.06		Supermarket White	0.74	(North American Philips Lamp)
Natural*	0.67		Optima 32	0.76	(Duro-Test)
Soft White (new-3000 K)	0.69	B	Super White	0.93	(Duro-Test)
Cool Green	0.83		Turquoise	0.96	(Duro-Test)
			Super Deluxe 45	0.79	(Duro-Test)
			Daylight 65	0.65	(Duro-Test)
			Vita-Lite, Vita-Lux	0.69	(Duro-Test)
			Ultralume 83, 84, 85	0.92	(North American Philips Lamp)

Lite White Deluxe	1.05	(GTE Sylvania)
3 K	1.05	(GTE Sylvania)
Color 86	1.05	(Philips Electronics, Canada)
High CRI, 5000 K	0.70	
Chroma 50 (GE), Colortone 50 (North American Philips) and Color-Matcher 50, Color Classer 50, Optima 50, and Magnalux (Duro-Test).		
High CRI, 7500 K	0.67	
Chroma 75 (GE), Colortone 75 (North American Philips Lamp), and Color Classer 75, and Color-Matcher 75 (Duro-Test).		
Lite White† (GE, North American Philips, and GTE Sylvania)	1.08	
Royal White	1.03	(GTE Sylvania)
Regal White	0.90	(GE)

*Formerly, Soft White/Natural.

† Currently available in energy saving fluorescent lamps only. Consult the lamp manufacturers for the availability and lumen output of colors in other lamp sizes.
The colors in groups A and B are made by most lamp manufacturers in several lamp sizes. The balance of the lamp colors are generally made by just one lamp company and in just a few lamp sizes. Practically all colors are made in the 40-watt preheat start/rapid start size. Consult the lamp manufacturers for the availability and lumen output of colors in other lamp sizes.

Fig. 24 Compact Fluorescent Lamps

Manufacturer's Lamp Designation	Lamp Watts	Maximum Over-All Length Millimeters	Maximum Over-All Length Inches	Bulb	Base	Approx. Lamp Current Amperes	Approx. Lamp Voltage Volts	Rated Life[a]	Initial Lumens[h]	Lamp Lumen Depreciation (LLD)	Color Rendering Index
Dulux[e]	5	105	4⅛	T-4[k]	G23	0.18	40	10000	250	—	86
F7TT/SPX27[b] / F7TT/27K[d]	7	135	5⁵/₁₆	T-4[k]	G23	0.18	45	10000	400	83	84
Dulux[e] / PL7/82[c]	7	133	5¼	T-4[m]	G23	0.18	45	10000	400	—	82
F9TT/SPX27[b] / F9TT/27K[d]	9	167	6⁹/₁₆	T-4[k]	G23	0.23	59	10000	600	85	84
Dulux[e] / PL9/82[d]	9	165	6½	T-4[m]	G23	0.18	59	10000	600	—	82
Dulux-D[e]	9	116	4⁹/₁₆	T-4[k]	G24	0.185	66	10000	600	—	85
PL13/82[c]	13	188	7⁷/₁₆	T-4[m]	GX23	0.30	60	10000	900	—	82
F13TT/27K[d]	13	188	7⁷/₁₆	T-4[k]	GX23	0.30	60	10000	900	—	82
Dulux[e]	13	177	6³¹/₃₂	T-4[k]	GX23	0.30	60	10000	900	—	86
Dulux-D[e]	13	146	5¾	T-4[k]	G24	0.175	92	10000	900	—	85
2D[g]	16	135	5⁵/₁₆	T-4[q]		0.20	97	10000	1050	—	82
Dulux-D[e]	17	170	6¹¹/₁₆	T-4[k]	G24	0.24	106	10000	1250	—	85
SL-18[c]	18	183	7³/₁₆	n	Med.	0.24	—	7500	1100	—	81
Dulux-T5[e]	18	251	9⅞	T-5	TBA	0.38	57	10000	1250	—	85
FBU19T8[f]	19	222	8¾	T-8[o]	Med.	—	—	7500	p	—	—
Dulux T5[e]	24	362	14¼	T-5	TBA	0.34	91	10000	2000	—	85
Dulux-D[e]	25	190	7½	T-4[k]	G24	0.30	114	10000	1800	—	85
Dulux T5[e]	35	443	17⁷/₁₆	T-5	TBA	0.43	108	10000	2900	—	85

[a] At three hours per start.
[b] General Electric.
[c] North American Philips Lighting.
[d] GTE Sylvania.
[e] Osram.
[f] Interlectric.
[g] Thorn EMI.
[h] 2700K color.
[k] U-shaped molded with square corners.
[m] Twin-tube with small connection near top.
[n] U-shaped bent into another U-shape and with enclosing plastic housing.
[o] U-shaped band.
[p] Bent to appear as 2 joined D shapes.
[q] Special recessed 2-pin base.
[r] Cool white 665, Warm white 675 and Tru-lite 520.

Fig. 25 Typical Cold-Cathode Instant-Starting Fluorescent Lamps

Manufacturers' designation ANSI designation	48T8 25-millimeter 45-Inch			72T8 25-Millimeter 69-Inch			96T8 25-Millimeter 93-Inch			U6-96T8 (Hairpin)[d] ——		
Lamp length[a]												
Millimeters		1125			1725			2325			1125[d]	
Inches		45			69			93			45[d]	
Lamp base		Ferrule			Ferrule			Ferrule			Ferrule	
Bulb		25-mm, T-8			25-mm, T-8			25-mm, T-8			25-mm, T-8	
Open-circuit volts (rms) for starting												
type L.P.		450			600			750			750	
type H.P.		600			750			835			835	
Lamp current (milliamperes)[b]	120	150	200	120	150	200	120	150	200	120	150	200
Lamp voltage—lamp type L.P.	250	240	—	330	310	—	420	400	—	420	400	—
Lamp voltage—lamp type H.P.	270	250	240	350	330	310	450	425	400	450	425	400
Lamp power (watts)												
type L.P.	26	30	—	34	40	—	42	49	—	42	49	—
type H.P.	28	33	40	37	43	52	46	54	65	46	54	65
Initial Lumens												
Warm white[c]	1100	1300	1600	1700	2000	2350	2300	2700	3400	2300	2700	3400
White (3500 K)[c]	1050	1250	1550	1650	1900	2300	2250	2650	3300	2225	2650	3300
White (4500 K)[c]	1000	1200	1500	1600	1850	2200	2200	2600	3200	2200	2600	3200
Daylight[c]	950	1150	1450	1550	1800	2150	2150	2550	3100	2150	2550	3100
Luminance (candelas per square meter)												
Warm white	4040	4800	5860	4350	5140	6100	4490	5240	6610	4490	5240	6610
White (3500 K)	3940	4590	5690	4250	5040	5930	4385	5140	6440	4385	5140	6440
White (4500 K)	3770	4385	5480	4145	4930	5650	4280	5070	5820	4280	5070	5820
Daylight	3730	4210	5310	4080	4830	5480	4210	4970	5650	4210	4970	5650
Rated lamp life (thousands of hours)[e]												
type L.P.	15	12.5	—	15	12.5	—	18	16	—	18	16	—
type H.P.	25	20	15	25	20	15	30	25	20	30	25	20

[a] Length of lamp without lamp holders.   [b] Lamps can be operated up to 200 mA.
[c] Initial rating after 100 hours for types LP and HP. Other color lamps are available.
[d] Extended lamp length 2325 millimeters (93 inches) formed to U shape with 180-degree, 150-millimeter (6-inch) arc.
[e] Life not affected by number of starts.

Fig. 26 Typical Hot Cathode

Nominal lamp watts	4	6	8	13	14	15
Nominal length (millimeters)	150	225	300	525	375	450
(inches)	6	9	12	21	15	18
Bulb	T-5 (16 mm)	T-5 (16 mm)	T-5 (16 mm)	T-5 (16 mm)	T-12 (38 mm)	T-8 (25 mm)
Base (bipin)	min.	min.	min.	min.	med.	med.
Approx. lamp current (amperes)	0.17	0.16	0.145	0.165	0.38	0.305
Approx. lamp volts	29	42	57	95	40	55
Approx. lamp watts	4.5	6.0	7.2	13	14	14
Preheat curent-max. amperes	0.25	0.25	0.25	0.27	0.65	0.65
Preheat current-min. amperes	0.16	0.16	0.16	0.18	0.44	0.44
Rated life (hours)[b]	6000	7500	7500	7500	9000	7500
Lamp lumen depreciation (LLD)[c]	67	67	75	72	82	79
Initial lumens[d]						
Cool white	138	293	400	833	687	873
Deluxe Cool White	105	205	280	—	—	613
Warm White	127	290	400	880	710	880
Deluxe Warm White	85	180	275	580	476	610
White	138	275	405	880	710	880
Daylight	115	230	333	750	593	750
Factors for Calculating Luminance[e]						
Candelas per Square Meter	78.9	44.4	31.4	15.2	10.3	12.0

  [a] The life and light output ratings of fluorescent lamps are based on their use with ballasts that provide operating characteristics. Ballasts that do not provide proper electrical values may substantially reduce either lamp life or light output, or both.
  [b] Rated life under specified test conditions at three hours per start. At longer burning intervals per start, longer life can be expected.
  [c] Per cent of initial light output at 70 per cent rated life at three hours per start. Average for cool white lamps. Approximate values.
  [d] At 100 hours. Where lamp is made by more than one manufacturer, light output is the average of manufacturers. For the lumen output of other colors of fluorescent lamps, multiply the cool white lumens by the relative light output value from the table below Fig. 23

Fig. 27 Typical Hot-Cathode

Nominal length (millimeters)	1050	1600	1800	2400	1200	1500
(inches)	42	64	72	96	48	60
Bulb	T-6	T-6	T-8	T-8	T-12	T-17
Base	Single Pin	Single Pin	Single Pin	Single Pin	Med. Bipin	Mog. Bipin
Approx. lamp current (amperes)	0.200[f]	0.200[f]	0.200[f]	0.200[f]	0.425	0.425
Approx. lamp volts	150	233	220	295	104	107
Approx. lamp watts	25.5	38.5	38	51	40.5	42
Rated life (hours)[b]	7500	7500	7500	7500	7500–12000	7500–9000
Lamp lumen depreciation (LLD)[c]	76	77	83	89	83	89
Initial lumens[d]						
Cool White	1835	3000	3030	4265	3100	2900
Deluxe Cool White	1265	2100	2100	2910		2020
Warm White	1875	3050	3015	4215	3150	2940
Deluxe Warm White	1275	2100				1990
White	1900	2945	3050	4225	3150	2940
Daylight	1605	2600	2650	3525	2565	2410
Factors for Calculating Luminance[e]						
Candelas per Square Meter	6.1	3.93	2.55	1.87	2.65	1.5

  [a] The life and light output ratings of fluorescent lamps are based on their use with ballasts that provide proper operating characteristics. Ballasts that do not provide proper electrical values may substantially reduce either lamp life or light output, or both.
  [b] Rated life under specified test conditions at three hours per start. At longer burning intervals per start, longer life can be expected.
  [c] Per cent of initial light output at 70 per cent rated life at three hours per start. Average for cool white lamps. Approximate values.
  [d] At 100 hours. Where lamp is made by more than one manufacturer, light output is the average of manufacturers. For the lumen output of other colors of fluorescent lamps, multiply the cool white lumens by the relative light output value from the table below Fig. 23

Fluorescent Lamps (Preheat Starting)[a]

15 450 18	20 600 24	25 700 28	25 825 33	30 900 36	40 1200 48	90 1500 60
T-12 (38 mm) med.	T-12 (38 mm) med.	T-12 (38 mm) med.	T-12 (38 mm) med.	T-8 (25 mm) med.	T-12 (38 mm) med.	T-12 or T-17 mog.
0.325	0.38	0.46	0.46	0.355	0.43	1.5
47	57	63	61	99	101	65
14.5	20.5	25	25.5	30.5	40	90
0.65	0.65	0.95	0.95	0.65	0.75	2.2
0.44	0.44	0.41	0.41	0.40	0.55	1.45
9000	9000	7500	7500	7500	20000	9000
81	85	79	79	79	82	85
793	1270	1725	1915	2200	3150	6400
537	858	—	—	1555	2200	—
800	1310	—	1935	2235	3175	6350
532	848	—	—	1505	2165	5710
800	1290	—	1925	2255	3185	6350
650	1050	1450	1635	1900	2615	5525
8.0	5.5	4.68	3.93	5.4	2.65	2.1, 1.5[f]

[a] To calculate approximate lamp luminance, multiply the lamp lumens of the lamp color desired by the appropriate factor.
[f] First factor is for the T-12 lamp; second factor is for the T-17 lamp.
Note: All electrical and lumen values apply only under standard photometric conditions.

Fluorescent Lamps (Instant Starting)[a]

600 24 T-12	900 36 T-12	1050 42 T-12	1200 48 T-12	1500 60 T-12	1600 64 T-12	1800 72 T-12	2100 84 T-12	2400 96 T-12
Single Pin	Single Pin	Single Pin	Single Pin	Single Pin	Single Pin	Single Pin	Single Pin	Single Pin
0.425	0.425	0.425	0.425	0.425	0.425	0.425	0.425	0.425
53	77	88	100	123	131	149	172	197
21.5	30	34.5	39	48	50.5	57	66.5	75
7500–9000	7500–9000	7500–9000	7500–12000	7500–12000	7500–12000	7500–12000	7500–12000	12000
81	82	80	82	78	78	89	91	89
1065	2000	2350	3000	3585	3865	4650	5400	6300
			2065			3175		4465
1150	2050	2500	3000		3950	4675		6500
			2050			3200		4365
			3000			4700		6400
1010	1715	2015	2500	3135	3250	3850	4450	5425
5.54	3.58	3.04	2.65	2.1	1.95	1.7	1.44	1.24

[a] To calculate approximate lamp luminance, multiply the lamp lumens of the lamp color desired by the appropriate factor.
[f] These lamps can also be operated at 0.120 and 0.300 amperes.
Note: All electrical and lumen values apply only under standard photometric conditions.

Fig. 28   Fluorescent Lamp Power Requirements

Lamp Designation	Nominal Length (millimeters)	(inches)	Minimum Required rms Voltage across lamp for reliable starting [a]	Operating Current (milliperes)	Approximate Watts Consumed [f] Lamp	Single Lamp Circuit Ballast	Single Lamp Circuit Total	Two Lamp Circuit Ballast	Two Lamp Circuit Total
Preheat Start									
4T5	150	6	b	170	4.5	2	6.5		
6T5	225	9	b	160	6	2	8		
8T5	300	12	b	145	7.2	2	9.2		
15T8	450	18	b	305	15	4.5	19.5	9	39
14T12	375	18	b	380	14	5	19	5	33
15T12	450	18	b	325	14.5	4.5	19	9	38
20T12	600	24	b	380	20.5	5	25.5	10	41
25T12	825	33	b	460	25.5	6	31.5		
13T5	525	21	180	165	13	5	18		
30T8	900	36	176	355	30.5	10.5	41	17	78
40T12	1200	48	176	430	40	12	52	16	96
90T17	1500	60	132	1500	90	20	110	24	204
Instant Start with Bipin Bases									
40T12	1200	48	385	425	40.5	23.5	64	25	106
40T12	1500	60	385	425	42	24	66		
40T17	1500	60	385	425	42	24	66	26	110
with Single Pin Bases (Slimline Lamps)									
42T6d	1050	42	405	200	25.5	13	38.5	21	72
64T6d	1600	64	540	200	38.5	16	54.5	22	99
72T8d	1800	72	540	200	38	16	54	31	107
96T8d	2400	96	675	200	51	16	67		
48T12 (lead-lag)	1200	48	385	425	39			26	104
(Series)	1200	48	385	425	39			17	95
72T12 (lead-lag)	1800	72	475	425	57			47	161
(Series)	1800	72	475	425	57			25	139
96T12 (lead-lag)	2400	96	565	425	75			40	190
(Series)	2400	96	565	425	75			22	172
96T12	2400	96	565	425	75				160
96T12	2400	96	565	440	60				128
Rapid Start [c]									
Lightly Loaded									
20T9e	165	6½	150	380	19	11.5	30.5		
22T9e	210	8¼	180	370	22.5	11.5	34		
25T8	900	36	170	265	25	5	30	8	58
28T12j	1200	48	231	330	28				60
32T8	1200	48	200	265	32	5	37	7	71
32T10e	305	12	200	425	33	12	45		
40T10e	406	16	205	415	41.5	12.5	54		
30T12	900	36	150	430	33.5	10.5	44		
"32"T12j	1200	48			32				76
40T8	1500	60	250	265	40	6	46	11	91
40T10	1200	48	200	420	41	13	54	13	95
40T12	1200	48	200	430	41	13	54	13	95
40W-U3f	600	24	200	420	41	13	54	13	95
40W-U6f	600	24	200	430	40.5	13	53.5	13	94
Medium Loaded									
24T12	600	24	b	800	37		70		115
48T12	1200	48	155	800	63		85		146
72T12	1800	72	260	800	87		106		200
84T12	2100	84	280	800	100		119		213
96T12	2400	96	296	800	113		140		252
Highly Loaded									
48T12, 48PG17	1200	48	160	1500	116		146		252
72T12, 72PG17	1800	72	225	1500	168		213		326
96T12, 96PG17	2400	96	300	1500	215		260		450

Continued on following page.

Fig. 28 *Continued*

Lamp Designation	Nominal Length		Minimum Required rms Voltage across lamp for reliable starting[a]	Operating Current (milli-peres)	Approximate Watts Consumed[i]				
	(milli-meters)	(inches)			Lamp	Single Lamp Circuit		Two Lamp Circuit	
						Ballast	Total	Ballast	Total
Cold Cathode									
96T8LP	2400	96	750	120	42	9	51	17	101
				150	49				
				200	59	19	78	30	148
96T8HP	2400	96	835	120	46	12	58	24	115
				150	54			28	136
				200	65			30	160
Using Energy Saving Ballasts[g]									
Rapid Start[c]									
40T12	1200	48	200	430	40	7	47	6	86
40T12[h]	1200	48	200	460	34–35		41		74
Slimline									
96T12	2400	96	565	425	75				160
96T12[h]	2400	96	565	440	60				128
Rapid Start[c] Medium Loaded									
96T12	2400	96	296	800	113				237
96T12[h]	2400	96	296	840	95				207

[a] Between 10°C (50°F) and 43°C (110°F). The voltage shown is that required across one lamp. A ballast to operate two rapid start lamps in series will require an open circuit voltage roughly 40 per cent higher than shown here. Consult lamp manufacturer for information.
[b] Suitable for operation on 120-watt ac lines with series reactor as ballast.
[c] Requires starting aid.
[d] T-6 and T-8 slimline lamps also operate at 120 and 300 milliamperes.
[e] Circular lamp-dimension given is nominal outside diameter.
[f] U-shaped lamp.
[g] Energy saving CBM rated ballasts are designed to operate standard fluorescent lamps at CBM rated light output with lower ballast losses than the usual fluorescent lamp ballasts. The data shown here are typical for this type of ballast for the three most common fluorescent lamps and the energy saving fluorescent lamp that can be used as a replacement.
[h] Energy saving fluorescent lamp.
[i] Lamp watts and light output will vary with lamp temperature.
[j] Cathode-cutout type.

Fig. 29 Representative Fluorescent Lamp Ballast Factors

Lamp	Ballast		
	Standard	Low Loss	High Performance
4-Foor Rapid Start System (F40T12)			
Standard	0.95	0.95	0.97
Reduced Wattage	0.89	0.87	0.95
8-Foot Slimline Systems (F96T12)			
Standard	0.94	0.93	0.97
Reduced Wattage	0.87	0.85	0.96
8-Foot High Output Systems			
Standard	0.98	—	1.03
Reduced Wattage	0.93	—	0.98

Fig. 30 Typical High-Intensity Discharge Lamps*
A. Mercury Lamps

ANSI Lamp Designation	Manufacturer's Lamp Designation	ANSI Ballast Number	Lamp Watts	Approximate Ballast Watts	Approximate Lumens** (vertical) Initial (100 hours)	Mean	Bulb	Outer Bulb Finish	Maximum Over-All Length Millimeters	Inches	Light Center Length Millimeters	Inches	Arc Length Millimeters	Inches	Base	Average Rated Life (hours)	Lamp Voltage and Current
H45AY-40/50 DX	H40/50 DX 45-46*	45	40	7-12	1140	910	E17, ED17	Phos. coat	130	5⅛	79	3⅛	—	—	Med.	16000	A
H45/46DL-40/50DX	H45/46 DL-40/50/DX[c]	46	50		1575	1260	E17	Phos. coat	130	5⅛	79	3⅛	—	—	Med.	16000	A
H45AY-40/50DX	H45AY-40/50DX[f]	45	40		1350	1070	B21	Phos. coat	165	6½	95	3¾	—	—	Med.	16000	A
		46	50		1680	1330	B21	Phos. coat	165	6½	95	3¾	—	—	Med.	16000	A
H46DL-40-50/DX	H46DL-40-50/DX[b]	45	40	7-12	1140	910	B17	Phos. coat	130	5⅛	79	3⅛	—	—	Med.	24000+	A
		46	50		1575	1260	B17	Phos. coat	130	5⅛	79	3⅛	—	—	Med.	24000+	A
H46DL-40-50/N	H46DL-40-50/N[b]	45	40		800	580	B17	Phos. coat	130	5⅛	79	3⅛	—	—	Med.	24000+	A
		46	50		1100	830	B17	Phos. coat	130	5⅛	79	3⅛	—	—	Med.	24000+	A
H43AZ-75	H43AZ-75[b]	43	75	8.0	2800	2400	B21	Clear	165	6½	95	3¾	27	1¹/₁₆	Med.	24000	B
H43AV-75/DX	H75DX43*	43	75	8.0	3000	2430	B17,21 ED17	Phos. coat	165	6½	95	3¾	—	—	Med.	24000	B
	H43AV-75/DX[b,c]																
H43AV-75/N	H43AV-75/N[b]	43	75	8.0	2800	2200	B21, ED17	Phos. coat	165	6½	95	3¾	—	—	Med.	24000+	B
H43/44-75/100/PFL	H75/100PFL43-44/4*	43	75	8.0	1780	1165	PAR38	Clear	110	4¹¹/₃₂	—	—	—	—	Med. side prong	16000	B
	H43/44KL-75-100[b] H43/44-75/100)/PFL[c]	44	100	10-35	2585	1755	PAR38	Clear	110	4¹¹/₃₃	—	—	—	—		16000	B
H38AV-100/DX/N	H38AV-100/DX/N[c]	38	100	10-35	4400	3560	ED17	Phos. coat	165	6½	95	3¾	—	—	Med.	24000+	C
H38BM-100	H100RFL38-4* H38BM-100[b]	38-4	100	10-35	2850	2250	R40	I.F. Refl.	178-191	7-7½	—	—	—	—	Med.	24000+	C
H38BP-100/DX	H100RDXFL38-4* H38BP-100/DX[b]	38-4	100	10-35	2865	2300	R40	Phos. Refl.	178	7	—	—	—	—	Med.	24000+	C
H38BP-100/N	H38BP-100/N[b]	38-4	100	10-35	2450	1950	R40	Phos. Refl.	191	7½	—	—	—	—	Med.	24000+	C

H38HT-100	H100A38-4[a] H38HT-100[b]	38–4	100	10–35	4040	3415	BT25 or E23½	Clear	191	7½	127	5	29	1⅛	Mog.	24000+	C
H38JA-100/C	H38JA-100/c[c]	38–4	100	10–35	4100	3230	BT25	Phos. coat	191	7½	127	5		—	Mog.	24000	C
H38JA-100/DX	H100DX38-4[a]	38–4	100	10–35	4425	3620	BT25 or E23½	Phos. coat	191	7½	127	5		—	Mog.	24000+	C
	H38JA-100/DX[b,c]																
H38JA-100/N	H38JA-100/N[b,c]	38–4	100	10–35	4600	3700	BT25 ED23½	Phos. coat	191	7½	127	5		—	Mog. Med.	24000+	C
H38JA-100/WDX	H100WDX38-4[a]	38–4	100	10–35	4000	3100	BT25 or E23½	Phos. coat	191	7½	127	5		—	Mog.	24000+	D
	H38JA-100/WDX[d]																
H38LL-100	H100A38-4/A23[a] H38LL-100[b]	38–4	100	10–35	3900	3225	A23	Clear	138	5 7/16	89	3½	29	1⅛	Med.	24000	C
H38MP-100/DX	H100DX38-4/A23[a]	38–4	100	10–35	4275	3500	A23	Phos. coat	138	5 7/16	89	3½		—	Med.	24000	C
	H38MP-100/DX[b,c]																
H38MP-100/N	H38MP-100/N[b,c]	38–4	100	10–35	4195	3185	A23	Phos. coat	138	5 7/16	89	3½		—	Med.	24000+	C
H44GS100	H100PSP44-4[a]	44–4	100	10–35	2585	1820	PAR38	Refl. spot	138	5 7/16		—		—	Admed.	16000	C
	H44GS-100[b,c]																
H44JM100	H100PFL44-4[a]	44–4	100	10–35	2585	1820	PAR38	Refl. Flood	138	5 7/16		—		—	Admed.	16000	C
	H44JM-100[b,c]																
H44AB100	H100A4/T[a] H44AB100[b]	4	100	10–35	3725	2740	T10	Clear	143	5⅝	87	3 7/16		—	Admed.	12000	C
H39BM-175	H175RFL39-22[a] H39BM-175[b]	39–22	175	15–35	6030	5075	R40	I.F. Refl.	178–191	7–7½		—		—	Med.	24000+	D
H39BN-175/DX	H39BM-175/DX[c]	39–22	175	15–35	5800	4900	R40	Phos. Refl.	191	7½		—		—	Med.	24000+	D
H39BP-175/DX	H175RDXFL39-22[a]	39–22	175	15–35	5715	4685	R40	Phos. I.F. Refl.	178–191	7–7½		—		—	Med.	24000+	D
	H39BP-175/DX[b,c]																
H39BP-175/N	H39BP-175/N[b]	39–22	175	15–35	4600	3650	R40	Phos. I.F. Refl.	191	7½		—		—	Med.	24000+	D
H39BS-175	H175RFL39-22/M[a] H39BS-175[b]	39–22	175	15–35	5830	5075	R40	I.F. Refl.	178–191	7–7½		—		—	Mog.	24000+	D
H39BV-175/DX	H39BV-175/DX[b]	39–22	175	15–35	5725	4800	R40	Phos. Refl.	191	7½				—	Mog.	24000+	D

Fig. 30 —Continued

ANSI Lamp Designation	Manufacturer's Lamp Designation	ANSI Ballast Number	Lamp Watts	Approximate Ballast Watts	Approximate Lumens (vertical) Initial (100 hours)	Approximate Lumens (vertical) Mean	Bulb	Outer Bulb Finish	Maximum Over-All Length Millimeters	Maximum Over-All Length Inches	Light Center Length Millimeters	Light Center Length Inches	Arc Length Millimeters	Arc Length Inches	Base	Average Rated Life (hours)	Lamp Voltage and Current
H39KB-175	H175A39–22* / H39KB-175^b,c,g	39–22	175	15–35	7975	7430	BT28 or E28	Clear	211	8⁹⁄₁₆	127	5	51	2	Mog.	24000+	D
H39KC-175/C	H39KC-175/C^c,b	39–22	175	15–35	7850	7140	BT28	Phos. coat	211	8⁹⁄₁₆	127	5	—	—	Mog.	24000	D
H39KC-175/DX	H175DX39–22* / H39KC-175/DX^b,c,g	39–22	175	15–35	8600	7640	BT28 or E28	Phos. coat		8⁹⁄₁₆	127	5	—	—	Mog.	24000+	D
H39KC175/N	H39KC-175/N^b,c	39–22	175	15–35	8800	7830	BT28	Phos. coat	211	8⁹⁄₁₆	127	5	—	—	Mog.	24000+	D
H39KC175/WDX	H175WDX39–22*	39–22	175	15–35	7650	6600	BT28 or E28	Phos. coat	211	8⁹⁄₁₆	127	5	—	—	Mog.	24000+	D
H37FS-250/DX	H37FS-250/DX^b	37–5	250	25–35	8000		R60	Phos. Refl.	257	10⅛	—	—	—	—	Mog.	24000	E
H37KB-250	H250A37-5* / H37KB-250^b,c,g	37–5	250	25–35	11825	10625	BT28 or E28	Clear	211	8⁹⁄₁₆	127	5	57	2¼	Mog.	24000+	E
H37KC-250/C	H37KC-250/C^c,b	37–5	250	25–35	11850	10540	BT28	Phos. coat	211	8⁹⁄₁₆	127	5	—	—	Mog.	24000+	E
H37KC-250/DX	H250DX37-5* / H37KC-250/DX^b,c,g	37–5	250	25–35	12775	10790	BT28 or E28	Phos. coat	211	8⁹⁄₁₆	127	5	—	—	Mog.	24000+	E
H37KC-250/N	H37KC-250/N^b,c	37–5	250	25–35	12345	9335	BT28	Phos. coat	211	8⁹⁄₁₆	127	5	—	—	Mog.	24000+	E
H37KC-250/WDX	H37KC-250/WDX^e	37–5	250	25–35	10750	8950	BT28 or E28	Phos. coat	211	8⁹⁄₁₆	127	5	—	—	Mog.	24000+	E
H33CD-300	H33CD-300^b	33	300		14000	—	BT37	Clear	292	11½	178	7	—	—	Mog.	16000+	—
H33GL-300/DX	H33GL-300/DX^b	33	300		15700	—	BT37	Phos. coat	292	11½	178	7	—	—	Mog.	16000+	—
H33GL-300/N	H33GL-300/N^b	33	300		1300	—	BT37	Phos. coat	279	11	178	7	—	—	Mog.	16000+	—
H33AR-400	H400A33-1/T16* / H33AR-400*	33–1	400	20–55	19750	18200	T16	Clear	279	11	178	7	70	2¾	Mog.	12000	F
H33CD-400	H400A33-1* / H33CD-400^b,g	33–1	400	20–55	21000	19150	BT37 or E37	Clear	292	11½	178	7	70	2¾	Mog.	24000+	F
H33DN400/C	H33DN400/C^c	33–1	400	20–55	21000	18800	R57	Phos. Refl.	324	12¾	—	—			Mog.	24000	F

H33DN400/DX	H400RDX33-1[a] H33DN400/DX[b,c]	33–1	400	20–55	22670	19370	R52 or R57	Phos. Refl.	298	11¾ 12¾		—			Mog.	24000+	F
H33FP-400	H400RSP33-1[a] H33FP-400[d]	33–1	400	20–55	15300	12200	R60	Clear Refl.	257	10⅛		—			Mog.	24000+	F
H33FS-400/DX	H400RDXFL33-1[a] H33FS-400/DX[b,c]	33–1	400	20–55	15775	12810	R60	Phos. Refl.	257	10⅛ 10⅞		—			Mog.	24000+	F
H33FY-400	H400R33–1[a] H33FY-400[b,c]	33–1	400	20–55	18270	16670	R52 or R57	I.F. Refl.	298	11¾ 12¾		—			Mog.	24000+	F
H33GL-400/C	H33GL-400/C[c]	33–1	400	20–55	20500	18570	BT37	Phos. coat	292	11½	178	7			Mog.	24000+	F
H33GL-400/DX	H400DX33-1[a] H33GL-400/DX[b,c,f,g]	33–1	400	20–55	23125	19840	BT37 or E37	Phos. coat	292	11½	178	7			Mog.	24000+	F
H33GL-400/DX/BT	H400DX33-1/BT[a]	33–1	400	20–55	22100	18800	BT37	Phos. coat	287	11 5/16	178	7			Mog.	24000+	F
H33GL-400/N	H33GL-400/N[b,c]	33–1	400	20–55	23000	20000	BT37	Phos. coat	292	11½	178	7			Mog.	24000+	F
H33GL-400/WDX	H400WDX33-1[a]	33–1	400	20–55	21500	18000	BT37 or E37	Phos. coat	292	11½	178	7			Mog.	24000+	F
H33HS-400 H33LN400	H33HS-400[b] H33LN-400[b,c]	33–1 33–1	400 400	20–55 20–55	17500 16660	15200 15300	R57 R60	I.F. Refl. Clear Refl.	324 276	12¾ 10⅞		— —			Mog. Mog.	24000+ 24000+	F F
H35NA-700	H700A35-18[a] H35NA-700[b,c]	35–18	700	35–65	40500	36250	BT46	Clear	368	14½	241	9½	127	5	Mog.	24000+	G
H35ND-700/C	H35ND-700/C[c]	35–18	700	35–65	41000	36490	BT46	Phos. coat	368	14½	241	9½		—	Mog.	24000+	G
H35ND-700/DX	H700DX35-18[a] H35ND-700/DX[b,c]	35–18	700	35–65	42750	36045	BT46	Phos. coat	368	14½	241	9½		—	Mog.	24000+	G
H35ND-700/N	H35ND-700/N[b]	35–18	700	35–65	36400	29850	BT46	Phos. coat	368	14½	241	9½		—		24000+	G
H34GW-1000/C	H346W-1000/C[c]	34–12	1000	40–100	57500	49450	BT56	Phos. coat	391	15⅜	241	9½		—	Mog.	24000	H
H34GW-1000/DX	H1000DX34-12[a] H34GW-1000/DX[b,c]	34–12	1000	40–100	61670	47670	BT56	Phos. coat	391	15⅜	241	9½		—	Mog.	24000+	H
H34GV-1000	H1000A34-12[a] H34GV-1000[b,c,g]	34–12	1000	40–100	55330	48480	BT56	Phos. coat	391	15⅜	241	9½		—	Mog.	24000+	H
H36FA-1000/DX	H1000RDXFL36-15[a]	36–15	1000	40–90	48330	30100	R80	Phos. Refl.	352	13⅞		—		—	Mog.	24000+	I

Fig. 30—Continued

ANSI Lamp Designation[i]	Manufacturer's Lamp Designation[j]	ANSI Ballast Number	Lamp Watts	Approximate Ballast Watts	Approximate Lumens** (vertical) Initial (100 hours)	Mean	Bulb[h]	Outer Bulb Finish	Maximum Over-All Length Millimeters	Inches	Light Center Length Millimeters	Inches	Arc Length Millimeters	Inches	Base	Average Rated Life (hours)	Lamp Voltage and Current
H36FB-1000	H36FA-1000/DX[b] H1000RFL36-15[a] H36FB-1000[b]	36–15	1000	40–90	43500	33130	R80	I.F. Refl.	352	13 7/8	—	—		—	Mog.	24000+	—
H36HR-1000/DX	H1000RSDX36-15[a]	36–15	1000	40–90	43000	25200	R40	Semi Phos. Refl.	352	17 7/8	—	—		—	Mog.	24000+	—
H36GV-1000	H1000A36-15[a] H36GV-1000[b,c,g]	36–15	1000	40–90	56150	48400	BT56	Clear	391	15 3/8	241	9 1/2	152	6	Mog.	24000+	—
H36GW-1000/C	H36GW-1000/C[c]	36–15	1000	40–90	55000	46200	BT56	Phos. goat	391	15 3/8	241	9 1/2		—	Mog.	24000+	—
H36GW-1000/DX	H1000DX36-15[a] H36GW-1000/DX[b,c,g]	36–15	1000	40–90	63000	48380	BT56	Phos. coat	391	15 3/8	241	9 1/2		—	Mog.	24000+	—
H36GW-1000/N	H36GW-1000/N[b]	36–15	1000	40–90	53500	38000	BT56	Phos. coat	391	15 3/8	241	9 1/2		—	Mog.	24000+	—
H36GW-1000/R	H36GW-1000/R[b]	36–15	1000	40–90	63000	44700	BT56	Phos. coat	391	15 3/8	241	9 1/2		—	Mog.	24000+	—
H36GW-1000/WDX	H1000WDX36-15[a]	36–15	1000	40–90	58000	39440	BT56	Phos. coat	383	15 1/16	241	9 1/2		—	Mog.	24000+	—
H36KY-1000/DX	H1000RDX36-15[a] H36KY-1000/D[b]	36–15	1000	40–90	58250	38750	BT56	Semi. Phos. Refl.	391	15 3/8	241	9 1/2		—	Mog.	24000+	—

* The life and light output ratings of mercury lamps are based on their use with ballasts that provide proper operating characteristics. Ballasts that do not provide proper electrical values may substantially reduce either lamp life or light output, or both. Unless otherwise noted, ratings apply to operation in ac circuits.

** Average of manufacturers' rating.

[a] General Electric. [b] North American Philips Lighting. [c] GTE Sylvania. [d] Osram. [e] Westron. [f] Action Tungsram.

[g] Duro-Test (T versions only).

[h] All bulbs are of heat resistant glass.

[i] The basic ANSI designation adds an R in front of the wattage designation for non-self extinguishing mercury lamps and a T for self extinguishing mercury lamps to comply with Federal Standard #21 CFR 1040.30. The system does not distinguish between different bulb coatings for the same lamp type. Each manufacturer adds a suffix to identify the coating.

[j] The R or T designation is added after the H for manufacturer[a], and in front of the wattage for manufacturer[b,c,d,g].

Lamp Voltage and Current for Mercury Lamps

	Lamp Voltage (volts)			Lamp Current (amperes)	
	Minimum for starting at −29°C (−20°F)		Operating	Starting	Operating
	rms	peak			
A	180	255	90	0.80	0.53
B	225	320	130	0.92	0.64
C	225	320	130	1.3	0.85
D	225	320	130	2.5	1.5
E	225	320	130	3.1	2.1
F	225	320	135	5.0	3.2
G	375	530	265	5.0	2.8
H	300	425	135	12.0	8.0
I	375	375	265	6.5	4.0

Fig. 30 —*Continued*
B. Metal Halide Lamps*

ANSI Lamp Designation[p]	Manufacturer's Lamp Designation[p]	ANSI Ballast Number	Lamp Watts	Approximate Lumens — Initial (100 hours) Vert.	Initial Horiz.	Mean Vert.	Mean Horiz.	Bulb	Outer Bulb Finish	Max Over-All Length Millimeters	Inches	Light Center Length Millimeters	Inches	Arc Length Millimeters	Inches	Base	Rated Life (hours)	Lamp Voltage and Current
M85PX-70	HQI-TS70W/WDL•	M85	75		5000		4000	T-6.3	Clear	114	4½	57	2¼	7	9/32		15000	A
M81PS-150	HQI-TS150W/NDL•	M81	150		11250		9000	T-7.2	Clear	138	5⅞	69	2 23/32	18	23/32	Mog.	15000	A
M57PE-175/X	MXR175/BU,BD•[a,g]	M57	175	16600		13300		E-23½	Clear	197	7¾	127	5	—	—	Mog.	10000	A
M57PF-175/X	MXR175/C/BU,BD•[a,g]	M57	175	15750		12600		E-23½	Diffuse	197	7¾	127	5	—	—	Mog.q	10000	A
—	MS175/3K/HOR[h,c]	M57	175		14000		10500	BT-28	Phos. Coat	211	8 5/16	127	5	—	—	Mog.q	10000	A
M57PE-175	MV175/U[b,x] MH175/4[b,x,g] M175BU,BD-only[c,h,g]	M57	175	14000		10800	8300	BT-28 or E-28	Clear	211	8 5/16	127	5	25	1	Mog.	7500	A
M57PF-175	MV175/C/U[b,x] MH175/C[b,x,g] M175/C/BU,BD-only[c,x,g]	M57	175	14000		10200		BT-28 or E-28	Phos Coat	211	8 5/16	127	5	—	—	Mog.	7500	A
—	MS175/HOR[c,h]	M57	175		15000		12000	BT-28	Clear	211	8 5/16	127	5	—	—	Mog.q	10000	A
—	MS175/C/HOR[c,h]	M57	175		15000		11300	BT-28	Phos. Coat	211	8 5/16	127	5	—	—	Mog.q	10000	A
M58PG-250	MV250/U[b,x] MH250/U[p] M250U[c,x]	M58	250	20500	19500	17000	14000	BT-28 or E-28	Clear	211	8 5/16	127	5	36	1.4	Mog.	10000	B
M58PH-250	MV250/C/U[b,x] MH250/C/U[p] M250/C/U[c,x]	M58	250	20500	19500	16000	13500	BT-28 or E-28	Phos. Coat	211	8 5/16	127	5	—	—	Mog.	10000	B

Fig. 30—*Continued*

B. Metal Halide Lamps*

ANSI Lamp Designation[p]	Manufacturer's Lamp Designation[p]	ANSI Ballast Number	Lamp Watts	Approximate Lumens: Initial (100 hours) Vert.	Approximate Lumens: Initial (100 hours) Horiz.	Approximate Lumens: Mean Vert.	Approximate Lumens: Mean Horiz.	Bulb	Outer Bulb Finish	Maximum Over-All Length: Millimeters	Maximum Over-All Length: Inches	Light Center Length: Millimeters	Light Center Length: Inches	Arc Length: Millimeters	Arc Length: Inches	Base	Rated Life (hours)	Lamp Voltage and Current
M80PR-250[e]	HQI-TS250W/D[e]	M80	250		19000		13300	T-8	Clear	165	6½	83	3¼	27	1 1/16	FC2[r,s]	15000	C
M88PN-250	HQI-T250W/D[e]	M88	250		19000		13300	T-14½	Clear	220	8 21/32	150	5 29/32	27	1 1/16	Mog.	15000	C
M80PR-250	HQI-TS250W/NDL[e]	M80	250		20000		16000	T-8	Clear	165	6½	83	3¼	27	1 1/16	FC2[r,s]	15000	C
—	913[d]	S50	250		19000			T-14½	Clear	222	8¾	146	5¾	—	—	Mog.	10000	
—	HGMI250/D[w]	—	250	20000				T-14	Clear	320	12 19/32	150	5 29/32	—	—	Mog.	12000	
—	HGMIL250/D[w]	—	250	14000				E-28	Phos. Coat	227	8 15/16	—	—	—	—	Mog.	12000	
—	MS250/3K/HOR[c,h]	M58	250		21500		16500	BT-28	Phos. Coat	211	8 5/16	127	5	36	1 13/32	Mog[q]	10000	
—	MS250/HOR[c,h]	M58	250		23000		18000	BT-28	Clear	211	8 5/16	127	5	36	1 11/32	Mog.[q]	10000	
—	MS250/C/HOR[c,h]	M58	250		23000		17000	BT-28	Phos. Coat	211	8 5/16	127	5	36	1 11/32	Mog.[q]	10000	
—	MH300BU,BD/4[b]	M59	300	24000				BT-37	Clear	292	11½	178	7	—	—	Mog.	10000	—
—	MH300BU,BD/C[b]	M59	300	24000				BT-37	Phos. Coat	292	11½	178	7	—	—	Mog.	10000	—
—	MV325/I/U/WM[a,j]	M33[i]	325	28000	26000	18200	16400	E-37	Clear	178	7	287	11 5/16	27	1 1/16	Mog.	15000	E
—	MV325/C/I/U/WM[a,j]	H33[i]	325	28000	26000	17600	15800	E-37	Phos. Coat	178	7	287	11 5/16	27	1 1/16	Mog.	15000	E
M86PZ-360	HQI-T400W/DH/DV[e]	M86	360	28000	25000	19600	17500	T-14½	Clear	285	11 7/32	175	6 7/8	48	1 29/32	Mog.	15000	F
M86PY-360	HQI-TS400W/D[e]	M86	360		25000		17500	T-9½	Clear	206	8⅛	103	4 1/16	48	1 29/32	FC2[r,s]	15000	F
M59PJ-400	{ MV400/U[a,j] MH400/U[b,j] M400/U[c,n] }	M59	400	34000	32000	25600	22600	BT-37 or E-37	Clear	178	7	292	11½	33	1.3	Mog.	15-20000	D
M59PK-400	{ MV400/C/U[a,j] MH400/C[b] M400/C/U[c,n] }	M59	400	34000	32000	24600	22600	BT-37 or E-37	Phos. Coat	178	7	292	11½	—	—	Mog.	15-20000	D
M59PL-400	MH400/E[b]	M59	400	34000		25600		E-18	Clear	248	9¾	146	5¾	—	—	Mog.	10000	D
—	MS400/HOR-only[c,h]	M59	400	[h]	40000		32000	BT-37	Clear	292	11½	178	7	—	—	Mog.[q]	20000	D
—	MS400/C/HOR-only[c,h]	M59	400	[h]	40000		31000	BT-37	Phos. Coat	292	11½	178	7	—	—	Mog.[q]	20000	D
M59PJ-400	MS400/BD,BU-only[c]	M59	400	32000				BT-37 or E-37	Clear	292	11½	178	7	—	—	Mog.	15000	D
—	MS400/C/BD,BU-only[c]	M59	400	31000				BT-37 or E-37	Phos. Coat	292	11½	178	7	—	—	Mog.	15000	D
M59PJ-400/ VBD,VBU	MV400/VBD,VBU[a,g] MS400/BU/BD-only[c,t]	M59	400	40000		32000		BT-37 or E-37	Clear	287	11 5/16	178	7	33	1.3	Mog.	2000	D
M59PK-400/ VBD,VBU	{ MV400/C/VBD,VBU[a,g] MS400/C/BU/BD-only[c,t] }	M59	400	40000		31000		BT-37 or E-37	Phos. Coat	287	11 5/16	178	7	—	—	Mog.	20000	D

M59PJ-400/I/U	MV400/I/U[a,l]	H33M59	400	34000	32000	20400	18400	BT-37 or E-37	Clear	287	11 5/16	178	7	33	1.3	Mog.	15000	D
M59PK-400/I/U	MV400/C/I/U[a,l]	H33M59	400	34000	32000	19600	17600	BT-37 or E-37	Phos. Coat	287	11 5/16	178	7	—	—	Mog.	15000	D
—	MV400/BD,BU/I[a,g]	H33[m]M59	400	34000		26500		BT-37 or E-37	Clear	287	11 5/16	178	7	—	—	Mog.	15000	D
	MS400/3K/HOR-only[c]	M59	400		37000		28000	BT-37	Phos. Coat	292	11 1/2	178	7	—	—	Mog.[q]	20000	D
M59PK-400	MS400/3K/BU-only[c]	M59	400	37000		28000		BT-37	Phos. Coat	292	11 1/2	178	7	—	—	Mog.[q]	20000	D
	914[d]	S51	400		28000			T-14 1/2	Clear	250	9 7/8	146	5 3/4	—	—	Mog.	10000	
	M750/BU-only[c,t]		750	83500		66800		BT-37	Clear	292	11 1/2	178	7	—	—	Mog.	5000	
M47PA-1000	MV1000/U[a,f] MV1000/VBU,VBD[a,g] MH1000/U[b] M1000/U[c,n]	M47	1000	11000		88000		BT-56	Clear	391	15 3/8	241	9 1/2	89	3 1/2	Mog.	10-12000	G
M47PB-1000	MV1000/C/U[a,l] MH1000/C/U[b] M1000/C/U[c,h]	M47	1000	10500		82000		BT-56	Phos. Coat	391	15 3/8	241	9 1/2			Mog.	10-12000	G
M47PA-1000/ BD,BU/I	MV1000/BD,BU[a,l]	H36,M47	1000	115000		92000		BT-56	Clear	383	15 1/16	238	9 3/8	89	3 1/2	Mog.	10000	H
—	MS1000/BD,BU-only[c,g]	M47	1000	125000		100000		BT-56	Clear	391	15 3/8	241	9 1/2	—	—	Mog.	10000	H
—	MS1000/C/BD,BU-only[c,g]	M47	1000	125000		95800		BT-56	Phos. Coat	391	15 3/8	241	9 1/2	—	—	Mog.	10000	H
—	HQI-T1000W/D[e]		1000		90000		63000	T-25	Clear	340	13 1/4	220	8 5/8	70	2 3/4	Mog.	10000	I
M48PC-1500/-	MV1500/HBU/E,HBD/E[a] MH1500/BD,BU[b] M1500/BD,BU-HOR[c,k]	M48	1500	15500		142000		BT-56	Clear	391	15 3/8	214	9 1/2	89	3 1/2	Mog.	3000	J
—	MW1500T-7/7H[c,k]	—	1500		150000		130000	T-7	Clear	256	10 1/16			178	7	RSC[s]	3000	—
—	MB1L[f]	—	1500	—	140000			T-8	Frosted	256	10.1	128	5			RSC[s]	4000	K
M82PT-2000	HQI-T2000W/D[e]	M82	2000		170000		119000	T-32	Clear	430	19 5/16	260	10 1/4	105	4 1/8	E40[r,s]	3000	L
M82PW-3500	HQI-T3500W/D[e]	M84	3500		300000		210000	T-32	Clear	430	19 5/16	260	10 1/4	150	5 15/16	E40[r,s]	2500	M

* The life and light output ratings of metal halide amps are based on their use with ballasts that provide proper operating characteristics. Ballasts that do not provide proper electrical values may substantially reduce either lamp life or light output, or both. Unless otherwise noted, ratings apply to operation in ac circuits.

[a] General Electric. [b] North American Philips Lighting. [c] GTE Sylvania. [d] Duro-Test. [e] Osram. [f] Thorn. [w] Action Tungsram.

[g] ±15° of vertical operation and enclosed luminaires only.

[h] ±15° horizontal operation and enclosed luminaires only.

[i] CW/CWA mercury ballasts only.

[j] Minimum starting volts at −18°C (0°F) and crest factor = 2.0.

[k] Operation restricted to enclosed luminaires only.

[l] Open of enclosed luminaires when operated within ±15° vertical, all other enclosed.

[m] Do not use on CW or CWA mercury ballasts, consult lamp manufacturer.

[n] Enclosed operation when operated horizontal or ±60° of horizontal.

[o] Base up to 15° below horizontal only (HBU), base down to 15° above horizontal only (HBD).

Fig. 30—Continued
B. Metal Halide Lamps*

p The basic ANSI designation adds R in front of the wattage designation for self-extinguishing metal halide lamps and a T for self-extinguishing metal halide lamps to comply with Federal standard #21 CFR 1040.30.

q Position oriented mogul screw base.

r European base.

s Double ended construction.

t Operation restricted to ±15° of vertical only.

	Lamp Voltage (volts)			Lamp Current (amperes)	
	Minimum for starting at −29°C (−20°F)		Operating	Starting	Operating
	rms	peak			
A	382	540	130	1.8	1.4
B	382	540	130	2.8	2.1
C	198	3500	100	6.0	3.0
D	382	540	135	5.0	3.2
E	190^i	380^i	111	5.0	3.5
F	198	3500	120	7.0	3.5
G	530	750	250	6.0	4.3
H	440	622	250	6.0	4.3
I	230	4500	125	19.0	9.5
J	530	750	270	9.0	6.0
K	340	2000	250	—	6.7
L	380	4500	230	22.0	10.3
M	380	4500	220	38.0	18.0

Fig. 30—Continued
C. High Pressure Sodium Lamps

ANSI Lamp Designation	Manufacturer's Lamp Designation	ANSI Ballast Number	Lamp Watts*	Approximate Lumens		Bulb	Outer Bulb Finish	Maximum Over-All Length		Light Center Length		Base	Average Rated Life (hours)^j	Lamp Voltage and Current
				Initial	Mean^i			(millimeters)	(inches)	(millimeters)	(inches)			
S76HA-35	LU35/MED^{q,c,t} C35S76/M^p	S76	35	2250	2025	E-17, B-17 or ED-17	Clear	138	5^7/16	87	3^7/16	Med.	16000	A
S76HB-35	LU35/D/MED^{q,c,t} C35S76/D/M^p	S76	35	2150	1935	E-17, B-17 or ED-17	Diffuse	138	5^7/16	87	3^7/16	Med.	16000	A

S68MS-50	{LU50[a,c] C50S68[b]}	S68	50	4000	3600	BT-25 or E-23½	Clear	197	7¾	127	5	Mog.	24000	A
S68MT-50	{LU50/D[a] C50S68/D[b]}	S68	50	3800	3420	BT-25 or E-23½	Diffuse	197	7¾	127	5	Mog.	24000	A
—	{LU50/MED[a,c,t] C50S68/M[b]}	S68	50	4000	3600	E-17 or B-17	Clear	138	5 7/16	87	3 7/16	Med.	24000	A
—	{LU50/D/MED[a,c,t] C50S68/D/M[b]}	S68	50	3800	3420	E-17 or B-17	Diffuse	138	5 7/16	87	3 7/16	Med.	24000	A
S62ME-70	{LU70[a,c] C70S62[b]}	S62	70	6000	5500	BT-25 or E-23½	Clear	197	7¾	127	5	Mog.	24000	B
S62MF-70	{LU70/D[a,c] C70S62/D[b]}	S62	70	5800	5000	BT-25 or E-23½	Diffuse	197	7¾	127	5	Mog.	24000	B
—	{LU70/(MED)[c,t] C70S62/M[b]}	S62	70	6000	5500	E-17 or B-17	Clear	138	5 7/16	87	3 7/16	Med.	24000	B
—	C70S62/PAR[b]	S62	70	4500	—	PAR-38		110	4 5/16		—	Med. Side Prong	7500	B
—	C70S62/PAR/M[b]	S62	70	4500	—	PAR-38	—	135	5 5/16	—	—	Med.	7500	B
S62LF-70	{LU70/D/MED[c,t] C70S62/D/M[b]}	S62	70	5985	5390	B-17	Diffuse	140	5½	—	—	Med.	24000	B
S54SB-100	{LU100[a,c,t] C100S54[b]}	S54	100	9500	8550	BT-25 or E-23½	Clear	197	7¾	127	5	Mog.	24000	C
S54MC-100	{LU100/D[a,c,t] C100S54/D[b]}	S54	100	8800	7920	BT-25 or E-23½	Diffuse	197	7¾	127	5	Mog.	24000	C
—	{LU100/MED[c] C100S54/M[b]}	S54	100	9500	8550	B-17	Clear	140	5½	87	3 7/16	Med.	24000	C
—	{LU100/D/MED[c] C100S54/D/M[b]}	S54	100	8800	7920	B-17	Diffuse	140	5½	87	3 7/16	Med.	24000	C
S55SC-150	{LU150/55[a,c,t] C150S55[b]}	S55	150	16000	14400	BT-25 or E-23½	Clear	194	7⅝	127	5	Mog.	24000	D
S55MD-150	{LU150/55/D[a,t] C150S55/D[b]}	S55	150	15000	13500	BT-25 or E-23½	Diffuse	194	7⅝	127	5	Mog.	24000	D
S56SD-150	{C150S56[b] LU150/100[c]}	S56	150	16000	14400	BT-28 or E-28	Clear	211	8 5/16	127	5	Mog.	24000	E
S56SE-150	C150S56/D[b]	S56	150	15000	13500	BT-28	Diffuse	211	8 5/16	127	5	Mog.	24000	E
—	150 SONDL-E[e,f]	S56[i]	150	13000	11700	E-28	Diffuse	227	18 15/16	142	5⅝	Mog.	12000	E
—	150 SONDL-T[e,f]	S56[i]	150	13500	12150	T-15	Clear	210	8 3/16	127	5	Mog.	12000	E

Fig. 30—Continued
C. High Pressure Sodium Lamps

ANSI Lamp Designation	Manufacturer's Lamp Designation	ANSI Ballast Number	Lamp Watts*	Approximate Lumens		Bulb	Outer Bulb Finish	Maximum Over-All Length		Light Center Length		Base	Average Rated Life (hours)	Lamp Voltage and Current
				Initial	Mean'			(millimeters)	(inches)	(millimeters)	(inches)			
S63MG-150	{LUH150/BU,BD/EZ* ULX150[c] C150S63/EL[b]}	H39**	150	13000	11700	BT-28 or E-28	Clear	211	8 9/16	127	5	Mog.	12000–16000	F
S63MH-150	{LUH150/D/BU,BD/EZ* ULX150/D[c]}	H39**	150	11330	10000	BT-28 or E-28	Diffuse	211	8 9/16	127	5	Mog.	12000–16000	F
—	{LU150/55/MED[c] C150S55/M[b]}	S55	150	16000	14400	B-17	Clear	140	5 1/2	95	3 3/4	Med.	24000	D
—	{LU150/55/D/MED[c] C150S55/D/M[b]}	S55	150	15000	13500	B-17	Diffuse	140	5 1/2	95	3 3/4	Med.	24000	D
S66MN-200	{LU200[a,c] C200S66[b]}	S66	200	22000	19800	E-18	Clear	248	9 3/4	146	5 3/4	Mog.	24000	G
S65ML-215	{LUH215/BU,BD/EZ C215S65/EL[b] ULX215[c]}	H37**	215	19750	17775	BT-28 or E-28	Clear	229	9	127	5	Mog.	12000	H
S65MM-215	{LUH215/D/BU,BD/EZ LU215H/D[b]}	H37**	215	18250	16350	E-28	Diffuse	229	9	127	5	Mog.	12000	H
S50VA-250	{LU250[a,c,l] C250S50[b]}	S50	250	27500	24750	E-18	Clear	248	9 3/4	146	5 3/4	Mog.	24000	I
S50VA-250/S	{LU250S[a,l,c] C250S50/S[b]}	S50	250	30000	27000	E-18	Clear	248	9 3/4	146	5 3/4	Mog.	24000	I
S50VB-250	250 SON-TD[e]	S50	250	27000'	24300'	T-7	Clear	191	7 1/2	95	3 1/2	RSC[a]	24000	I
—	250 SONDL-E[e,l] 250 SONDL-T[e,l]	S50' S50'	250 250	23000 23500	20700 21150	E-28 T-15	Diffuse Clear	227 257	8 15/16 10 1/8	142 158	5 5/8 6 1/4	Mog. Mog.	12000 12000	— —
—	LU250/T7/RSC[c,a]	S50	250	25500	23400	T-7	Clear	256	10 1/16	130	5 1/8	RSC[a]	24000	I
S50VC-250	{LU250/D[a,c] C250S50/D/28[b]}	S50	250	26000	23400	BT-28 or E-28	Diffuse	229	9	127	5	Mog.	24000	I
S67MR-310	{LU310[c] C310S67[b]}	S67	310	37000	33300	E-18	Clear	248	9 3/4	146	5 3/4	Mog.	24000	J
S64MJ-360	{ULX360[c] C360S64/EL[b]}	H33**	360	38000	35000	BT-37 or E-37	Clear	292	11 1/2	178	7	Mog.	16000	I
S64MK-360	ULX360/D[c]	H33**	360	36000	32500	E-37	Diffuse	292	11 1/2	178	7	Mog.	16000	I

			Watts			Bulb	Finish					Base	Life	Color
S51WA-400	LU400^a,c,i / C400S51^b	S51	400	50000	45000	E-18	Clear	248	9¾	146	5¾	Mog.	24000	K
S51WB-400	LU400/D^a,c / C400S51/D^b	S51	400	47500	42750	BT-37 or E-37	Diffuse	292	11½	178	7	Mog.	24000	K
S51WC-400	400 SON-TD^e	S51	400	50000	45000	T-7	Clear	256	10 1/16	128	5	RSC*	24000	K
—	400 SONDL-E^a,f	S51^i	400	39000	35100	E-37	Diffuse	286	11¼	173	6 13/16	Mog.	12000	K
—	400 SONDL-T^a,f	S51^i	400	41000	36900	T-15	Clear	285	11¼	173	6 13/16	Mog.	12000	K
—	LU400/T7/RSC^c,k	S50	400	46000	41400	T-7	Clear	256	10 1/16	130	5⅛	RSC*	24000	—
—	C400S51/DE^b	S51	400	45000	40500	T-6	Clear	256	10 1/16	—	—	RSC*	24000	K
—	ULX880^c	H36**	880	102000	91800	E-25	Clear	383	15 1/16	241	9½	Mog.	16000	—
S52XB-1000	LU1000^a,c / C1000S52^b	S52	1000	140000	126000	E-25	Clear	383	15 1/16	222	8¾	Mog.	24000	L
S52XE-1000	1000 SON-TD^e	S52	1000	140000	126000	T-7	Clear	334	13⅜	167	6 9/16	RSC*	24000	L

* Actual lamp watts may vary depending on the ballast characteristic curve. Use only on ballasts that provide proper electrical values.
** Do not use on CW- or CWA-type mercury ballasts. Consult lamp manufacturer.
a General Electric. b North American Philips Lighting. c GTE Sylvania. e Thorn. f Action Tungsram.
f Deluxe color CRI ≥ 70.
g Horizontal operation ±20°.
i At 10 hours per start.
k Double ended construction.
l Ballasts are restricted to certain types; consult lamp manufacturer.

Lamp Voltage and Current for High Pressure Sodium Lamps

	Lamp Volts (Operating)	Ballast Open-Circuit Volts (minimum)†	Lamp Current (amperes)	
			Starting	Operating
A	52	110	1.85	1.18
B	52	110	2.4	1.6
C	55	110	3.2	2.1
D	55	110	5.0	3.3
E	100	—	—	1.8
F	130	—	—	1.5
G	100	195	3.5	2.3
H	130	—	2.1	—
I	100	195	4.5	3.0
J	100	190	5.5	3.6
K	100	195	7.0	4.7
L	250	456	7.0	4.7

† Requires a 2500- to 5000-volt pulse to start.

Fig. 31 Self-Ballasted Mercury Lamps Used for General Lighting

Watts	Bulb	Volts[h]	Bulb Finish	Approximate Lumens[i]		Lamp Current (amperes)[j]		Rated Life (hours)[j]	Base	Max. Over-All Length		Light Center Length	
				Initial (100 hours)	Mean	Starting	Operating			(millimeters)	(inches)	(millimeters)	(inches)
100	B-21[b,f,k]	Std.[l]	Phos. Coat	1650		1.1	.8	12000	Med.	159	6¼		
110	PAR-38[b,f,m]	Std.	I. M. Refl.			1.3	.9	12000	Med. Skt.	165	6½		
110	R-40[b,f,m]	Std.	I. M. Refl.			1.3	.9	12000	Med. Skt.	191	7½		
160	B-21[d]	Std.	Phos. Coat	2800		1.6	1.3	12000	Med.	178	7	122	4 13/16
	E-23[f,d]	Std.	Phos. Coat	2800–3125	1700	1.6	1.3	12000	Med.	173	6 13/16		
	R-40[b,f,m]	High	I. M. Refl.			.9	.6	12000	Med. Skt. or Mog.	178–191	7–7½		5½
	PS-30[a,b,f,g]	Std.	Phos. Coat	2100–3125		1.6	1.3	8–16000	Med.	184	7¼	140	4 13/16
	B-21[d]	High	Phos. Coat	3000		.9	.8	12000	Med.	178	7	122	
	E-23[b,f]	High	Phos. Coat	3520		.9	.8	12000	Med.	173	6 13/16		
	PAR-38[b,d,f,g]	Std.	Clear Flood	1970		1.6	1.3	12–14000	Med. Skt.	165	6½	100	3 15/16
	R-40[b,d,f,g]	Std.	I. F. Refl.	1850		1.6	1.3	12000	Med.	167	6 9/16		
250	E-28[a,b,d,f]	Std.	Phos. Coat	5000–5990		2.8	2.2	12000	Med. or Mog.	213	8⅜	146	5¾
	E-28[b,d,f]	High	Phos. Coat	5000–6450		1.3	1.1	12–14000	Med.	213	8⅜	146	5¾
	E-28[b,f,k]	High	Phos. Coat	6250		1.5	1.1	12000	Med. or Mog.	213	8⅜		
	PS-30[g]	Std.	Phos. Coat	4800		2.8	2.2	10–12000	Med.	205	8 1/16	152	6
	PS-35[b,f,g]	Std.	Phos. Coat	4700–5750		2.8	2.2	10–14000	Med. or Mog.	229–238	9–9⅜	165–178	6½–7
	PS-35[b,g]	High	Phos. Coat	5900		1.3	1.1	12–14000	Mog.	238	9⅜	178	7
	R-40[b,d,f,g]	Std.	I. F. Refl.	2650–3810		2.8	2.2	12–14000	Med. or Mog.	178	7		
300	E-28[b,f,m]	Std.	Phos. Coat	6750		3.5	2.9	14000	Med. Skt.	229	9		
	E-28[b,f,m]	Std.	Phos. Coat	6750		3.5	2.9	14000	Mog.	216	8½		
	R-40[b,f]	Std.	I. F. Refl.			3.5	2.9	14000	Med. Skt., or Mog.	178–191	7–7½		
	E-28[f]	High	Phos. Coat	7400		1.6	1.1	14000	Mog.	211	8 5/16		
450	BT-37[d]	Std.	Clear	9100	8280	6	4	16000	Mog.	292	11½	178	7
	BT-37[a,d]	Std.	Phos. Coat	8200–9500	6970–7125	6	4	16000	Mog.	290–292	11⅓–11½	178–187	7–7⅜
	BT-37[d,g]	High	Clear	9850–11800	9050–9700	3.5	2.3	14–16000	Mog.	292	11½	178	7
	BT-37[d,g]	High	Phos. Coat	8900–12200	7750–8500	3.5	2.3	14–16000	Mog.	292	11½	178	7
	E-37[b,f,m]	Std.	Phos. Coat	10400		5.5	3.8	16000	Mog.	260	10¼		
	E-37[b,f,m]	High	Phos. Coat	13000		3.4	2.2	16000	Mog.	260	10¼		
	PS-40[g]	Std.	Clear	9500		6	4	14–16000	Mog.	248	9¾	178	7
	PS-40[g]	Std.	Phos. Coat	10200		6	4	14–16000	Mog.	248	9¾	178	7
	PS-40[g]	High	Phos. Coat	12000–12200		3.5	2.3	14–16000	Mog.	248	9¾	178	7
	R-57[g]	Std.	Phos. Refl.			6	4	14–16000	Mog.	324	12¾	213	8⅜
	R-60[d,g]	Std.	I. F. Refl.	8170		6	4	14–16000	Mog.	260	10¼		
	R-60[g]	Std.	Phos. Refl.			6	4	14–16000	Mog.	260	10¼		
	R-60[g]	High	I. F. Refl.			3.5	2.3	14–16000	Mog.	260	10¼		
	R-60[g]	High	Phos. Refl.			3.5	2.3	14–16000	Mog.	260	10¼		
500	E-37[b,d,f]	Std.	Phos. Coat	10850–11780	8670	5.6	4.2	16000	Mog.	298	11¾	178	7
	E-37[b,d,f]	High	Phos. Coat	13590–14750	10880	3.3	2.2	16000	Mog.	286	11¼		
	R-57[d,f]	Std.	I. F. Refl.			5.6	4.2	16000	Mog.	324	12¾		
	R-57[b,f]	Std.	Phos. Refl.			5.6	4.2	16000	Mog.	324	12¾		

	R-57[b,f]	High	I. F. Refl.			3.3	2.2	16000	Mog.	257	10⅛		
	R-57[d]	High	Phos. Refl.			3.3	2.2	16000	Mog.	324	12¾		
	R-60[d]	Std.	I. F. Refl.	9160		5.6	4.2	16000	Mog.	257	10⅛		
	R-60[d]	Std.	Phos. Refl.	7800		5.6	4.2	16000	Mog.	257	10⅛		
750	BT-46[b,d,f]	Std.		18900	15120	8.8	6.6	16000	Mog.	368	14½		
	BT-46[b,f,m]	High	Phos. Coat	22000		4.8	3.4	16000	Mog.	368	14½		
	R-57[g]	Std.	Clear	18500		8.8	6.6	14–16000	Mog.	324	12¾	213	8⅜
	R-57[b,f,g]	Std.	I. F. Refl.	14000	11200	8.8	6.6	14–16000	Mog.	324	12¾		
	R-57[a,b,f,g]	Std.	Phos. Refl.	14000–19200	10500	8.8	6.3	14–16000	Mog.	324	12¾		
	R-57[g]	High	Clear	22200		5.1	3.6	18–20000	Mog.	324	12¾		
	R-57[b,f,g]	High	I. F. Refl.			5.1	3.6	18–20000	Mog.	324	12¾		
	R-57[a,b,f]	High	Phos. Refl.	14500–23000	12325	5.1	3.6	16–20000	Mog.	324	12¾		
	R-60[d]	Std.	I. F. Refl.	12120		8.8	6.6	16000	Mog.	257	10⅛		
	R-60[d]	Std.	Phos. Refl.	8160–11500		8.8	6.6	16000	Mog.	276	10⅞		
	R-60[d]	High	I. F. Refl.	12310–15000		5.1	3.6	16000	Mog.	257–276	10⅛–10⅞		
	R-60[d]	High	Phos. Refl.	8160–11250		5.1	3.6	16000	Mog.	257	10⅛		
1250	BT-56[d,g]	High	Clear	38000		10	6.4	14–16000	Mog.	381–391	15–15⅜	241	9½
	BT-56[b,d,f,g]	High	Phos. Coat	37780–41000	30230	10	6.4	14–16000	Mog.	381–391	15–15⅜	241	9½
	R-57[f]	High	I. F. Refl.			10	6.4	8000	Mog.	324	12¾		
	R-80[b,f]	High	I. F. Refl.			10	6.4	16000	Mog.	352	13⅞		

[a] General Electric.
[b] Public Service Lamp.
[d] North American Philips Lighting.
[f] Westron.
[g] Duro-Test.
[h] Std. = 120 volts ac nominal. High = 208, 220, 240 or 277 volts ac nominal.
[i] Values apply to lower voltage where two different voltage lamps were indicated.
[j] Life expectancy dependent on line voltage.
[k] For indoor use. Outdoors, use enclosed luminaire.
[l] Can be used in voltage range 115 volts to 130 volts.
[m] Hard glass. For indoor or outdoor use.
[o] Action Tungsram.

Fig. 32 Typical Low Pressure Sodium Lamps†

Manufacturers' Designations	ANSI Designations	Type**	Rated Watts	Nominal Volts	Lamp Current (amperes)	Bulb Diameter (millimeters)	Length (millimeters)	Initial Lumens*	Rated Average Life (hours)
SOX10W	L77RG-10	"U" Tube	10	55	0.20	40	150	1000	10000
SOX18W	L69RA-18	"U" Tube	18	57	0.35	54	216	1800	14000
SOX35W	L70RB-35	"U" Tube	35	70	0.60	54	311	4800	18000
SOX55W	L71RC-55	"U" Tube	55	109	0.59	54	425	8000	18000
SOX90W	L72RD-90	"U" Tube	90	112	0.94	68	528	13500	18000
SOX135W	L73RE-135	"U" Tube	135	164	0.95	68	775	22500	18000
SOX180W	L74RF-180	"U" Tube	180	240	0.91	68	1120	33000	18000

* After 100 burning hours.
** Linear types are still manufactured for replacement purposes.
† All have BY22d bases—an IEC designation for a specific D.C. bayonet base.

Fig. 33 Color and Color Rendering Characteristics of Common Light Sources.*

Test Lamp Designation	CIE Chromaticity Coordinates		Correlated Color Temperature (Kelvins)	CIE General Color Rendering Index	CIE Special Color Rendering Indices, R_i													
	x	y		R_a	R_1	R_2	R_3	R_4	R_5	R_6	R_7	R_8	R_9	R_{10}	R_{11}	R_{12}	R_{13}	R_{14}
Fluorescent Lamps																		
Warm White	.436	.406	3020	52	43	70	90	40	42	55	66	13	−111	31	21	27	48	94
Warm White Deluxe	.440	.403	2940	73	72	80	81	71	69	67	83	64	14	49	60	43	73	88
White	.410	.398	3450	57	48	72	90	47	49	61	68	20	−104	36	32	38	52	94
Cool White	.373	.385	4250	62	52	74	90	54	56	64	74	31	−94	39	42	48	57	93
Cool White Deluxe	.376	.368	4050	89	91	91	85	89	90	86	90	88	70	74	88	78	91	90
Daylight	.316	.345	6250	74	67	82	92	70	72	78	82	51	−56	59	64	72	71	95
Three-Component A	.376	.374	4100	83	98	94	48	89	89	78	88	82	32	46	73	53	95	65
Three-Component B	.370	.381	4310	82	84	93	66	65	28	94	83	85	44	69	62	68	90	76
Three-Component C	.448	.408	2857	80	78	96	75	61	81	87	89	74	18	89	51	69	79	80
Simulated D_{50}	.342	.359	5150	95	93	96	98	95	94	95	98	92	76	91	94	93	94	99
Simulated D_{55}	.333	.352	5480	98	99	98	96	99	99	98	98	96	91	95	98	97	98	98
Simulated D_{65}	.313	.325	6520	91	93	91	85	91	93	88	90	92	89	76	91	86	92	91
Simulated D_{70}	.307	.314	6980	93	97	93	87	92	97	91	91	94	95	82	95	93	94	93
Simulated D_{75}	.299	.315	7500	93	93	94	91	93	93	91	94	91	73	83	92	90	93	95
Mercury, clear	.326	.390	5710	15	−15	32	59	2	3	7	45	−15	−327	−55	−22	−25	−3	75
Mercury, improved color	.373	.415	4430	32	10	43	60	20	18	14	60	31	−108	−32	−7	−23	17	77
Metal Halide, clear	.396	.390	3720	60	52	84	81	54	60	83	59	5	−142	68	55	78	62	88
Xenon, high pressure arc	.324	.324	5920	94	94	91	90	96	95	92	95	96	81	81	97	93	92	95
High pressure sodium	.519	.418	2100	21	11	65	52	−9	10	55	32	−52	−212	45	−34	32	18	69
Low pressure sodium	.569	.421	1740	−44	−68	44	−2	−101	−67	29	−23	−165	−492	20	−128	−21	−39	31
DXW Tungsten Halogen	.424	.399	3190	100	100	100	100	100	100	99	100	100	100	99	100	100	100	100

* Lamps representative of the industry are listed. Variations from manufacturer to manufacturer are likely. A high positive value of R_i indicates a small color difference for sample *i*. A low value of R_i indicates a large color difference.

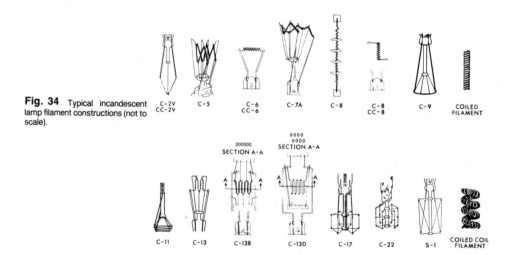

Fig. 34 Typical incandescent lamp filament constructions (not to scale).

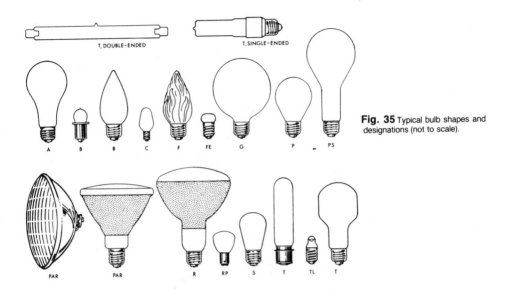

Fig. 35 Typical bulb shapes and designations (not to scale).

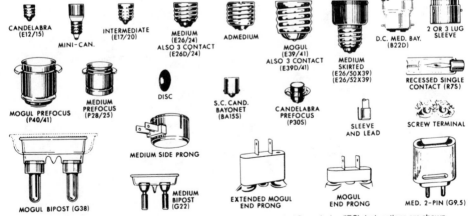

Fig. 36 Common lamp bases (not to scale). International Electrotechnical Commission (IEC) designations are shown, where available.

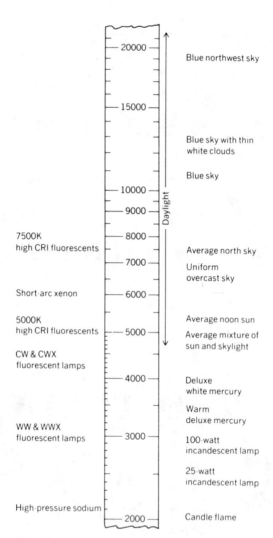

Fig. 37   Correlated color temperature in kelvins, of several electric light and daylight sources.

Reflectance
and
Transmittance Data

(From IES Lighting Handbook—*1984 Reference Volume*
and *1987 Application Volume*)

Fig. 38 Transmittance Data of Glass and Plastic Materials

Material	Approximate Transmittance (per cent)
Polished Plate/Float Glass	80–90
Sheet Glass	85–91
Heat Absorbing Plate Glass	70–80
Heat Absorbing Sheet Glass	70–85
Tinted Polished Plate	40–50
Figure Glass	70–90
Corrugated Glass	80–85
Glass Block	60–80
Clear Plastic Sheet	80–92
Tinted Plastic Sheet	90–42
Colorless Patterned Plastic	80–90
White Translucent Plastic	10–80
Glass Fiber Reinforced Plastic	5–80
Double Glazed—2 Lights Clear Glass	77
Tinted Plus Clear	37–45
Reflective Glass*	5–60

* Includes single glass, double glazed units and laminated assemblies. Consult manufacturer's material for specific values.

Fig. 39 Reflectances of Building Materials and Outside Surfaces

Material	Reflectance (per cent)	Material	Reflectance (per cent)
Bluestone, sandstone	18	Asphalt (free from dirt)	7
Brick		Earth (moist culti-	7
light buff	48	vated)	
dark buff	40	Granolite pavement	17
dark red glazed	30	Grass (dark green)	6
Cement	27	Gravel	13
Concrete	40	Macadam	18
Marble (white)	45	Slate (dark clay)	8
Paint (white)		Snow	
new	75	new	74
old	55	old	64
Glass		Vegetation (mean)	25
clear	7		
reflective	20–30		
tinted	7		

Fig. 40 Properties of Lighting Materials
(Per Cent Reflectance (R) and Transmittance (T) at Selected Wavelengths)

Material	Visible Wavelengths						Near Infrared Wavelengths						Far Infrared Wavelengths							
	400 nm		500 nm		600 nm		1000 nm		2000 nm		4000 nm		7000 nm		10,000 nm		12,000 nm		15,000 nm	
	R	T	R	T	R	T	R	T	R	T	R	T	R	T	R	T	R	T	R	T
Specular aluminum	87	0	82	0	86	0	97	0	94	0	88	0	84	0	27	0	16	0	14	0
Diffuse aluminum	79	0	75	0	84	0	86	0	95	0	88	0	81	0	68	0	49	0	44	0
White synthetic enamel	48	0	85	0	84	0	90	0	45	0	8	0	4	0	4	0	2	0	9	0
White porcelain enamel	56	0	84	0	83	0	76	0	38	0	4	0	2	0	22	0	8	0	9	0
Clear glass-3.2 millimeters (.125 inch)	8	91	8	92	7	92	5	92	23	90	2	0	0	0	24	0	6	0	5	0
Opal glass-3.9 millimeters (.155 inch)	28	36	26	39	24	42	12	59	16	71	2	0	0	0	24	0	6	0	5	0
Clear acrylic-3.1 millimeters (.120 inch)	7	92	7	92	7	92	4	90	8	53	3	0	2	0	2	0	3	0	3	0
Clear polystyrene-3.1 millimeters (.120 inch)	9	87	9	89	8	90	6	90	11	61	4	0	4	0	4	0	4	0	5	0
White acrylic-3.2 millimeters (.125 inch)	18	15	34	32	30	34	13	59	6	40	2	0	3	0	3	0	3	0	3	0
White polystyrene-3.1 millimeters (.120 inch)	26	18	32	29	30	30	22	48	9	35	3	0	3	0	3	0	3	0	4	0
White vinyl-0.76 millimeters (.030 inch)	8	72	8	78	8	76	6	85	17	75	3	0	2	0	3	0	3	0	3	0

Note: (a) Measurements in visible range made with General Electric Recording Spectrophotometer. Reflectance with black velvet backing for samples (b) Measurements at 1000 nm and 2000 nm made with Beckman DK2-R Spectrophotometer. (c) Measurements at wavelengths greater than 2000 nm made with Perkin-Elmer Spectrophotometer. (d) Reflectances in infrared relative to evaporative aluminum on glass.

Fig. 41 Reflecting and Transmitting Materials

Material	Reflectance * or Trans- mittance† (per cent)	Characteristics
	Reflecting	
Specular		
Mirrored and optical coated glass	80 to 99	Provide directional control of light and brightness at specific viewing angles. Effective as efficient reflectors and for special decorative lighting effects.
Metallized and optical coated plastic	75 to 97	
Processed anodized and optical coated aluminum	75 to 95	
Polished aluminum	60 to 70	
Chromium	60 to 65	
Stainless steel	55 to 65	
Black structural glass	5	
Spread		
Processed aluminum (diffuse)	70 to 80	General diffuse reflection with a high specular surface reflection of from 5 to 10 per cent of the light.
Etched aluminum	70 to 85	
Satin chromium	50 to 55	
Brushed aluminum	55 to 58	
Aluminum paint	60 to 70	
Diffuse		
White plaster	90 to 92	Diffuse reflection results in uniform surface brightness at all viewing angles. Materials of this type are good reflecting backgrounds for coves and luminous forms.
White paint**	75 to 90	
Porcelain enamel**	65 to 90	
White terra-cotta**	65 to 80	
White structural glass	75 to 80	
Limestone	35 to 65	
	Transmitting	
Glass		
Clear and optical coated	80 to 99	Low absorption; no diffusion; high concentrated transmission. Used as protective cover plates for concealed light sources.
Configurated, obscure, etched, ground, sandblasted, and frosted	70 to 85	Low absorption; high transmission; poor diffusion. Used only when backed by good diffusing glass or when light sources are placed at edges of panel to light the background.
Opalescent and alabaster	55 to 80	Lower transmission than above glasses; fair diffusion. Used for favorable appearance when indirectly lighted.
Flashed (cased) opal	30 to 65	Low absorption; excellent diffusion. used for panels of uniform brightness with good efficiency.
Solid opal glass	15 to 40	Higher absorption than flashed opal glass; excellent diffusion. Used in place of flashed opal where a white appearance is required.
Plastics		
Clear prismatic lens	70 to 92	Low absorption; no diffusion; high concentrated transmission. Used as shielding for fluorescent luminaires, outdoor signs and luminaires.
White	30 to 70	High absorption; excellent diffusion. Used to diffuse lamp images and provide even appearance in fluorescent luminaires.
Colors	0 to 90	Available in any color for special color rendering lighting requirements or esthetic reasons.
Marble (impregnated)	5 to 30	High absorption; excellent diffusion; used for panels of low brightness. Seldom used in producing general illumination because of the low efficiency.
Alabaster	20 to 50	High absorption; good diffusion. Used for favorable appearance when directly lighted.

* Specular and diffuse reflectance.
** These provide compound diffuse-specular reflection unless matte finished.
† Inasmuch as the amount of light transmitted depends upon the thickness of the material and angle of incidence of light, the figures given are based on thicknesses generally used in lighting applications and on near normal angles of incidence.

Fig. 42 Relationship Between Munsell Value
and Luminous Reflectance

Munsell Value	Luminous Reflectance* (per cent)
10.0	100.0
9.5	87.8
9.0	76.7
8.5	66.7
8.0	57.6
7.5	49.4
7.0	42.0
6.5	35.3
6.0	29.3
5.5	24.0
5.0	19.3
4.5	15.2
4.0	11.7
3.5	8.8
3.0	6.4
2.5	4.5
2.0	3.0
1.5	2.0
1.0	1.2
0	0

* Relative to a perfect diffuser.

Illuminance Selection

(From IES Lighting Handbook—*1987 Application Volume*)

Fig. 43 Currently Recommended Illuminance Categories and Illuminance Values for Lighting Design—
Targeted Maintained Levels

The tabulation that follows is a consolidated listing of the Society's current illuminance recommendations. The listing is intended to guide the lighting designer in selecting an appropriate illuminance for design and evaluation of lighting systems.

Guidance is provided in two forms: (1), in parts I, II and III as an *Illuminance Category*, representing a range of illuminances (see page 2-3* for a method of selecting a value within each illuminance range); and (2), in parts IV, V and VI as an Illuminance Value. Illuminance Categories are represented by letter designations A through I. Illuminance Values are given in *lux* with an approximate equivalence in *footcandles* and as such are intended as target (nominal) values with deviations expected. These target values also represent *maintained* values (see page 2-23*).

This table has been divided into the six parts for ease of use. Part I provides a listing of both Illuminance Categories and Illuminance Values for generic types of interior activities and normally is to be used when illuminance Categories cannot be found in parts II and III. Parts IV, V and VI provide target maintained Illuminance Values for outdoor facilities, sports and recreational areas, and transportation vehicles where special considerations apply as discussed on page 2-4.*

In all cases the recommendations in this table are based on the assumption that the lighting will be properly designed to take into account the visual characteristics of the task. See the design information in the particular application sections of this Application Handbook for further recommendations.

I. Illuminance Categories and Illuminance Values for Generic Types of Activities in Interiors

Type of Activity	Illuminance Category	Ranges of Illuminances		Reference Work-Plane
		Lux	Footcandles	
Public spaces with dark surroundings	A	20-30-50	2-3-5	
Simple orientation for short temporary visits	B	50-75-100	5-7.5-10	General lighting throughout spaces
Working spaces where visual tasks are only occasionally performed	C	100-150-200	10-15-20	
Performance of visual tasks of high contrast or large size	D	200-300-500	20-30-50	
Performance of visual tasks of medium contrast or small size	E	500-750-1000	50-75-100	Illuminance on task
Performance of visual tasks of low contrast or very small size	F	1000-1500-2000	100-150-200	
Performance of visual tasks of low contrast and very small size over a prolonged period	G	2000-3000-5000	200-300-500	
Performance of very prolonged and exacting visual task	H	5000-7500-10000	500-750-1000	Illuminance on task, obtained by a combination of general and local (supplementary lighting)
Performance of very special visual tasks of extremely low contrast and small size	I	10000-15000-20000	1000-1500-2000	

II. Commercial, Institutional, Residential and Public Assembly Interiors

Area/Activity	Illuminance Category	Area/Activity	Illuminance Category
Accounting (see **Reading**)		**Churches and synagogues** .. (see Fig. 45)[4]	
Air terminals (see **Transportation terminals**)		**Club and lodge rooms**	
Armories	C[1]	Lounge and reading	D
Art galleries (see **Museums**)		**Conference rooms**	
Auditoriums		Conferring...............................	D
Assembly	C[1]	Critical seeing (refer to individual task)	
Social activity	B	**Court rooms**	
Banks (also see **Reading**)		Seating area............................	C
Lobby		Court activity area	E[3]
General	C	**Dance halls and discotheques**	B
Writing area	D		
Tellers' stations	E[3]	**Depots, terminals and stations**	
Barber shops and beauty parlors	E	(see **Transportation terminals**)	

For footnotes, see page 102.
*In IES Lighting Handbook, *1987 Application Volume.* Also see Fig. 56 and Fig. 75 (inside back cover).

Fig. 43 *Continued*

II. *Continued*

Area/Activity	Illuminance Category	Area/Activity	Illuminance Category
Drafting		**Health care facilities**	
Mylar		Ambulance (local)	E
High contrast media; India ink, plastic		Anesthetizing	E
leads, soft graphite leads	E^3	Autopsy and morgue[17, 18]	
Low contrast media; hard graphite leads	F^3	Autopsy, general	E
Vellum		Autopsy table	G
High contrast	E^3	Morgue, general	D
Low contrast	F^3	Museum	E
Tracing paper		Cardiac function lab	E
High contrast	E^3	Central sterile supply	
Low contrast	F^3	Inspection, general	E
Overlays[5]		Inspection	F
Light table	C	At sinks	E
Prints		Work areas, general	D
Blue line	E	Processed storage	D
Blueprints	E	Corridors[17]	
Sepia prints	F	Nursing areas—day	C
		Nursing areas—night	B
Educational facilities		Operating areas, delivery, recovery, and	
Classrooms		laboratory suites and service	E
General (see **Reading**)		Critical care areas[17]	
Drafting (see **Drafting**)		General	C
Home economics (see **Residences**)		Examination	E
Science laboratories	E	Surgical task lighting	H
Lecture rooms		Handwashing	F
Audience (see **Reading**)		Cystoscopy room[17, 18]	E
Demonstration	F	Dental suite[17]	
Music rooms (see **Reading**)		General	D
Shops (see Part III, Industrial Group)		Instrument tray	E
Sight saving rooms	F	Oral cavity	H
Study halls (see **Reading**)		Prosthetic laboratory, general	D
Typing (see **Reading**)		Prosthetic laboratory, work bench	E
Sports facilities (see Part V, Sports		Prosthetic laboratory, local	F
and Recreational Areas)		Recovery room, general	C
Cafeterias (see **Food service facilities**)		Recovery room, emergency examination	E
Dormitories (see **Residences**)		Dialysis unit, medical[17]	F
		Elevators	C
Elevators, freight and passenger	C	EKG and specimen room[17]	
		General	B
Exhibition halls	C^1	On equipment	C
		Emergency outpatient[17]	
Filing (refer to individual task)		General	E
		Local	F
Financial facilities (see **Banks**)		Endoscopy rooms[17, 18]	
		General	E
Fire halls (see **Municipal buildings**)		Peritoneoscopy	D
		Culdoscopy	D
Food service facilities		Examination and treatment rooms[17]	
Dining areas		General	D
Cashier	D	Local	E
Cleaning	C	Eye surgery[17, 18]	F
Dining	B^6	Fracture room[17]	
Food displays (see **Merchandising spaces**)		General	E
Kitchen	E	Local	F
		Inhalation therapy	D
Garages—parking (see Fig. 55)		Laboratories[17]	
		Specimen collecting	E
Gasoline stations (see **Service stations**)		Tissue laboratories	F
		Microscopic reading room	D
Graphic design and material		Gross specimen review	F
Color selection	F^{11}		
Charting and mapping	F		
Graphs	E		
Keylining	F		
Layout and artwork	F		
Photographs, moderate detail	E^{13}		

For footnotes, see page 102. For illuminance ranges for each Illuminance Category, see page 87.

Fig. 43 *Continued*

II *Continued*

Area/Activity	Illuminance Category	Area/Activity	Illuminance Category
Chemistry rooms	E	Radiological suite[17]	
Bacteriology rooms		Diagnostic section	
General	E	General[18]	A
Reading culture plates	F	Waiting area	A
Hematology	E	Radiographic/fluoroscopic room	A
Linens		Film sorting	F
Sorting soiled linen	D	Barium kitchen	E
Central (clean) linen room	D	Radation therapy section	
Sewing room, general	D	General[18]	B
Sewing room, work area	E	Waiting area	B
Linen closet	B	Isotope kitchen, general	E
Lobby	C	Isotope kitchen, benches	E
Locker rooms	C	Computerized radiotomography section	
Medical illustration studio[17, 18]	F	Scanning room	B
Medical records	E	Equipment maintenance room	E
Nurseries[17]		Solarium	
General[18]	C	General	C
Observation and treatment	E	Local for reading	D
Nursing stations[17]		Stairways	C
General	D	Surgical suite[17]	
Desk	E	Operating room, general[18]	F
Corridors, day	C	Operating table	(see page 7-15) *
Corridors, night	A	Scrub room[18]	F
Medication station	E	Instruments and sterile supply room	D
Obstetric delivery suite[17]		Clean up room, instruments	E
Labor rooms		Anesthesia storage	C
General	C	Substerilizing room	C
Local	E	Surgical induction room[17, 18]	E
Birthing room	F[7]	Surgical holding area[17, 18]	E
Delivery area		Toilets	C
Scrub, general	F	Utility room	D
General	G	Waiting areas[17]	
Delivery table	(see page 7-19) *	General	C
Resuscitation	G	Local for reading	D
Postdelivery recovery area	E	**Homes** (see **Residences**)	
Substerilizing room	B	**Hospitality facilities** (see **Hotels, Food service facilities**)	
Occupational therapy[17]			
Work area, general	D	**Hospitals** (see **Health care facilities**)	
Work tables or benches	E	**Hotels**	
Patients' rooms[17]		Bathrooms, for grooming	D
General[18]	B	Bedrooms, for reading	D
Observation	A	Corridors, elevators and stairs	C
Critical examination	E	Front desk	E[3]
Reading	D	Linen room	
Toilets	D	Sewing	F
Pharmacy[17]		General	C
General	E	Lobby	
Alcohol vault	D	General lighting	C
Laminar flow bench	F	Reading and working areas	D
Night light	A	Canopy (see Part IV, Outdoor Facilities)	
Parenteral solution room	D	**Houses of worship** (see Fig. 45)	
Physical therapy departments		**Kitchens** (see **Food service facilities or Residences**)	
Gymnasiums	D		
Tank rooms	D	**Libraries**	
Treatment cubicles	D	Reading areas (see **Reading**)	
Postanesthetic recovery room[17]		Book stacks (vertical 760 millimeters (30 inches) above floor)	
General[18]	E	Active stacks	D
Local	H	Inactive stacks	B
Pulmonary function laboratories[17]	E		

For footnotes, see page 102. For illuminance ranges for each Illuminance Category, see page 87.
*In IES Lighting Handbook, *1987 Application Volume.*

Fig. 43 *Continued*

II. *Continued*

Area/Activity	Illuminance Category
Book repair and binding	D
Cataloging	D[3]
Card files	E
Carrels, individual study areas (see **Reading**)	
Circulation desks	D
Map, picture and print rooms (see **Graphic design and material**)	
Audiovisual areas	D
Audio listening areas	D
Microform areas (see **Reading**)	
Locker rooms	C
Merchandising spaces	
Alteration room	F
Fitting room	
Dressing areas	D
Fitting areas	F
Locker rooms	C
Stock rooms, wrapping and packaging	D
Sales transaction area (see **Reading**)	
Circulation	(see Fig. 47)[8]
Merchandise	(see Fig. 47)[8]
Feature display	(see Fig. 47)[8]
Show windows	(see Fig. 47)[8]
Motels (see Hotels)	
Minicipal buildings—fire and police	
Police	
Identification records	F
Jail cells and interrogation rooms	D
Fire hall	D
Museums	
Displays of non-sensitive materials	D
Displays of sensitive materials	(see Fig. 46)[2]
Lobbies, general gallery areas, corridors	C
Restoration or conservation shops and laboratories	E
Nursing homes (see Health care facilities)	
Offices	
Accounting (see **Reading**)	
Audio-visual areas	D
Conference areas (see **Conference rooms**)	
Drafting (see **Drafting**)	
General and private offices (see **Reading**)	
Libraries (see **Libraries**)	
Lobbies, lounges and reception areas	C
Mail sorting	E
Off-set printing and duplicating area	D
Spaces with VDTs	(see page 5-13) *
Parking facilities	(see Fig. 55)
Post offices (see Offices)	
Reading	
Copied tasks	
Ditto copy	E[3]
Micro-fiche reader	B[12, 13]
Mimeograph	D
Photograph, moderate detail	E[13]

Area/Activity	Illuminance Category
Thermal copy, poor copy	F[3]
Xerograph	D
Xerography, 3rd generation and greater	E
Electronic data processing tasks	
CRT screens	B[12, 13]
Impact printer	
good ribbon	D
poor ribbon	E
2nd carbon and greater	E
Ink jet printer	D
Keyboard reading	D
Machine rooms	
Active operations	D
Tape storage	D
Machine area	C
Equipment service	E[10]
Thermal print	E
Handwritten tasks	
#2 pencil and softer leads	D[3]
#3 pencil	E[3]
#4 pencil and harder leads	F[3]
Ball-point pen	D[3]
Felt-tip pen	D
Handwritten carbon copies	E
Non photographically reproducible colors	F[3]
Chalkboards	E[3]
Printed tasks	
6 point type	E[3]
8 and 10 point type	D[3]
Glossy magazines	D[13]
Maps	E
Newsprint	D
Typed originals	D
Typed 2nd carbon and later	E
Telephone books	E
Residences	
General lighting	
Conversation, relaxation and entertainment	B
Passage areas	B
Specific visual tasks [20]	
Dining	C
Grooming	
Makeup and shaving	D
Full-length mirror	D
Handcrafts and hobbies	
Workbench hobbies	
Ordinary tasks	D
Difficult tasks	E
Critical tasks	F
Easel hobbies	E
Ironing	D
Kitchen duties	
Kitchen counter	
Critical seeing	E
Noncritical	D
Kitchen range	
Difficult seeing	E
Noncritical	D[3]

For footnotes, see page 102. For illuminance ranges for each Illuminance Category, see page 87.
*In IES Lighting Handbook, *1987 Application Volume.*

Fig. 43 *Continued*

II. *Continued*

Area/Activity	Illuminance Category	Area/Activity	Illuminance Category
Kitchen sink		Safety . . . . (see Fig. 44)	
Difficult seeing	E		
Noncritical.	D	Schools (see Educational facilities)	
Laundry		Service spaces (see also Storage rooms)	
Preparation and tubs	D	Stairways, corridors	C
Washer and dryer	D	Elevators, freight and passenger	C
Music study (piano or organ)		Toilets and washrooms	C
Simple scores	D		
Advanced scores	E	Service stations	
Substand size scores	F	Service bays (see Part III, Industrial Group)	
Reading		Sales room (see Merchandising spaces)	
In a chair			
Books, magazines and newspapers	D	Show windows . . . . (see Fig. 47)	
Handwriting, reproductions and poor			
copies	E	Stairways (see Service spaces)	
In bed		Storage rooms (see Part III, Industrial Group)	
Normal	D		
Prolonged serious or critical	E	Stores (see Merchandising spaces and Show	
Desk		windows)	
Primary task plane, casual	D	Television . . . . (see Section 11) [*]	
Primary task plane, study	E		
Sewing		Theatre and motion picture	
Hand sewing		houses . . . . (see Section 11) [*]	
Dark fabrics, low contrast	F	Toilets and washrooms	C
Light to medium fabrics	E		
Occasional, high contrast	D	Transportation terminals	
Machine sewing		Waiting room and lounge	C
Dark fabrics, low contrast	F	Ticket counters	E
Light to medium fabrics	E	Baggage checking	D
Occasional, high contrast	D	Rest rooms	C
Table games	D	Concourse	B
Restaurants (see **Food service facilities**)		Boarding area	C

III. Industrial Group

Area/Activity	Illuminance Category	Area/Activity	Illuminance Category
		Mechanical	D
Aircraft maintenance . . . . (see Fig. 48)[21]		Hand	E
		Scales and thermometers	D
Aircraft manufacturing . . . . (see Fig. 48)[21]		Wrapping	D
Assembly		**Book binding**	
Simple	D	Folding, assembling, pasting	D
Moderately difficult	E	Cutting, punching, stitching	E
Difficult	F	Embossing and inspection	F
Very difficult	G		
Exacting	H	**Breweries**	
		Brew house	D
Automobile manufacturing . . . . (see Fig. 49)[21]		Boiling and keg washing	D
Bakeries		Filling (bottles, cans, kegs)	D
Mixing room	D		
Face of shelves	D	**Building construction** (see Part IV, Outdoor	
Inside of mixing bowl	D	Facilities)	
Fermentation room	D		
Make-up room		**Building exteriors** (see Part IV, Outdoor Facilities)	
Bread	D		
Sweet yeast-raised products	D	**Candy making**	
Proofing room	D	Box department	D
Oven room	D	Chocolate department	
Fillings and other ingredients	D	Husking, winnowing, fat extraction, crushing	
Decorating and icing		and refining, feeding	D

For footnotes, see page 102. For illuminance ranges for each Illuminance Category, see page 87.
*In IES Lighting Handbook, *1987 Application Volume.*

Fig. 43 *Continued*

III. *Continued*

Area/Activity	Illuminance Category	Area/Activity	Illuminance Category
Bean cleaning, sorting, dipping, packing, wrapping	D	Examining (perching)	I
Milling	E	Sponging, decating, winding, measuring	D
Cream making		Piling up and marking	E
Mixing, cooking, molding	D	Cutting	G
Gum drops and jellied forms	D	Pattern making, preparation of trimming, piping, canvas and shoulder pads	E
Hand decorating	D	Fitting, bundling, shading, stitching	D
Hard candy		Shops	F
Mixing, cooking, molding	D	Inspection	G
Die cutting and sorting	E	Pressing	F
Kiss making and wrapping	E	Sewing	G
Canning and preserving		**Control rooms** (see **Electric generating stations—interior**)	
Initial grading raw material samples	D		
Tomatoes	E	**Corridors** (see **Service spaces**)	
Color grading and cutting rooms	F	**Cotton gin industry**	
Preparation		Overhead equipment—separators, driers, grid cleaners, stick machines, conveyers, feeders and catwalks	D
Preliminary sorting			
Apricots and peaches	D	Gin stand	D
Tomatoes	E	Control console	D
Olives	F	Lint cleaner	D
Cutting and pitting	E	Bale press	D
Final sorting	E	**Dairy farms** (see **Farms**)	
Canning		**Dairy products**	
Continuous-belt canning	E	Fluid milk industry	
Sink canning	E	Boiler room	D
Hand packing	D	Bottle storage	D
Olives	E	Bottle sorting	E
Examination of canned samples	F	Bottle washers	E[22]
Container handling		Can washers	D
Inspection	F	Cooling equipment	D
Can unscramblers	E	Filling: inspection	E
Labeling and cartoning	D	Gauges (on face)	E
Casting (see **Foundries**)		Laboratories	E
		Meter panels (on face)	E
Central stations (see **Electric generating stations**)		Pasteurizers	D
		Separators	D
Chemical plants (see **Petroleum and chemical plants**)		Storage refrigerator	D
		Tanks, vats	
Clay and concrete products		Light interiors	C
Grinding, filter presses, kiln rooms	C	Dark interiors	E
Molding, pressing, cleaning, trimming	D	Thermometer (on face)	E
Enameling	E	Weighing room	D
Color and glazing—rough work	E	Scales	E
Color and glazing—fine work	F	**Dispatch boards** (see **Electric generating stations—interior**)	
Cleaning and pressing industry			
Checking and sorting	E	**Dredging** (see Part IV, Outdoor Facilities)	
Dry and wet cleaning and steaming	E		
Inspection and spotting	G	**Electrical equipment manufacturing**	
Pressing	F	Impregnating	D
Repair and alteration	F	Insulating: coil winding	E
Cloth products		**Electric generating stations—interior** (see also **Nuclear power plants**)	
Cloth inspection	I		
Cutting	G	Air-conditioning equipment, air preheater and fan floor, ash sluicing	B
Sewing	G		
Pressing	F	Auxiliaries, pumps, tanks, compressors, gauge area	C
Clothing manufacture (men's) (see also **Sewn Products**)			
Receiving, opening, storing, shipping	D		

For footnotes, see page 102. For illuminance ranges for each Illuminance Category, see page 87.

Fig. 43 *Continued*

III. *Continued*

Area/Activity	Illuminance Category
Battery rooms	D
Boiler platforms	B
Burner platforms	C
Cable room	B
Coal handling systems	B
Coal pulverizer	C
Condensers, deaerator floor, evaporator floor, heater floors	B
Control rooms	
Main control boards	D[23]
Auxiliary control panels	D[23]
Operator's station	E[23]
Maintenance and wiring areas	D
Emergency operating lighting	C
Gauge reading	D
Hydrogen and carbon dioxide manifold area	C
Laboratory	E
Precipitators	B
Screen house	C
Soot or slag blower platform	C
Steam headers and throttles	B
Switchgear and motor control centers	D
Telephone and communication equipment rooms	D
Tunnels or galleries, piping and electrical	B
Turbine building	
Operating floor	D
Below operating floor	C
Visitor's gallery	C
Water treating area	D

Electric generating stations—exterior (see Part IV, Outdoor Facilities)

Elevators (see **Service spaces**)

Explosives manufacturing

Area/Activity	Illuminance Category
Hand furnaces, boiling tanks, stationary driers, stationary and gravity crystallizers	D
Mechanical furnace, generators and stills, mechanical driers, evaporators, filtration, mechanical crystallizers	D
Tanks for cooking, extractors, percolators, nitrators	D

Farms—dairy

Area/Activity	Illuminance Category
Milking operation area (milking parlor and stall barn)	
General	C
Cow's udder	D
Milk handling equipment and storage area (milk house or milk room)	
General	C
Washing area	E
Bulk tank interior	E
Loading platform	C
Feeding area (stall barn feed alley, pens, loose housing feed area)	C
Feed storage area—forage	
Haymow	A
Hay inspection area	C
Ladders and stairs	C

Area/Activity	Illuminance Category
Silo	A
Silo room	C
Feed storage area—rain and concentrate	
Grain bin	A
Concentrate storage area	B
Feed processing area	B
Livestock housing area (community, maternity, individual calf pens, and loose housing holding and resting areas)	B
Machine storage area (garage and machine shed)	B
Farm shop area	
Active storage area	B
General shop area (machinery repair, rough sawing)	D
Rough bench and machine work (painting, fine storage, ordinary sheet metal work, welding, medium benchwork)	D
Medium bench and machine work (fine woodworking, drill press, metal lathe, grinder)	E
Miscellaneous areas	
Farm office (see **Reading**)	
Restrooms (see **Service spaces**)	
Pumphouse	C

Farms—poultry (see **Poultry industry**)

Flour mills

Area/Activity	Illuminance Category
Rolling, sifting, purifying	E
Packing	D
Product control	F
Cleaning, screens, man lifts, aisleways and walkways, bin checking	D

Forge shops .... E

Foundries

Area/Activity	Illuminance Category
Annealing (furnaces)	D
Cleaning	D
Core making	
Fine	F
Medium	E
Grinding and chipping	F
Inspection	
Fine	G
Medium	F
Molding	
Medium	F
Large	E
Pouring	E
Sorting	E
Cupola	C
Shakeout	D

Garages—service

Area/Activity	Illuminance Category
Repairs	E
Active traffic areas	C
Write-up	D

Glass works

Area/Activity	Illuminance Category
Mix and furnace rooms, pressing and lehr, glass-blowing machines	C

For footnotes, see page 102. For illuminance ranges for each Illuminance Category, see page 87.

Fig. 43 *Continued*

III. *Continued*

Area/Activity	Illuminance Category
Grinding, cutting, silvering	D
Fine grinding, beveling, polishing	E
Inspection, etching and decorating	F
Glove manufacturing	
Pressing	G
Knitting	F
Sorting	F
Cutting	G
Sewing and inspection	G
Hangars (see **Aircraft manufacturing**)	
Hat manufacturing	
Dyeing, stiffening, braiding, cleaning, refining	E
Forming, sizing, pouncing, flanging, finishing, ironing	F
Sewing	G
Inspection	
Simple	D
Moderately difficult	E
Difficult	F
Very difficult	G
Exacting	H
Iron and steel manufacturing (see Fig. 51)[21]	
Jewelry and watch manufacturing	G
Laundries	
Washing	D
Flat work ironing, weighing, listing, marking	D
Machine and press finishing, sorting	E
Fine hand ironing	E
Leather manufacturing	
Cleaning, tanning and stretching, vats	D
Cutting, fleshing and stuffing	D
Finishing and scarfing	E
Leather working	
Pressing, winding, glazing	F
Grading, matching, cutting, scarfing, sewing	G
Loading and unloading platforms (see Part IV, Outdoor Facilities)	
Locker rooms	C
Logging (see Part IV, Outdoor Facilities)	
Lumber yards (see Part IV, Outdoor Facilities)	
Machine shops	
Rough bench or machine work	D
Medium bench or machine work, ordinary automatic machines, rough grinding, medium buffing and polishing	E
Fine bench or machine work, fine automatic machines medium grinding, fine buffing and polishing	G
Extra-fine bench or machine work, grinding, fine work	H
Materials handling	
Wrapping, packing, labeling	D
Picking stock, classifying	D
Loading, inside truck bodies and freight cars	C

Area/Activity	Illuminance Category
Meat packing	
Slaughtering	D
Cleaning, cutting, cooking, grinding, canning, packing	D
Nuclear power plants (see also **Electric generating stations**)	
Auxiliary building, uncontrolled access areas	C
Controlled access areas	
Count room	E[23]
Laboratory	E
Health physics office	F
Medical aid room	F
Hot laundry	D
Storage room	C
Engineered safety features equipment	D
Diesel generator building	D
Fuel handling building	
Operating floor	D
Below operating floor	C
Off gas building	C
Radwaste building	D
Reactor building	
Operating floor	D
Below operating floor	C
Packing and boxing (see **Materials handling**)	
Paint manufacturing	
Processing	D
Mix comparison	F
Paint shops	
Dipping, simple spraying, firing	D
Rubbing, ordinary hand painting and finishing art, stencil and special spraying	D
Fine hand painting and finishing	E
Extra-fine hand painting and finishing	G
Paper-box manufacturing	E
Paper manufacturing	
Beaters, grinding, calendering	D
Finishing, cutting, trimming, papermaking machines	E
Hand counting, wet end of paper machine	E
Paper machine reel, paper inspection, and laboratories	F
Rewinder	F
Parking facilities (see Fig. 55)	
Petroleum and chemical plants (see Fig. 52)[21]	
Plating	D
Polishing and burnishing (see **Machine shops**)	
Power plants (see **Electric generating stations**)	
Poultry industry (see also **Farm—dairy**)	
Brooding, production, and laying houses	
Feeding, inspection, cleaning	C
Charts and records	D
Thermometers, thermostats, time clocks	D
Hatcheries	
General area and loading platform	C

For footnotes, see page 102. For illuminance ranges for each Illuminance Category, see page 87.

Fig. 43 *Continued*

III. *Continued*

Area/Activity	Illuminance Category
Inside incubators	D
Dubbing station	F
Sexing	H
Egg handling, packing, and shipping	
General cleanliness	E
Egg quality inspection	E
Loading platform, egg storage area, etc.	C
Egg processing	
General lighting	E
Fowl processing plant	
General (excluding killing and unloading area)	E
Government inspection station and grading stations	E
Unloading and killing area	C
Feed storage	
Grain, feed rations	C
Processing	C
Charts and records	D
Machine storage area (garage and machine shed)	B
Printing industries	
Type foundries	
Matrix making, dressing type	E
Font assembly—sorting	D
Casting	E
Printing plants	
Color inspection and appraisal	F
Machine composition	E
Composing room	E
Presses	E
Imposing stones	F
Proofreading	F
Electrotyping	F
Molding, routing, finishing, leveling molds, trimming	E
Blocking, tinning	D
Electroplating, washing, backing	D
Photoengraving	
Etching, staging, blocking	D
Routing, finishing, proofing	E
Tint laying, masking	E
Quality control (see **Inspection**)	
Receiving and shipping (see **Materials handling**)	
Railroad yards (see Part IV, Outdoor Facilities)	
Rubber goods—mechanical (see Fig. 50)[21]	
Rubber tire manufacturing (see Fig. 50)[21]	
Safety (see Fig. 44)	
Sawmills	
Secondary log deck	B
Head saw (cutting area viewed by sawyer)	E
Head saw outfeed	B
Machine in-feeds (bull edger, resaws, edgers, trim, hula saws, planers)	B
Main mill floor (base lighting)	A
Sorting tables	D

Area/Activity	Illuminance Category
Rough lumber grading	D
Finished lumber grading	F
Dry lumber warehouse (planer)	C
Dry kiln colling shed	B
Chipper infeed	B
Basement areas	
Active	A
Inactive	A
Filing room (work areas)	E
Service spaces (see also **Storage rooms**)	
Stairways, corridors	B
Elevators, freight and passenger	B
Toilets and wash rooms	C
Sewn products	
Receiving, packing, shipping	E
Opening, raw goods storage	E
Designing, pattern-drafting, pattern grading and marker-making	F
Computerized designing, pattern-making and grading digitizing, marker-making, and plotting	B
Cloth inspection and perching	I
Spreading and cutting (includes computerized cutting)	F[27]
Fitting, sorting and bundling, shading, stitch marking	G
Sewing	G
Pressing	F
In-process and final inspection	G
Finished goods storage and picking orders	F[28]
Trim preparation, piping, canvas and shoulder pads	F
Machine repair shops	G
Knitting	F
Sponging, decating, rewinding, measuring	E
Hat manufacture (see **Hat manufacture**)	
Leather working (see **Leather working**)	
Shoe manufacturing (see **Shoe manufacturing**)	
Sheet metal works	
Miscellaneous machines, ordinary bench work	E
Presses, shears, stamps, spinning, medium bench work	E
Punches	E
Tin plate inspection, galvanized	F
Scribing	F
Shoe manufacturing—leather	
Cutting and stitching	
Cutting tables	G
Marking, buttonholing, skiving, sorting, vamping, counting	G
Stitching, dark materials	G
Making and finishing, nailers, sole layers, welt beaters and scarfers, trimmers, welters, lasters, edge setters, sluggers, randers, wheelers, treers, cleaning, spraying, buffing, polishing, embossing	F
Shoe manufacturing—rubber	
Washing, coating, mill run compounding	D

For footnotes, see page 102. For illuminance ranges for each Illuminance Category, see page 87.

Fig. 43 *Continued*

III. *Continued*

Area/Activity	Illuminance Category
Varnishing, vulcanizing, calendering, upper and sole cutting	D
Sole rolling, lining, making and finishing processes	E
Soap manufacturing	
Kettle houses, cutting, soap chip and powder	D
Stamping, wrapping and packing, filling and packing soap powder	D
Stairways (see **Service spaces**)	
Steel (see **Iron and steel**)	
Storage battery manufacturing	D
Storage rooms or warehouses	
Inactive	B
Active	
Rough, bulky items	C
Small items	D
Storage yards (see Part IV, Outdoor Facilities)	
Structural steel fabrication	E
Sugar refining	
Grading	E
Color inspection	F
Testing	
General	D
Exacting tests, extra-fine instruments, scales, etc.	F
Textile mills	
Staple fiber preparation	
Stock dyeing, tinting	D
Sorting and grading (wood and cotton)	E[16]
Yarn manufacturing	
Opening and picking (chute feed)	D
Carding (nonwoven web formation)	D[24]

Area/Activity	Illuminance Category
Drawing (gilling, pin drafting)	D
Combing	D[24]
Roving (slubbing, fly frame)	E
Spinning (cap spinning, twisting, texturing)	E
Yarn preparation	
Winding, quilling, twisting	E
Warping (beaming, sizing)	F[16]
Warp tie-in or drawing-in (automatic)	E
Fabric production	
Weaving, knitting, tufting	F
Inspection	G[16]
Finishing	
Fabric preparation (desizing, scouring, bleaching, singeing, and mercerization)	D
Fabric dyeing (printing)	D
Fabric finishing (calendaring, sanforizing, sueding, chemical treatment)	E[16]
Inspection	G[16, 25]
Tobacco products	
Drying, stripping	D
Grading and sorting	F
Toilets and wash rooms (see **Service spaces**)	
Upholstering	F
Warehouse (see **Storage rooms**)	
Welding	
Orientation	D
Precision manual arc-welding	H
Woodworking	
Rough sawing and bench work	D
Sizing, planing, rough sanding, medium quality machine and bench work, gluing, veneering, cooperage	D
Fine bench and machine work, fine sanding and finishing	E

IV. Outdoor Facilities

Area/Activity	Lux	Footcandles	Area/Activity	Lux	Footcandles
Advertising Signs (see **Bulletin and poster boards**			Medium light surfaces	200	20
			Medium dark surfaces	300	30
Bikeways (see page 14-16) *			Dark surfaces	500	50
			Dark surroundings		
Building (construction)			Light surfaces	50	5
General construction	100	10	Medium light surfaces	100	10
Excavation work	20	2	Medium dark surfaces	150	15
			Dark surfaces	200	20
Building exteriors					
Entrances			**Bulletin and poster boards**		
Active (pedestrian and/or conveyance)	50	5	Bright surroundings		
			Light surfaces	500	50
Inactive (normally locked, infrequently used)	10	1	Dark surfaces	1000	100
			Dark surroundings		
Vital locations or structures	50	5	Light surfaces	200	20
Building surrounds	10	1	Dark surfaces	500	50
Buildings and monuments, floodlighted					
Bright surroundings			**Central station** (see **Electric generating stations—exterior**)		
Light surfaces	150	15			

For footnotes, see page 102. For illuminance ranges for each Illuminance Category, see page 87.
*In IES Lighting Handbook, *1987 Application Volume.*

Fig. 43 Continued

IV. Continued

Area/Activity	Lux	Footcandles	Area/Activity	Lux	Footcandles
Coal yards (protective)	2	0.2	Focal points, large	100	10
Dredging	20	2	Focal points, small	200	20
Electric generating stations— exterior			**Gasoline station** (see **Service stations** in Part II)		
Boiler areas			**Highways** (see Fig. 54)		
Catwalks, general areas .	20	2	**Loading and unloading**		
Stairs and platforms	50	5	Platforms	200	20
Ground level areas including precipitators,			Freight car interiors	100	10
FD and ID fans,			**Logging** (see also **Sawmills**)		
bottom ash hoppers	50	5	Yarding	30	3
Cooling towers			Log loading and unloading .	50	5
Fan deck, platforms,			Log stowing (water)	5	0.5
stairs, valve areas .	50	5	Active log storage area (land)	5	0.5
Pump areas	20	2	Log booming area (water)—		
Fuel handling			foot traffic	10	1
Barge unloading, car dumper, unloading hoppers,			Active log handling area		
truck unloading,			(water)	20	2
pumps, gas metering	50	5	Log grading—water or land	50	5
Conveyors	20	2	Log bins (land)	20	2
Storage tanks	10	1	**Lumber yards**	10	1
Coal storage piles, ash			**Parking areas** (see Fig. 55)		
dumps	2	0.2	**Piers**		
Hydroelectric			Freight	200	20
Powerhouse roof, stairs,			Passenger	200	20
platform and intake			Active shipping area		
decks	50	5	surrounds	50	5
Inlet and discharge water			**Prison yards**	50	5
area	2	0.2	**Quarries**	50	5
Intake structures			**Railroad yards**		
Deck and laydown area .	50	5	Retarder classification yards		
Value pits	20	2	Receiving yard		
Inlet water area	2	0.2	Switch points	20	2
Parking areas			Body of yard	10	1
Main plant parking	20	2	Hump area (vertical)	200	20
Secondary parking	10	1	Control tower and retarder		
Substation			area (vertical)	100	10
Horizontal general area .	20	2	Head end	50	5
Vertical tasks	50	5	Body	10	1
Transformer yards			Pull-out end	20	2
Horizontal general area .	20	2	Dispatch or forwarding		
Vertical tasks	50	5	yard	10	1
Turbine areas			Hump and car rider classification yard		
Building surrounds	20	2	Receiving yard		
Turbine and heater decks,			Switch points	20	2
unloading bays	50	5	Body of yard	10	1
Entrances, stairs and			Hump area	50	5
platforms	50[9]	5[9]	Flat switching yards		
Flags, floodlighted (see **Bulletin and poster boards**)			Side of cars (vertical) . . .	50	5
Gardens[19]			Switch points	20	2
General lighting	5	0.5	Trailer-on-flatcars		
Path, steps, away from			Horizontal surface of		
house	10	1	flatcar	50	5
Backgrounds—fences, walls,			Hold-down points (vertical)	50	5
trees, shrubbery . . .	20	2	Container-on-flatcars	30	3
Flower beds, rock gardens .	50	5	**Roadways** (see Fig. 54)		
Trees, shrubbery, when					
emphasized	50	5			

For footnotes, see page 102.

Fig. 43 Continued

IV. Continued

Area/Activity	Lux	Footcandles	Area/Activity	Lux	Footcandles
Sawmills (see also **Logging**)			Service areas	70	7
Cut-off saw	100	10	Landscape highlights ...	50	5
Log haul	20	2	**Ship yards**		
Log hoist (side lift)........	20	2	General	50	5
Primary log deck	100	10	Ways..................	100	10
Barker in-feed	300	30	Fabrication areas	300	30
Green chain	200 to 300[26]	20 to 30[26]	**Signs**		
Lumber strapping	150 to 200[26]	15 to 20[26]	Advertising (see **Bulletin and**		
Lumber handling areas	20	2	**poster boards**)		
Lumber loading areas	50	5	Externally lighted roadway		
Wood chip storage piles ...	5	0.5	(see page14-25)		
Service station (at grade)			**Smokestacks with advertising**		
Dark surrounding			**messages** (see		
Approach	15	1.5	**Bulletin and**		
Driveway	15	1.5	**poster boards**)		
Pump island area	200	20	**Storage yards**		
Building faces (exclusive			Active	200	20
of glass)	100[14]	10[14]	Inactive................	10	1
Service areas	30	3	**Streets** (See Fig. 54)		
Landscape highlights ...	20	2			
Light surrounding			**Water tanks with advertising**		
Approach	30	3	**messages** (see **Bulle-**		
Driveway	50	5	**tin and poster boards**)		
Pump island area	300	30			
Building faces (exclusive					
of glass)	300[14]	30[14]			

V. Sports and Recreational Areas

Area/Activity	Lux	Footcandles	Area/Activity	Lux	Footcandles
Archery (indoor)			Junior league (Class I and		
Target, tournament	500[14]	50[14]	Class II)		
Target, recreational........	300[14]	30[14]	Infield	300	30
Shooting line, tournament ...	200	20	Outfield................	200	20
Shooting line, recrational	100	10	On seat during game.......	20	2
Archery (outdoor)			On seats before & after game	50	5
Target, tournament	100[14]	10[14]	**Basketball**		
Target, recreational........	50[14]	5[14]	College and professional	500	50
Shooting line, tournament ...	100	10	College intramural and high		
Shooting line, recreational ...	50	5	school	300	30
Badminton			Recreational (outdoor)	100	10
Tournament	300	30	**Bathing beaches**		
Club	200	20	On land................	10	1
Recreational	100	10	150 feet from shore	30[14]	3[14]
Baseball			**Billiards** (on table)		
Major league			Tournament..............	500	50
Infield	1500	150	Recreational	300	30
Outfield................	1000	100	**Bowling**		
AA and AAA league			Tournament		
Infield	700	70	Approaches	100	10
Outfield................	500	50	Lanes	200	20
A and B league			Pins	500[14]	50[14]
Infield	500	50	Recreational		
Outfield................	300	30	Approaches	100	10
C and D league			Lanes	100	10
Infield	300	30	Pins	300[14]	30[14]
Outfield................	200	20	**Bowling on the green**		
Semi-pro and municipal league			Tournament	100	10
Infield	200	20	Recreational	50	5
Outfield................	150	15	**Boxing or wrestling (ring)**		
Recreational			Championship	5000	500
Infield	150	15			
Outfield................	100	10			

For footnotes, see page 102.

Fig. 43 Continued

V. Continued

Area/Activity	Lux	Footcandles	Area/Activity	Lux	Footcandles
Professional	2000	200	**Golf**		
Amateur	1000	100	Tee	50	5
Seats during bout	20	2	Fairway	10,30[14]	1,3[14]
Seats before and after bout	50	5	Green	50	5
Casting—bait, dry-fly, wet-fly			Driving range		
Pier or dock	100	10	At 180 meters [200 yards]	50[14]	5[14]
Target (at 24 meters [80 feet]			Over tee area	100	10
for bait casting and 15			Miniature	100	10
meters [50 feet] for wet or			Practice putting green	100	10
dry-fly casting)	50[14]	5[14]	**Gymnasiums** (refer to individual		
Combination (outdoor)			sports listed)		
Baseball/football			General exercising and		
Infield	200	20	recreation	300	30
Outfield and football	150	15	**Handball**		
Industrial softball/football			Tournament	500	50
Infield	200	20	Club		
Outfield and football	150	15	Indoor—four-wall or squash	300	30
Industrial softball/6-man foot-			Outdoor—two-court	200	20
ball			Recreational		
Infield	200	20	Indoor—four-wall or squash	200	20
Outfield and football	150	15	Outdoor—two-court	100	10
Croquet or Roque			**Hockey, field**	200	20
Tournament	100	10	**Hockey, ice (indoor)**		
Recreational	50	5	College or professional	1000	100
Curling			Amateur	500	50
Tournament			Recreational	200	20
Tees	500	50	**Hockey, ice (outdoor)**		
Rink	300	30	College or professional	500	50
Recreational			Amateur	200	20
Tees	200	20	Recreational	100	10
Rink	100	10	**Horse shoes**		
Fencing			Tournament	100	10
Exhibitions	500	50	Recreational	50	5
Recreational	300	30	**Horse shows**	200	20
Football			**Jai-alai**		
Distance from nearest side-			Professional	1000	100
line to the farthest row			Amateur	700	70
of spectators			**Lacrosse**	200	20
Class I Over 30 meters [100			**Playgrounds**	50	5
feet]	1000	100	**Quoits**	50	5
Class II 15 to 30 meters [50			**Racing (outdoor)**		
to 100 feet]	500	50	Auto	200	20
Class III 9 to 15 meters [30			Bicycle		
to 50 feet	300	30	Tournament	300	30
Class IV Under 9 meters			Competitive	200	20
[30 feet]	200	20	Recreational	100	10
Class V No fixed seating			Dog	300	30
facilities	100	10	Dragstrip		

It is generally conceded that the distance between the spectators and the play is the first consideration in determining the class and lighting requirements. However, the potential seating capacity of the stands should also be considered and the following ratio is suggested: Class I for over 30,000 spectators; Class II for 10,000 to 30,000; Class III for 5000 to 10,000; and Class IV for under 5000 spectators.

Football, Canadian—rugby
   (see **Football**)

Football, six-man

Area/Activity	Lux	Footcandles
High school or college	200	20
Jr. high and recreational	100	10

Area/Activity	Lux	Footcandles
Staging area	100	10
Acceleration, 400 meters [1320 feet]	200	20
Deceleration, first 200 meters [660 feet]	150	15
Deceleration, second 200 meters [660 feet]	100	10
Shutdown, 250 meters [820 feet]	50	5
Horse	200	20
Motor (midget of motorcycle)	200	20

For footnotes, see page 102.

Fig. 43 *Continued*

V. *Continued*

Area/Activity	Lux	Footcandles	Area/Activity	Lux	Footcandles
Racquetball (see Handball)			Infield	300	30
Rifle 45 meters [50 yards]—out-			Outfield	2000	20
door)			Industrial league		
On targets	500[14]	50[14]	Infield	200	20
Firing point	100	10	Outfield	150	15
Range	50	5	Recreational (6-pole)		
Rifle and pistol range (indoor)			Infield	100	10
On targets	1000[14]	100[14]	Outfield	70	7
Firing point	200	20	Slow pitch, tournament—see		
Range	100	10	industrial league		
Rodeo			Slow pitch, recreational		
Arena			(6-pole)—see recreational		
Professional	500	50	(6-pole)		
Amateur	300	30	**Squash (see Handball)**		
Recreational	100	10	**Swimming (indoor)**		
Pens and chutes	50	5	Exhibitions	500	50
Roque (see Croquet)			Recreational	300	30
Shuffleboard (indoor)			Underwater—1000 [100] lamp		
Tournament	300	30	lumens per square meter		
Recreational	200	20	[foot] of surface area		
Shuffleboard (outdoor)			**Swimming (outdoor)**		
Tournament	100	10	Exhibitions	200	20
Recreational	50	5	Recreational	100	10
Skating			Underwater—600 [60] lamp		
Roller rink	100	10	lumens per square meter		
Ice rink, indoor	100	10	[foot] of surface area		
Ice rink, outdoor	50	5	**Tennis (indoor)**		
Lagoon, pond, or flooded area	10	1	Tournament	1000	100
Skeet			Club	750	75
Targets at 18 meters [60 feet]	300[14]	30[14]	Recreational	500	50
Firing points	50	5	**Tennis (outdoor)**		
Skeet and trap (combination)			Tournament	300	30
Targets at 30 meters [100 feet]			Club	200	20
for trap, 18 meters [60 feet]			Recreational	100	10
for skeet	300[14]	30[14]	**Tennis, platform**	500	50
Firing points	50	5	**Tennis, table**		
Ski slope	10	1	Tournament	500	50
Soccer (see Football)			Club	300	30
Softball			Recreational	200	20
Professional and championship			**Trap**		
Infield	500	50	Targets at 30 meters [100 feet]	300[14]	30[14]
Outfield	300	30	Firing points	50	5
Semi-professional			**Volley ball**		
			Tournaments	200	20
			Recreational	100	10

VI. Transportation Vehicles

Area/Activity	Lux	Footcandles	Area/Activity	Lux	Footcandles
Aircraft			Fare box (rapid transit train)	150	15
Passenger compartment			Vestibule (commuter and in-		
General	50	5	tercity trains)	100	10
Reading (at seat)	200	20	Aisles	100	10
Airports			Advertising cards (rapid tran-		
Hangar apron	10	1	sit and commuter trains)	300	30
Terminal building apron			Back-lighted advertising		
Parking area	5	0.5	cards (rapid transit and		
Loading area	20[14]	2[14]	commuter trains)—860		
Rail conveyances			cd/m^2 (80 cd/ft^2) average		
Boarding or exiting	100	10	maximum.		
			Reading	300[3]	30[3]

For footnotes, see page 102.

Fig. 43 *Continued*

VI. *Continued*

Area/Activity	Lux	Footcandles
Rest room (inter-city train) . . .	200	20
Dining area (inter-city train) . .	500	50
Food preparation (inter-city train)	700	70
Lounge (inter-city train)		
General lighting	200	20
Table games	300	30
Sleeping car		
General lighting	100	10
Normal reading	300^3	30^3
Prolonged seeing	700^3	70^3
Road conveyances		
Step well and adjacent ground area	100	10
Fare box	150	15
General lighting (for seat selection and movement)		
City and inter-city buses at city stop	100	10
Inter-city bus at country stop	20	2
School bus while moving .	150	15
School bus at stops	300	30
Advertising cards	300	30
Back-lighted advertising cards (see **Rail conveyances**)		
Reading	300^3	30^3
Emergency exit (school bus) .	50	5
Ships		
Living Areas		
Staterooms and Cabins		
General lighting	100	10
Reading and writing. . . .	$300^{15, 3}$	$30^{15, 3}$
Prolonged seeing	$700^{16, 3}$	$70^{16, 3}$
Baths (general lighting) .	100	10
Mirrors (personal grooming)	500	50
Barber shop and beauty parlor	500	50
On subject	1000	100
Day rooms		
General lighting	200^{15}	20^{15}
Desks	$500^{16, 3}$	$50^{16, 3}$
Dining rooms and mess-rooms	200	20
Enclosed promenades		
General lighting	100	10
Entrances and passageways		
General	100	10
Daytime embarkation . . .	300	30
Gymnasiums		
General lighting	300	30
Hospital		
Dispensary (general lighting)	300^{16}	30^{16}
Operating room		
General lighting	500^{16}	50^{16}
Doctor's office	300^{16}	30^{16}
Operating table	20000	2000
Wards		
General lighting	100	10
Reading	300	30

Area/Activity	Lux	Footcandles
Toilets	200	20
Libraries and lounges		
General lighting	200	20
Reading	$300^{16, 3}$	$30^{16, 3}$
Prolonged seeing	$700^{16, 3}$	$70^{16, 3}$
Purser's office	200^{16}	20^{16}
Shopping areas	200	20
Smoking rooms	150	15
Stairs and foyers	200	20
Recreation areas		
Ball rooms	150^{15}	15^{15}
Cocktail lounges	150^{15}	15^{15}
Swimming pools		
General	150^{15}	15^{15}
Underwater		
Outdoors—600 [60] lamps lumens/square meter [foot] of surface area		
Indoors—1000 [100] lamp lumens/square meter [foot] of surface area		
Theatre		
Auditorium		
General	100^{15}	10^{15}
During picture	1	0.1
Navigating Areas		
Chart room		
General	100	10
On chart table	$500^{16, 3}$	$50^{16, 3}$
Gyro room	200	20
Radar room	200	20
Radio room	100^{16}	10^{16}
Radio room, passenger		
Foyer	100	10
Ship's offices		
General	200^{16}	20^{16}
On desks and work tables	$500^{16, 3}$	$50^{16, 3}$
Wheelhouse	100	10
Service Areas		
Food preparation		
General	200^{16}	20^{16}
Butcher shop	200^{16}	20^{16}
Galley	300^{16}	30^{16}
Pantry	200^{16}	20^{16}
Thaw room	200^{16}	20^{16}
Sculleries	200^{16}	20^{16}
Food storage nonrefrigerated)	100	10
Refrigerated spaces (ship's stores)	50	5
Laundries		
General	200^{16}	20^{16}
Machine and press finishing, Sorting. . . .	500	50
Lockers	50	5
Offices		
General	200	20
Reading	$500^{16, 3}$	$50^{16, 3}$
Passenger counter	$500^{16, 3}$	$50^{16, 3}$
Storerooms	50	5
Telephone exchange	200	20
Operating Areas		
Access and casing	100	10
Battery room	100	10
Boiler rooms	200^{16}	20^{16}

For footnotes, see page 102.

VI. *Continued*					
Area/Activity	Lux	Footcandles	Area/Activity	Lux	Footcandles
Cargo handling (weather deck)	50^{16}	5^{16}	Motor generator rooms (cargo handling)	100	10
Control stations (except navigating areas)			Pump room	100	10
General			Shaft alley	100	10
Control consoles	200	20	Shaft alley escape	30	3
Gauge and control	300	30	Steering gear room	200	20
boards	300	30	Windlass rooms	100	10
Switchboards	300	30	Workshops		
Engine rooms	200^{16}	20^{16}	General	300^{16}	30^{16}
Generator and switchboard rooms	200^{16}	20^{16}	On top of work bench	500^{16}	50^{16}
Fan rooms (ventilation & air conditioning)	100	10	Tailor shop	500^{16}	50^{16}
Motor rooms	200	20	Cargo holds		
			Permanent luminaires	30^{16}	3^{16}
			Passageways and trunks	100	10

[1]Include provisions for higher levels for exhibitions.

[2]Specific limits are provided to minimize deterioration effects.

[3]Task subject to veiling reflections. Illuminance listed is not an ESI value. Currently, insufficient experience in the use of ESI target values precludes the direct use of Equivalent Sphere Illumination in the present consensus approach to recommend illuminance values. Equivalent Sphere Illumination may be used as a tool in determining the effectiveness of controlling veiling reflections and as a part of the evaluation of lighting systems.

[4]Illuminance values are listed based on experience and consensus. Values relate to needs during various religious ceremonies.

[5]Degradation factors: Overlays—add 1 weighting factor for each overlay; Used material—estimate additional factors.

[6]Provide higher level over food service or selection areas.

[7]Supplementary illumination as in delivery room must be available.

[8]Illuminance values developed for various degrees of store area activity.

[9]Or not less than 1/5 the level in the adjacent areas.

[10]Only when actual equipment service is in process. May be achieved by a general lighting system or by localized or portable equipment.

[11]For color matching, the spectral quality of the color of the light source is important.

[12]Veiling reflections may be produced on glass surfaces. It may be necessary to treat plus weighting factors as minus in order to obtain proper illuminance.

[13]Especially subject to veiling reflections. It may be necessary to shield the task or to reorient it.

[14]Vertical

[15]Illuminance values may vary widely, depending upon the effect desired, the decorative scheme, and the use made of the room.

[16]Supplementary lighting should be provided in this space to produce the higher levels required for specific seeing tasks involved.

[17]Good to high color rendering capability should be considered in these areas. As lamps of higher luminous efficacy and higher color rendering capability become available and economically feasible, they should be applied in all areas of health care facilities.

[18]Variable (dimming or switching).

[19]Values based on a 25 percent reflectance, which is average for vegetation and typical outdoor surfaces. These figures must be adjusted to specific reflectances of materials lighted for equivalent brightness. Levels give satisfactory brightness patterns when viewed from dimly lighted terraces or interiors. When viewed from dark areas they may be reduced by at least 1/2; or they may be doubled when a high key is desired.

[20]General lighting should not be less than 1/3 of visual task illuminance nor less than 200 lux [20 footcandles].

[21]Industry representatives have established a table of single illuminance values which, in their opinion, can be used in preference to employing reference 6. Illuminance values for specific operations can also be determined using illuminance categories of similar tasks and activities found in this table and the application of the appropriate weighting factors in Fig. 56.

[22]Special lighting such that (1) the luminous area is large enough to cover the surface which is being inspected and (2) the luminance is within the limits necessary to obtain comfortable contrast conditions. This involves the use of sources of large area and relatively low luminance in which the source luminance is the principal factor rather than the illuminance produced at a given point.

[23]Maximum levels—controlled system.

[24]Additional lighting needs to be provided for maintenance only.

[25]Color temperature of the light source is important for color matching.

[26]Select upper level for high speed conveyor systems. For grading redwood lumber 3000 lux [300 footcandles] is required.

[27]Higher levels from local lighting may be required for manually operated cutting machines.

[28]If color matching is critical, use illuminance category G.

Fig. 44 Illuminance Levels for Safety *

Hazards Requiring Visual Detection	Slight		High	
Normal† Activity Level	Low	High	Low	High
Illuminance Levels				
Lux	5.4	11	22	54
Footcandles	0.5	1	2	5

* Minimum illuminance for safety of people, absolute minimum at any time and at any location on any plane where safety is related to seeing conditions.

† Special conditions may require different illuminance levels. In some cases higher levels may be required as for example where security is a factor. In some other cases greatly reduced levels, including total darkness, may be necessary, specifically in situations involving manufacturing, handling, use, or processing of light-sensitive materials (notably in connection with photographic products). In these situations alternate methods of insuring safe operations must be relied upon.

Note: See specific application reports of the IES for guidelines to minimum illuminances for safety by area.

Fig. 45 Illuminances Currently Recommended in Main Worship Areas of Houses of Worship

Area/Activity	Illuminances	
	Lux	Footcandles
Reading in seated areas In architecturally rich interiors, for more liturgical groups	100*	10*
In modern, simple interiors, for groups with religious fervor	300*	30*
Accent lighting	a	a
Architectural lighting	b	b

* Maintained target design values on tasks, dimmable.
a Approximately three times the *reading* target illuminance
b Approximately 25 percent or less of the reading target illuminance.

Fig. 46 Recommended Total Exposure Limits in Terms of Illuminance-Hours Per Year to Limit Damage to Light-Susceptible Museum and Art Gallery Objects

Objects	Lux-Hours Per Annum	Footcandle-Hours Per Annum
Highly susceptible displayed materials—silk, art on paper, antique documents, lace, fugitive dyes	120,000*	12,000*
Moderately susceptible displayed materials—cotton, wool, other textiles where the dye is stable, certain wood finishes, leather	180,000**	18,000**

* Approximately 50 lux (5 footcandles) x 8 hours per day x 300 days per year.
** Approximately 75 lux (7.5 footcandles) x 8 hours per day x 300 days per year.
NOTE: These illuminances, if carefully applied, will not result in worse than just perceptible fading in the stated materials in ten years exposure. All wavelengths shorter than 400 nanometers should be rigidly excluded.

Fig. 47 Currently Recommended Illuminances for Merchandising Areas

Areas or Tasks	Description	Type of Activity Area*	Illuminance**	
			Lux	Footcandles
Circulation	Area not used for display or appraisal of merchandise or for sales transactions	High activity	300	30
		Medium activity	200	20
		Low activity	100	10
Merchandise† (including showcases & wall displays)	That plane area, horizontal to vertical, where merchandise is displayed and readily accessible for customer examination	High activity	1000	100
		Medium activity	750	75
		Low activity	300	30
Feature displays†	Single item or items requiring special highlighting to visually attract and set apart from the surround	High activity	5000	500
		Medium activity	3000	300
		Low activity	1000	150
Show windows				
Daytime lighting				
General			2000	200
Feature			10000	1000
Nighttime lighting				
Main business districts-highly competitive				
General			2000	200
Feature			10000	1000
Secondary business districts or small towns				
General			1000	100
Feature			5000	500

* One store may encompass all three types within the building
High activity area— Where merchandise displayed has readily recognizable usage. Evaluation and viewing time is rapid, and merchandise is shown to attract and stimulate the impulse buying decision.
Medium activity — Where merchandise is familiar in type or usage, but the customer may require time and/or help in evaluation of quality, usage or for the decision to buy.
Low activity — Where merchandise is displayed that is purchased less frequently by the customer, who may be unfamiliar with the inherent quality, design, value or usage. Where assistance and time is necessary to reach a buying decision.
** Maintained on the task or in the area at any time
† Lighting levels to be measured in the plane of the merchandise

Fig. 48 Illuminance Values Currently Recommended by Industry Representatives for Aircraft Maintenance and Manufacturing[1] (Maintained on Tasks)

Area and Task	Illuminance on Task		Area and Task	Illuminance on Task	
	Lux	Footcan-dles		Lux	Footcan-dles
Aircraft Maintenance					
Close up			Check, operate, pre-inspect, record	750	75
Install plates, panels, fairings, cowls, etc.	750	75	Install safety devices (lockpins-sleeves, etc.)	750	75
Seal plates	750	75	Drain tanks, relieve struts	500	50
Paint (exterior or interior of aircraft) where plates, panels, fairings, cowls, etc., must be in place before accomplishing	750	75	Remove any plates, doors, cowls, fairings, etc. required for precleaning	750	75
			Install protective covers and masking	750	75
Stencils, decals, seals, etc., where final paint coat needed before applying	750	75	Strip paint	750	75
			Clean	750	75
Final "fly-away" outfitting (trays, loose gear, certs, includes final cleaning)	750	75	Install personnel protective devices (sharp edge covers, people barriers, etc.)	300	30
Docking			**Preparation for dedock**		
Position doors and control surfaces for docking	300	30	Remove shoring	750	75
Move aircraft into position in dock	500	50	Remove workstands, ladders, etc.	750	75
Attach grounding wires and other safety equipment	300	30	Close aircraft doors and position control surfaces	300	30
Jack and level aircraft	750	75	Let aircraft down off jacks	750	75
Shore aircraft	750	75	Dedock	750	75
Position ramps, walk-overs and other work facilities and equipment	750	75	**Preparation for maintenance and modification**		
Systems deactivation and safety locks installed	750	75	Check, operations, recordings—required in dock prior to power shutdown	750	75
Removals—prior to power shutdown	750	75	Draining	750	75
Reposition doors, flaps, etc. after docking	750	75	Shutdown aircraft power systems	750	75
Maintenance, modification and repairs to airframe structures			Remove plates, panels, cowls, fairings, linings, etc., for accessibility	750	75
Jacking and shoring not accomplished during docking phase	750	75	Vent, purge, flush, etc., any tanks, lines, systems, etc., drain systems, drain and cap off not previously accomplished	750	75
Remove any carrier or energy transmission portions or systems	750	75	Precleaning prior to removals	750	75
Remove any linings, insulation, blankets, etc., to expose structure	750	75	Disconnect lines, cables, ducts, linkages, etc., required for accessibility	750	75
Remove any minor structures (brackets, clips, angles, boxes, shelves, etc.) that attach to, obstruct or cover up major structure to be replaced, modified or repaired	750	75	Remove components	750	75
			Install protective covers, masking, or devices	750	75
Remove any sealant necessary to expose structures	750	75	Dock cleaning and/or stripping required for inspection, later modification, maintenance and/or painting	750	75
Remove any major structural members (spars, stringers, longerons, circumferentials, etc.) that will be replaced with new ones	750	75	Sand painted areas	750	75
			Area inspection		
Install new structural members	750	75	Ordinary	500	50
Sealant installation after structural member adjustment, modification or repair	1000	100	Difficult	1000	100
			Highly difficult	2000	200
Install any linings, insulation, blankets, etc.	750	75	**Specialty shops**		
			Instruments, radio	1500	150
Prime paint exterior	750	75	Electrical	1500	150
Top coat paint exterior	1000	100	Hydraulic and pneumatic	1000	100
Modifications or repairs to systems			Components	1000	100
Install those carrier or energy transmission portions of systems previously removed which do not require modifications. (Elec. wires, hyd. lines, ducts, fuel lines, cables, etc.)	750	75	Upholstery, chairs, rugs	1000	100
			Sheet metal fabrication, repairs, welding	1000	100
			Paint	1000	100
			Parts inspection	1000	100
Modify any energy transmission or carrier portion of a system or add new ones previously nonexistent (Electrical, mechanical, liquid, pneumatic)	750	75	Plastics	1500	150
			System operations and functional checks requiring aircraft power systems activation to perform		
Repair any carrier or energy transmitting portions of any systems	750	75	Activate any aircraft power system	300	30
Post overhaul—ramp	50	5	Block areas for operationals	750	75
Predocking			Functional check of any system that prohibits other operations or actions within that system	750	75
Convert hanger to fit incoming plane	300	30	Operational or functional check of any system not a part of or requiring sequential accomplishment	750	75
			Test sequentially required operation of systems	750	75

Fig. 48 *Continued*

Area and Task	Illuminance on Task		Area and Task	Illuminance on Task	
	Lux	Footcandles		Lux	Footcandles
Aircraft Maintenance					
Release areas after operationals	300	30	Re-install components of systems that do not require a system check or ring-out before component installation	750	75
Nonpressure lube after operations	300	30			
Cleaning after operations	750	75			
System repairs after operations and close up preparation			Install components requiring preliminary checks and ring-outs	750	75
Repairs after system operationals	750	75	Hook up systems (wires, lines, pipes, ducts, cables, etc.) other than rigging	750	75
Corrosion treatment	750	75			
Apply masking	750	75	Physically block areas for dangerous operations	750	75
Painting and/or chromating	750	75			
Removing masking	500	50	Rig cable systems that do not require sequential rigging	750	75
Final inspections prior to close			Rig cable systems in step sequences	750	75
Ordinary	500	50	Operate any system for checking that can be operated from power source other than aircraft power	750	75
Difficult	1000	100			
Highly difficult	2000	200			
System restoration or new system component installation			Clear blocked area	300	30
			Reconnect lines to aircraft systems	750	75
Hook up any lines, cables, ducts, panels and insulation to be covered by later component installation	750	75	Install cavity or tank covers or plates necessary to filling	750	75
			Precheck before filling	500	50
Install any components previously removed that must be in place for others to attach to, or subsequent components which when installed would obstruct or cover	750	75	Fill tanks. Service or lube tanks, struts, accumulators, etc.	500	50
			Static leak checks	500	50
Paint	750	75	Pressure check systems from pressure sources external to the aircraft	750	75
Paint preparation and clean up	500	50			
Aircraft Manufacturing					
Fabrication (preparation for assembly)			General		
Rough bench work and sheet metal operations such as shears, presses, punches, countersinking, spinning	500	50	Rough easy seeing	300	30
			Rough difficult seeing	500	50
			Medium	1000	100
Drilling, riveting, screw fastening	750	75	Fine	5000[a]	500[a]
Medium bench work and machining such as ordinary automatic machines, rough grinding, medium buffing and polishing	1000	100	Extra fine	10000[a]	1000[a]
			First manufacturing operations (first cut)		
			Marking, shearing, sawing	500	50
Fine bench work and machining such as ordinary automatic machines, rough grinding, medium buffing and polishing	5000[a]	500[a]	**Flight test and delivery area**		
			On the horizontal plane	50	5
			On the vertical plane	20	2
Extra fine bench and machine work	10000[a]	1000[a]	**General warehousing**		
Layout and template work, shaping and smoothing of small parts for fuselage, wing sections, cowling etc.	1000[a]	100[a]	High activity	100	10
			Rough bulky	100	10
Scribing	2000[a]	200[a]	Medium	200	20
Plating	300	30	Fine	500	50
			Low activity	50	5
Final assembly such as placing of motors, propellers, wing sections, landing gear	1000	100	**Outdoor receiving and storage areas**		
			Unloading	200	20
			Storage		
			High activity	200	20
			Low activity	10	1

[a] Obtained with a combination of general lighting plus specialized supplementary lighting. Care should be taken to keep within the recommended luminance ratios(see page9-3). * These seeing tasks generally involve the discrimination of fine detail for long periods of time and under conditions of poor contrast. The design and installation of the combination system must not only provide a sufficient amount of light, but also the proper direction of light, diffusion, color and eye protection. As far as possible it should eliminate direct and reflected glare as well as objectionable shadows.

*In *IES Lighting Handbook, 1987 Application Volume.*

Fig. 49 Illuminance Values Currently Recommended by Industry Representatives for Automotive Industry Facilities (Maintained on Tasks)

Activity	Illuminance	
	Lux	Footcandles
Coal yards, oil storage	5	0.5
Exterior inactive storage, railroad switching points, outdoor substations, parking areas	15	1.5
Inactive interior storage areas, exterior pedestrian entrances, truck maneuvering areas	50	5
Elevators, steel furnace areas, locker rooms, exterior active storage areas	200	20
Waste treatment facilities (interior), clay mold and kiln rooms, casting furnace area, glass furnace rooms, HVAC and substation rooms, sheet steel rolling, loading docks, general paint manufacturing, plating, toilets and washrooms	300	30
Frame assembly, powerhouse, forgings, quick service dining, casting pouring and sorting, service garages, active storage areas, press rooms, battery manufacturing, welding area	500	50
Control and dispatch rooms, kitchens, large casting core and molding areas (engines), machining operations (engine and parts)	750	75
Chassis, body and component assembly, clay enamel and glazing, medium casting core and molding areas (crankshaft), grinding and chipping, glass cutting and inspection, hospital examination and treatment rooms, ordinary inspection, maintenance and machine repair areas, polishing and burnishing, upholstering	1000	100
Parts inspection stations	1500	150
Final assembly, body finishing and assembly, difficult inspection, paint color comparison	2000	200
Fine difficult inspection (casting cracks)	5000	500

Fig. 50 Illuminance Values Currently Recommended by Industry Representatives for the Manufacture of Rubber Tires and Mechanical Rubber Goods (Maintained on Tasks)

Area/Activity	Illuminance	
	Lux	Foot-candles
Rubber tire manufacturing		
Banbury	300	30
Tread stock		
General	500	50
Booking and inspection, extruder, check weighing, width measuring	1000[a]	100[a]
Calendering		
General	300	30
Letoff and windup	500	50
Stock cutting		
General	300	30
Cutters and splicers	1000[a]	100[a]
Bead Building	500	50
Tire Building		
General	500	50
At machines	1500[b]	150[b]
In-process stock	300	30
Curing		
General	300	30
At molds	750[b]	75[b]
Inspection		
General	1000	100
At tires	3000[a]	300[a]
Storage	200	20
Rubber goods—mechanical		
Stock preparation		
Plasticating, milling, Banbury	300	30
Calendering	500	50
Fabric preparation, stock cutting, hose looms	500	50
Extruded products	500	50
Molded products and curing	500	50
Inspection	2000[b]	200[b]

[a] Localized general lighting.
[b] Obtained with a combination of general lighting plus specialized supplementary lighting. Care should be taken to keep within the recommended luminance ratios.

Fig. 51 Illuminance Values Currently Recommended by Representatives from the Iron and Steel Industry (Maintained on Tasks)

Area/Activity	Illuminance		Area/Activity	Illuminance	
	Lux	Foot-candles		Lux	Foot-candles
Open hearth			Rolling mills		
Stock yard	100	10	Blooming, slabbing, hot strip, hot sheet	300	30
Charging floor	200	20	Cold strip, plate	300	30
Pouring slide			Pipe, rod, tube, wire drawing	500	50
Slag pits	200	20	Merchant and sheared plate	300	30
Control platforms	300	30	Tin plate mills		
Mold yard	50	5	Tinning and galvanizing	500	50
Hot top	300	30	Cold strip rolling	500	50
Hot top storage	100	10	Motor room, machine room	300	30
Checker cellar	100	10	Inspection		
Buggy and door repair	300	30	Black plate, bloom and billet chipping	1000	100
Stripping yard	200	20	Tin plate and other bright surfaces	2000*	200*
Scrap stockyard	100	10			
Mixer building	300	30			
Calcining building	100	10			
Skull cracker	100	10			

* The specular surface of the material may necessitate special consideration in selection and placement of lighting equipment, or orientation of work.

Fig. 52. Illuminances Currently Recommended by the Petroleum, Chemical and Petrochemical Industry Representatives.[a]

Area or Activity	Illuminance Lux (Foot-candles)	Elevation Millimeter (Inches)	Area or Activity	Illuminance Lux (Foot-candles)	Elevation Millimeter (Inches)
I. Process areas			Substation operating aisles	150 (15)	Floor
A. General process units			General substation (indoor)	50 (5)	Floor
Pump rows, valves, manifolds	50 (5)	Ground	Switch racks	50 (5)[b]	1200 (48)
Heat exchangers	30 (3)	Ground	G. Plant road lighting (where lighting is required[d]		
Maintenance platforms	10 (1)	Floor	Frequent use (trucking)	4 (0.4)	Ground
Operating platforms	50 (5)	Floor	Infrequent use	2 (0.2)	Ground
Cooling towers (equipment areas)	50 (5)	Ground	H. Plant parking lots[d]	1 (0.1)	Ground
Furnaces	30 (3)	Ground	I. Aircraft obstruction lighting[e]		
Ladders and stairs (inactive)	10 (1)	Floor	**III. Buildings[d]**		
Ladders and stairs (active)	50 (5)	Floor	A. Offices (See Part II)		
Gage glasses	50 (5)[b]	Eye level	B. Laboratories		
Instruments (on process units)	50 (5)[b]	Eye level	Qualitative, quantitative and physical test	500 (50)	900 (36)
Compressor houses	200 (20)	Floor	Research, experimental	500 (50)	900 (36)
Separators	50 (5)	Top of bay	Pilot plant, process and specialty	300 (30)	Floor
General area	10 (1)	Ground	ASTM equipment knock test	300 (30)	Floor
B. Control rooms and houses			Glassware, washrooms	300 (30)	900 (36)
Ordinary control house	300 (30)	Floor	Fume hoods	300 (30)	900 (36)
Instrument panel	300 (30)[b]	1700 (66)	Stock rooms	150 (15)	Floor
Console	300 (30)[b]	760 (30)	C. Warehouses and stock rooms[d]		
Back of panel	100 (10)[b]	760 (30)	Indoor bulk storage	50 (5)	Floor
Central control house	500 (50)	Floor	Outdoor bulk storage	5 (0.5)	Ground
Instrument panel	500 (50)[b]	1700 (66)	Large bin storage	50 (5)	760 (30)
Console	500 (50)[b]	760 (30)	Small bin storage	100 (10)[a]	760 (30)
Back of panel	100 (10)[b]	900 (36)	Small parts storage	200 (20)[a]	760 (30)
C. Specialty process units			Counter tops	300 (30)	1200 (48)
Electrolytic cell room	50 (5)	Floor	D. Repair shop[d]		
Electric furnace	50 (5)	Floor	Large fabrication	200 (20)	Floor
Conveyors	20 (2)	Surface	Bench and machine work	500 (50)	760 (30)
Conveyor transfer points	50 (5)	Surface	Craneway, aisles	150 (15)	Floor
Kilns (operating area)	50 (5)	Floor	Small machine	300 (30)	760 (30)
Extruders and mixers	200 (20)	Floor	Sheet metal	200 (20)	760 (30)
II. Nonprocess areas			Electrical	200 (20)	760 (30)
A. Loading, unloading, and cooling water pump houses			Instrument	300 (30)	760 (30)
Pump area	50 (5)	Ground	E. Change house[d]		
General control area	150 (15)	Floor	Locker room, shower	100 (10)	Floor
Control panel	200 (20)[b]	1100 (45)	Lavatory	100 (10)	Floor
B. Boiler and air compressor plants			F. Clock house and entrance gatehouse[d]		
Indoor equipment	200 (20)	Floor	Card rack and clock area	100 (10)	Floor
Outdoor equipment	50 (5)	Ground	Entrance gate, inspection	150 (15)	Floor
C. Tank fields (where lighting is required)			General	50 (5)	Floor
Ladders and stairs	5 (0.5)	Floor	G. Cafeteria		
Gaging area	10 (1)	Ground	Eating	300 (30)	760 (30)
Manifold area	5 (0.5)	Floor	Serving area	300 (30)	900 (36)
D. Loading racks			Food preparation	300 (30)	900 (36)
General area	50 (5)	Floor	General, halls, etc.	100 (10)	Floor
Tank car	100 (10)	Point	H. Garage and firehouse		
Tank trucks, loading point	100 (10)	Point	Storage and minor repairs	100 (10)	Floor
E. Tanker dock facilities[c]			I. First aid room[d]	700 (70)	760 (30)
F. Electrical substations and switch yards[d]					
Outdoor switch yards	20 (2)	Ground			
General substation (outdoor)	20 (2)	Ground			

[a] These illumination values are not intended to be mandatory by enactment into law. They are a recommended practice to be considered in the design of new facilities. For minimum levels for safety, see Fig. 44. All illumination values are average maintained levels.
[b] Indicates vertical illumination.
[c] Refer to local Coast Guard, Port Authority, or governing body for required navigational lights.
[d] The use of many areas in petroleum and chemical plants is often different from what the designation may infer. Generally, the areas are small, occupancy low (restricted to plant personnel), occupancy infrequent and only by personnel trained to conduct themselves safely under unusual conditions. For these reasons, illuminances may be different from those recommended for other industries, commercial areas, educational areas or public areas.
[e] Refer to local FAA regulations for required navigational and obstruction lighting and marking.

Fig. 53 Recommended Illuminances for
Acclimatized Plants (14 Hours of Light Per Day)

A. Trees 1.5 to 3 Meters (5 to 10 Feet) Tall

Tree	Illuminances	
	Lux	Footcandles
Araucaria excelsa (Norfolk Island Pine)	above 2000	above 200
Eriobotrya japonica (Chinese Loquator, Japan Plum)	above 2000	above 200
Ficus benjamina 'Exotica' (Weeping Java Fig)	750–2000	75–200
Ficus lyrata (Fiddleleaf Fig)	750–2000	75–200
Ficus retusa nitida (Indian Laurel)	750–2000	75–200
Ligustrum lucidum (Waxleaf)	750–2000	75–200

B. Floor plants 0.6 to 1.8 Meters (2 to 6 Feet) Tall

Plant	Illuminances	
	Lux	Footcandles
Brassaia actinophylla (Schefflera)	750–2000	75–200
Chamaedorea elegans 'bella' (Neanthe Bella Palm)	250–750	25–75
Chamaedorea erumpens (Bamboo Palm)	250–750	25–75
Chamaerops humilis (European Fanpalm)	above 2000	above 200
Dieffenbachia amoena (Giant Dumb Cane)	750–2000	75–200
Dizygotheca elegantissima (False Aralia)	above 2000	above 200
Dracaena deremensis 'Janet Craig' (Green Drasena)	750–200	75–200
Dracaena fragrans massangeana (Corn Plant)	250–750	25–75
Dracaena marginata (Dwarf Dragon Tree)	750–2000	75–200
Ficus elastica 'Decora' (Rubber Plant)	750–2000	75–200
Ficus philippinensis (Philippine Fig)	750–2000	75–200
Howeia forsteriana (Kentia Palm)	250–750	25–75
Philodendron x evansii (Selfheading Philodendron)	750–2000	75–200
Phoenix roebelenii (Pigmy Date Palm)	750–2000	75–200
Pittosporum tobira (Mock Orange)	above 2000	above 200
Podocarpus macrophylla Maki (Podocarpus)	above 2000	above 200
Polyscias guilfoylei (Parsley Aralia)	750–2000	75–200
Rhapis exclesa (Lady Palm)	750–2000	75–200
Yucca elephantipes (Palm-Lily)	above 2000	above 200

C. Table or desk plants

Plant	Illuminances	
	Lux	Footcandles
Aechmea fasciata (Bromeliad)	750–2000	75–200
Aglaonema commutatum (Variegated Chinese Evergreen)	250–750	25–75
Agalonema 'Pseudobacteatum' (Golden Aglaonema)	250–750	25–75
Aglaonema roebelinii (Peuter Plant)	250–750	25–75
Asparagus sprengeri (Asparagus Fern)	750–2000	75–200
Ciccus antarctiva (Kangaroo Vine)	above 2000	above 200
Cissus rhombifolia (Grape Ivy)	750–2000	75–200
Citrus mitis (Calamondin)	above 2000	above 200
Dieffenbachia 'Exoctica' (Dumb Cane)	750–2000	75–200
Dracaena deremensis 'Warneckei' (White Striped Dracaena)	750–2000	75–200
Dracaena fragrans massangeana (Corn Plant)	250–750	25–75
Hoya carnosa (Wax plant)	750–2000	75–200
Maranta leuconeura (Prayer Plant)	750–2000	75–200
Nephrolepsis exaltata bostoniensis (Boston Fern)	750–2000	75–200
Peperomia caperata (Emerald Ripple)	250–750	25–75
Philodendron oxycardium (cordatum) (Common Philodendron)	250–750	25–75
Spathiphyllum 'Mauna Loa' (White Flag)	750–2000	75–200

Fig. 54 Recommended Maintained Luminance and Illuminance Values for Roadways

(a) Maintained Luminance Values (L_{avg}) in Candelas per Square Meter*

Road and Area Classification		Average Luminance L_{avg}	Luminance Uniformity		Veiling Luminance Ratio (maximum) L_v to L_{avg}
			L_{avg} to L_{min}	L_{max} to L_{min}	
Freeway Class A		0.6	3.5 to 1	6 to 1	0.3 to 1
Freeway Class B		0.4	3.5 to 1	6 to 1	0.3 to 1
Expressway	Commercial	1.0	3 to 1	5 to 1	0.3 to 1
	Intermediate	0.8	3 to 1	5 to 1	
	Residential	0.6	3.5 to 1	6 to 1	
Major	Commercial	1.2	3 to 1	5 to 1	0.3 to 1
	Intermediate	0.9	3 to 1	5 to 1	
	Residential	0.6	3.5 to 1	6 to 1	
Collector	Commercial	0.8	3 to 1	5 to 1	0.4 to 1
	Intermediate	0.6	3.5 to 1	6 to 1	
	Residential	0.4	4 to 1	8 to 1	
Local	Commercial	0.6	6 to 1	10 to 1	0.4 to 1
	Intermediate	0.5	6 to 1	10 to 1	
	Residential	0.3	6 to 1	10 to 1	

(b) Average Maintained Illuminance Values (E_{avg})†

Road and Area Classification		Pavement Classification			Illuminance Uniformity Ratio E_{avg} to E_{min}
		R1	R2 and R3	R4	
Freeway Class A		6	9	8	3 to 1
Freeway Class B		4	6	5	
Expressway	Commercial	10	14	13	3 to 1
	Intermediate	8	12	10	
	Residential	6	9	8	
Major	Commercial	12	17	15	3 to 1
	Intermediate	9	13	11	
	Residential	6	9	8	
Collector	Commercial	8	12	10	4 to 1
	Intermediate	6	9	8	
	Residential	4	6	5	
Local	Commercial	6	9	8	6 to 1
	Intermediate	5	7	6	
	Residential	3	4	4	

Notes
L_v = veiling luminance
1. These tables do not apply to high mast interchange lighting systems, *e.g.*, mounting heights over 20 meters. See Fig. 14-12.*
2. The relationship between individual and respective luminance and illuminance values is derived from general conditions for dry paving and straight road sections. This relationship does not apply to averages.
3. For divided highways, where the lighting on one roadway may differ from that on the other, calculations should be made on each roadway independently.
4. For freeways, the recommended values apply to both mainline and ramp roadways.
5. The recommended values shown are meaningful only when designed in conjunction with other elements. The most critical elements as described in this section are: *
(a) Lighting System Depreciation   (f) Luminaire Selection
(b) Quality   (g) Traffic Conflict Area
(c) Uniformity   (h) Lighting Termination
(d) Luminaire Mounting Height   (i) Alleys
(e) Luminaire Spacing
* For approximate values in candelas per square foot, multiply by 0.1.
† For approximate values in footcandles, multiply by 0.1.

*In IES Lighting Handbook, *1987 Application Volume.*

Fig. 55 Recommended Maintained Horizontal Illuminances for Parking Facilities

(a) Open Parking Facilities

Level of Activity	General Parking and Pedestrian Area			Vehicle Use Area (only)		
	Lux (Minimum on Pavement)	Footcandles (Minimum on Pavement)	Uniformity Ratio (Average:Minimum)	Lux (Average on Pavement)	Footcandles (Average on Pavement)	Uniformity Ratio (Average:Minimum)
High	10	0.9	4:1	22	2	3:1
Medium	6	0.6	4:1	11	1	3:1
Low*	2	0.2	4:1	5	0.5	4:1

(b) Covered Parking Facilities

Areas	Day		Night		
	Lux (Average on Pavement)†	Footcandles (Average on Pavement)†	Lux (Average on on Pavement)	Footcandles (Average (Average:Minimum)	Uniformity Ratio
General parking and Pedestrian areas	54	5	54	5	4:1
Ramps and corners	110	10	54	5	4:1
Entrance areas	540	50	54	5	4:1

	Range of Illuminances	
	Lux	Footcandles
Stairways‡	100-150-200	10-15-20

* This recommendation is based on the requirement to maintain security at any time in areas where there is a low level of nighttime activity.
† Sum of electric lighting and daylight.
‡ See Fig. 43.

Fig. 56. Weighting Factors to be Considered in Selecting Specific Illuminance Within Ranges of Values for Each Category.

a. For Illuminance Categories A through C

Room and Occupant Characteristics	Weighting Factor		
	−1	0	+1
Occupants ages	Under 40	40–55	Over 55
Room surface reflectances*	Greater than 70 per cent	30 to 70 per cent	Less than 30 per cent

b. For Illuminance Categories D through I

Task and Worker Characteristics	Weighting Factor		
	−1	0	+1
Workers ages	Under 40	40–55	Over 55
Speed and/or accuracy**	Not important	Important	Critical
Reflectance of task background***	Greater than 70 per cent	30 to 70	Less than 30 per cent

* Average weighted surface reflectances, including wall, floor and ceiling reflectances, if they encompass a large portion of the task area or visual surround. For instance, in an elevator lobby, where the ceiling height is 7.6 meters (25 feet), neither the task nor the visual surround encompass the ceiling, so only the floor and wall reflectances would be considered.

** In determining whether speed and/or accuracy is not important, important or critical, the following questions need to be answered: What are the time limitations? How important is it to perform the task rapidly? Will errors produce an unsafe condition or product? Will errors reduce productivity and be costly? For example, in reading for leisure there are no time limitations and it is not important to read rapidly. Errors will not be costly and will not be related to safety. Thus, speed and/or accuracy is not important. If however, prescription notes are to be read by a pharmacist, accuracy is critical because errors could produce an unsafe condition and time is important for customer relations.

*** The task background is that portion of the task upon which the meaningful visual display is exhibited. For example, on this page the meaningful visual display includes each letter which combines with other letters to form words and phrases. The display medium, or task background, is the paper, which has a reflectance of approximately 85 per cent.

Lighting Calculation Data

(From IES Lighting Handbook—*1984 Reference Volume*)

Fig. 57. Average Illuminance Calculation Sheet

GENERAL INFORMATION

Project identification: _____

(*Give name of area and/or building and room number*)

Average maintained illuminance for design:___ lux or

.___ footcandles

Lamp data:

Type and color:_____

Luminaire data:

Manufacturer: _____

Catalog number: _____

Number per luminaire:_____

Total lumens per luminaire: _____

SELECTION OF COEFFICIENT OF UTILIZATION

Step 1: Fill in sketch at right.

Step 2: Determine Cavity Ratios from Fig. 58 or by
formulas.

Room Cavity Ratio, RCR = _____

Ceiling Cavity Ratio, CCR = _____

Floor Cavity Ratio, FCR = _____

$\rho=$___% $h_{CC}=$_____

$\rho=$___% $L=$_____

$\rho=$___% $h_{RC}=$_____ $W=$_____

—WORK–PLANE—

$\rho=$___% $\rho=$___% $h_{FC}=$_____

Step 3: Obtain Effective Ceiling Cavity Reflectance (ρ_{CC}) from Fig. 59

$\rho_{CC}=$ ____

Step 4: Obtain Effective Floor Cavity Reflectance (ρ_{FC}) from Fig. 59

$\rho_{FC}=$ ____

Step 5: Obtain Coefficient of Utilization (CU) from Manufacturer's Data.

$CU=$ ____

SELECTION OF LIGHT LOSS FACTORS

Nonrecoverable

Luminaire ambient temperature
(See page 9–5.) * _____

Voltage to luminaire
(See page 9–5.) * _____

Ballast factor
(See page 9–5.) * _____

Luminaire surface depreciation
(See page 9–7.) * _____

Recoverable

Room surface dirt depreciation
RSDD (See page 9–9.) * _____

Lamp lumen depreciation
LLD (See page 9–8.) * _____

Lamp burnouts factor
LBO (See page 9–9.) * _____

Luminaire dirt depreciation
LDD (See page 9–8.) * _____

Total light loss factor, LLF (product of individual factors above) = _____

CALCULATIONS

(Average Maintained Illuminance)

$$\text{Number of Luminaires} = \frac{(\text{Illuminance}) \times (\text{Area})}{(\text{Lumens per Luminaire}) \times (CU) \times (LLF)}$$

$$= \underline{\hspace{6cm}} = \underline{\hspace{2cm}}$$

$$\text{Illuminance} = \frac{(\text{Number of Luminaires}) \times (\text{Lumens per Luminaire}) \times (CU) \times (LLF)}{(\text{Area})}$$

$$= \underline{\hspace{7cm}} = \underline{\hspace{2cm}}$$

Calculated by: _____ Date: _____

*IES Lighting Handbook, *1984 Reference Volume*

Fig. 58. Cavity Ratios

For room and cavity dimensions other than those below, the cavity ratio can be calculated by the formulas on page 9–26.* If smaller room and cavity dimensions are required divide width, length and cavity depth by 10.

Room Dimensions		Cavity Depth																		
Width	Length	1	1.5	2	2.5	3	3.5	4	5	6	7	8	9	10	11	12	14	16	20	25	30
8	8	1.2	1.9	2.5	3.1	3.7	4.4	5.0	6.2	7.5	8.8	10.0	11.2	12.5	—	—	—	—	—	—	—
	10	1.1	1.7	2.2	2.8	3.4	3.9	4.5	5.6	6.7	7.9	9.0	10.1	11.3	12.4	—	—	—	—	—	—
	14	1.0	1.5	2.0	2.5	3.0	3.4	3.9	4.9	5.9	6.9	7.8	8.8	9.7	10.7	11.7	—	—	—	—	—
	20	0.9	1.3	1.7	2.2	2.6	3.1	3.5	4.4	5.2	6.1	7.0	7.9	8.8	9.6	10.5	12.2	—	—	—	—
	30	0.8	1.2	1.6	2.0	2.4	2.8	3.2	4.0	4.7	5.5	6.3	7.1	7.9	8.7	9.5	11.0	—	—	—	—
	40	0.7	1.1	1.5	1.9	2.3	2.6	3.0	3.7	4.5	5.3	5.9	6.5	7.4	8.1	8.8	10.3	11.8	—	—	—
10	10	1.0	1.5	2.0	2.5	3.0	3.5	4.0	5.0	6.0	7.0	8.0	9.0	10.0	11.0	12.0	—	—	—	—	—
	14	0.9	1.3	1.7	2.1	2.6	3.0	3.4	4.3	5.1	6.0	6.9	7.8	8.6	9.5	10.4	12.0	—	—	—	—
	20	0.7	1.1	1.5	1.9	2.3	2.6	3.0	3.7	4.5	5.3	6.0	6.8	7.5	8.3	9.0	10.5	12.0	—	—	—
	30	0.7	1.0	1.3	1.7	2.0	2.3	2.7	3.3	4.0	4.7	5.3	6.0	6.7	7.3	8.0	9.4	10.6	—	—	—
	40	0.6	0.9	1.2	1.6	1.9	2.2	2.5	3.1	3.7	4.4	5.0	5.6	6.2	6.9	7.5	8.7	10.0	12.5	—	—
	60	0.6	0.9	1.2	1.5	1.7	2.0	2.3	2.9	3.5	4.1	4.7	5.3	5.9	6.5	7.1	8.2	9.4	11.7	—	—
12	12	0.8	1.2	1.7	2.1	2.5	2.9	3.3	4.2	5.0	5.8	6.7	7.5	8.4	9.2	10.0	11.7	—	—	—	—
	16	0.7	1.1	1.5	1.8	2.2	2.5	2.9	3.6	4.4	5.1	5.8	6.5	7.2	8.0	8.7	10.2	11.6	—	—	—
	24	0.6	0.9	1.2	1.6	1.9	2.2	2.5	3.1	3.7	4.4	5.0	5.6	6.2	6.9	7.5	8.7	10.0	12.5	—	—
	36	0.6	0.8	1.1	1.4	1.7	1.9	2.2	2.8	3.3	3.9	4.4	5.0	5.5	6.0	6.6	7.8	8.8	11.0	—	—
	50	0.5	0.8	1.0	1.3	1.5	1.8	2.1	2.6	3.1	3.6	4.1	4.6	5.1	5.6	6.2	7.2	8.2	10.2	—	—
	70	0.5	0.7	1.0	1.2	1.5	1.7	2.0	2.4	2.9	3.4	3.9	4.4	4.9	5.4	5.8	6.8	7.8	9.7	12.2	—
14	14	0.7	1.1	1.4	1.8	2.1	2.5	2.9	3.6	4.3	5.0	5.7	6.4	7.1	7.8	8.5	10.0	11.4	—	—	—
	20	0.6	0.9	1.2	1.5	1.8	2.1	2.4	3.0	3.6	4.2	4.9	5.5	6.1	6.7	7.3	8.6	9.8	12.3	—	—
	30	0.5	0.8	1.0	1.3	1.6	1.8	2.1	2.6	3.1	3.7	4.2	4.7	5.2	5.8	6.3	7.3	8.4	10.5	—	—
	42	0.5	0.7	1.0	1.2	1.4	1.7	1.9	2.4	2.9	3.3	3.8	4.3	4.7	5.2	5.7	6.7	7.6	9.5	11.9	—
	60	0.4	0.7	0.9	1.1	1.3	1.5	1.8	2.2	2.6	3.1	3.5	3.9	4.4	4.8	5.2	6.1	7.0	8.8	10.9	—
	90	0.4	0.6	0.8	1.0	1.2	1.4	1.6	2.0	2.5	2.9	3.3	3.7	4.1	4.5	5.0	5.8	6.6	8.3	10.3	12.4
17	17	0.6	0.9	1.2	1.5	1.8	2.1	2.3	2.9	3.5	4.1	4.7	5.3	5.9	6.5	7.0	8.2	9.4	11.7	—	—
	25	0.5	0.7	1.0	1.2	1.5	1.7	2.0	2.5	3.0	3.5	4.0	4.5	5.0	5.5	6.0	7.0	8.0	10.0	12.5	—
	35	0.4	0.7	0.9	1.1	1.3	1.5	1.7	2.2	2.6	3.1	3.5	3.9	4.4	4.8	5.2	6.1	7.0	8.7	10.9	—
	50	0.4	0.6	0.8	1.0	1.2	1.4	1.6	2.0	2.4	2.8	3.1	3.5	3.9	4.3	4.5	5.4	6.2	7.7	9.7	11.6
	80	0.4	0.5	0.7	0.9	1.1	1.2	1.4	1.8	2.1	2.5	2.9	3.3	3.6	4.0	4.3	5.1	5.8	7.2	9.0	10.9
	120	0.3	0.5	0.7	0.8	1.0	1.2	1.3	1.7	2.0	2.3	2.7	3.0	3.4	3.7	4.0	4.7	5.4	6.7	8.4	10.1
20	20	0.5	0.7	1.0	1.2	1.5	1.7	2.0	2.5	3.0	3.5	4.0	4.5	5.0	5.5	6.0	7.0	8.0	10.0	12.5	—
	30	0.4	0.6	0.8	1.0	1.2	1.5	1.7	2.1	2.5	2.9	3.3	3.7	4.1	4.5	4.9	5.8	6.6	8.2	10.3	12.4
	45	0.4	0.5	0.7	0.9	1.1	1.3	1.4	1.8	2.2	2.5	2.9	3.3	3.6	4.0	4.3	5.1	5.8	7.2	9.1	10.9
	60	0.3	0.5	0.7	0.8	1.0	1.2	1.3	1.7	2.0	2.3	2.7	3.0	3.4	3.7	4.0	4.7	5.4	6.7	8.4	10.1
	90	0.3	0.5	0.6	0.8	0.9	1.1	1.2	1.5	1.8	2.1	2.4	2.7	3.0	3.3	3.6	4.2	4.8	6.0	7.5	9.0
	150	0.3	0.4	0.6	0.7	0.8	1.0	1.1	1.4	1.7	2.0	2.3	2.6	2.9	3.2	3.4	4.0	4.6	5.7	7.2	8.6
24	24	0.4	0.6	0.8	1.0	1.2	1.5	1.7	2.1	2.5	2.9	3.3	3.7	4.1	4.5	5.0	5.8	6.7	8.2	10.3	12.4
	32	0.4	0.5	0.7	0.9	1.1	1.3	1.5	1.8	2.2	2.6	2.9	3.3	3.6	4.0	4.3	5.1	5.8	7.2	9.0	11.0
	50	0.3	0.5	0.6	0.8	0.9	1.1	1.2	1.5	1.8	2.2	2.5	2.8	3.1	3.4	3.7	4.4	5.0	6.2	7.8	9.4
	70	0.3	0.4	0.6	0.7	0.8	1.0	1.1	1.4	1.7	2.0	2.2	2.5	2.8	3.0	3.3	3.8	4.4	5.5	6.9	8.2
	100	0.3	0.4	0.5	0.6	0.8	0.9	1.0	1.3	1.6	1.8	2.1	2.4	2.6	2.9	3.1	3.7	4.2	5.2	6.5	7.9
	160	0.2	0.4	0.5	0.6	0.7	0.8	1.0	1.2	1.4	1.7	1.9	2.1	2.4	2.6	2.8	3.3	3.8	4.7	5.9	7.1
30	30	0.3	0.5	0.7	0.8	1.0	1.2	1.3	1.7	2.0	2.3	2.7	3.0	3.3	3.7	4.0	4.7	5.4	6.7	8.4	10.0
	45	0.3	0.4	0.6	0.7	0.8	1.0	1.1	1.4	1.7	1.9	2.2	2.5	2.7	3.0	3.3	3.8	4.4	5.5	6.9	8.2
	60	0.3	0.4	0.5	0.6	0.7	0.9	1.0	1.2	1.5	1.7	2.0	2.2	2.5	2.7	3.0	3.5	4.0	5.0	6.2	7.4
	90	0.2	0.3	0.4	0.6	0.7	0.8	0.9	1.1	1.3	1.6	1.8	2.0	2.2	2.5	2.7	3.1	3.6	4.5	5.6	6.7
	150	0.2	0.3	0.4	0.5	0.6	0.7	0.8	1.0	1.2	1.4	1.6	1.8	2.0	2.2	2.4	2.8	3.2	4.0	5.0	5.9
	200	0.2	0.3	0.4	0.5	0.6	0.7	0.8	1.0	1.1	1.3	1.5	1.7	1.9	2.0	2.2	2.6	3.0	3.7	4.7	5.6
36	36	0.3	0.4	0.6	0.7	0.8	1.0	1.1	1.4	1.7	1.9	2.2	2.5	2.8	3.0	3.3	3.9	4.4	5.5	6.9	8.3
	50	0.2	0.4	0.5	0.6	0.7	0.8	1.0	1.2	1.4	1.7	1.9	2.1	2.3	2.6	2.9	3.3	3.8	4.8	5.9	7.2
	75	0.2	0.3	0.4	0.5	0.6	0.7	0.8	1.0	1.2	1.4	1.6	1.8	2.0	2.3	2.5	2.9	3.3	4.1	5.1	6.1
	100	0.2	0.3	0.4	0.5	0.6	0.7	0.8	0.9	1.1	1.3	1.5	1.7	1.9	2.1	2.3	2.6	3.0	3.8	4.7	5.7
	150	0.2	0.3	0.3	0.4	0.5	0.6	0.7	0.9	1.0	1.2	1.4	1.6	1.7	1.9	2.1	2.4	2.8	3.5	4.3	5.2
	200	0.2	0.2	0.3	0.4	0.5	0.6	0.7	0.8	1.0	1.1	1.3	1.5	1.6	1.8	2.0	2.3	2.6	3.3	4.1	4.9
42	42	0.2	0.4	0.5	0.6	0.7	0.8	1.0	1.2	1.4	1.6	1.9	2.1	2.4	2.6	2.8	3.3	3.8	4.7	5.9	7.1
	60	0.2	0.3	0.4	0.5	0.6	0.7	0.8	1.0	1.2	1.4	1.6	1.8	2.0	2.2	2.4	2.8	3.2	4.0	5.0	6.0
	90	0.2	0.3	0.3	0.4	0.5	0.6	0.7	0.9	1.0	1.2	1.4	1.6	1.7	1.9	2.1	2.4	2.8	3.5	4.4	5.2
	140	0.2	0.2	0.3	0.4	0.5	0.5	0.6	0.8	0.9	1.1	1.2	1.4	1.5	1.7	1.9	2.2	2.5	3.1	3.9	4.6
	200	0.1	0.2	0.3	0.4	0.4	0.5	0.6	0.7	0.9	1.0	1.1	1.3	1.4	1.6	1.7	2.0	2.3	2.9	3.6	4.3
	300	0.1	0.2	0.3	0.3	0.4	0.5	0.5	0.7	0.8	0.9	1.1	1.3	1.4	1.5	1.7	1.9	2.2	2.8	3.5	4.2

*IES Lighting Handbook, *1984 Reference Volume*

Fig. 58 *Continued*

Room Dimensions		Cavity Depth																			
Width	Length	1	1.5	2	2.5	3	3.5	4	5	6	7	8	9	10	11	12	14	16	20	25	30
50	50	0.2	0.3	0.4	0.5	0.6	0.7	0.8	1.0	1.2	1.4	1.6	1.8	2.0	2.2	2.4	2.8	3.2	4.0	5.0	6.0
	70	0.2	0.3	0.3	0.4	0.5	0.6	0.7	0.9	1.0	1.2	1.4	1.5	1.7	1.9	2.0	2.4	2.7	3.4	4.3	5.1
	100	0.1	0.2	0.3	0.4	0.4	0.5	0.6	0.7	0.9	1.0	1.2	1.3	1.5	1.6	1.8	2.1	2.4	3.0	3.7	4.5
	150	0.1	0.2	0.3	0.3	0.4	0.5	0.5	0.7	0.8	0.9	1.1	1.2	1.3	1.5	1.6	1.9	2.1	2.7	3.3	4.0
	300	0.1	0.2	0.2	0.3	0.3	0.4	0.5	0.6	0.7	0.8	0.9	1.0	1.1	1.3	1.4	1.6	1.9	2.3	2.9	3.5
60	60	0.2	0.2	0.3	0.4	0.5	0.6	0.7	0.8	1.0	1.2	1.3	1.5	1.7	1.8	2.0	2.3	2.7	3.3	4.2	5.0
	100	0.1	0.2	0.3	0.3	0.4	0.5	0.5	0.7	0.8	0.9	1.1	1.2	1.3	1.5	1.6	1.9	2.1	2.7	3.3	4.0
	150	0.1	0.2	0.2	0.3	0.3	0.4	0.5	0.6	0.7	0.8	0.9	1.0	1.2	1.3	1.4	1.6	1.9	2.3	2.9	3.5
	300	0.1	0.1	0.2	0.2	0.3	0.3	0.4	0.5	0.6	0.7	0.8	0.9	1.0	1.1	1.2	1.4	1.6	2.0	2.5	3.0
75	75	0.1	0.2	0.3	0.3	0.4	0.5	0.5	0.7	0.8	0.9	1.1	1.2	1.3	1.5	1.6	1.9	2.1	2.7	3.3	4.0
	120	0.1	0.2	0.2	0.3	0.3	0.4	0.4	0.5	0.6	0.8	0.9	1.0	1.1	1.2	1.3	1.5	1.7	2.2	2.7	3.3
	200	0.1	0.1	0.2	0.2	0.3	0.3	0.4	0.5	0.5	0.6	0.7	0.8	0.9	1.0	1.1	1.3	1.5	1.8	2.3	2.7
	300	0.1	0.1	0.2	0.2	0.2	0.3	0.3	0.4	0.5	0.6	0.7	0.7	0.8	0.9	1.0	1.2	1.3	1.7	2.1	2.5
100	100	0.1	0.1	0.2	0.2	0.3	0.3	0.4	0.5	0.6	0.7	0.8	0.9	1.0	1.1	1.2	1.4	1.6	2.0	2.5	3.0
	200	0.1	0.1	0.1	0.2	0.2	0.3	0.3	0.4	0.4	0.5	0.6	0.7	0.7	0.8	0.9	1.0	1.2	1.5	1.9	2.2
	300	0.1	0.1	0.1	0.2	0.2	0.2	0.3	0.3	0.4	0.5	0.5	0.6	0.7	0.7	0.8	0.9	1.1	1.3	1.7	2.0
150	150	0.1	0.1	0.1	0.2	0.2	0.2	0.3	0.3	0.4	0.5	0.5	0.6	0.7	0.7	0.8	0.9	1.1	1.3	1.7	2.0
	300	—	0.1	0.1	0.1	0.1	0.2	0.2	0.2	0.3	0.3	0.4	0.5	0.5	0.6	0.6	0.7	0.8	1.0	1.2	1.5
200	200	—	0.1	0.1	0.1	0.1	0.2	0.2	0.2	0.3	0.3	0.4	0.5	0.5	0.6	0.6	0.7	0.8	1.0	1.2	1.5
	300	—	0.1	0.1	0.1	0.1	0.1	0.2	0.2	0.2	0.3	0.3	0.4	0.4	0.5	0.5	0.6	0.7	0.8	1.0	1.2
300	300	—	—	0.1	0.1	0.1	0.1	0.1	0.2	0.2	0.2	0.3	0.3	0.3	0.4	0.4	0.5	0.5	0.6	0.7	0.8
500	500	—	—	—	—	0.1	0.1	0.1	0.1	0.1	0.1	0.2	0.2	0.2	0.2	0.2	0.3	0.3	0.4	0.5	0.6

Fig. 59. Per Cent Effective Ceiling or Floor Cavity Reflectances for Various Reflectance Combinations

See Page 9–26 ‡ for Algorithms for Calculation of Effective Cavity Reflectance*

Per Cent Base† Reflectance	90										80										70										60										50									
Per Cent Wall Reflectance	90	80	70	60	50	40	30	20	10	0	90	80	70	60	50	40	30	20	10	0	90	80	70	60	50	40	30	20	10	0	90	80	70	60	50	40	30	20	10	0	90	80	70	60	50	40	30	20	10	0
Cavity Ratio																																																		
0.2	89	88	88	87	86	85	84	84	84	82	79	78	78	77	77	76	76	75	74	72	70	69	68	68	67	67	66	66	65	64	60	59	59	58	57	56	56	55	55	53	50	49	49	48	48	47	46	46	44	44
0.4	88	87	86	85	84	83	81	80	79	76	79	77	76	75	74	73	72	71	70	68	69	68	67	66	64	64	63	62	61	58	60	59	57	56	55	54	53	52	52	50	50	48	48	47	46	45	45	44	42	42
0.6	87	86	84	82	80	77	76	75	74	73	78	77	75	73	71	70	68	66	65	63	69	67	65	64	63	61	59	59	57	54	60	58	56	54	53	51	51	50	48	46	50	48	47	46	45	44	43	41	41	38
0.8	87	85	82	80	77	75	73	71	69	67	78	75	73	71	69	67	65	63	61	57	68	66	64	62	60	58	56	55	53	50	59	57	56	54	51	48	48	45	44	43	50	48	47	45	44	43	42	40	38	36
1.0	86	83	80	77	75	72	69	66	64	62	77	74	72	69	67	65	62	60	57	55	68	65	62	60	58	55	53	52	48	47	59	57	55	53	51	48	45	43	41	41	50	48	46	44	43	41	38	37	36	34
1.2	85	82	78	75	72	69	66	63	60	57	76	73	70	67	64	61	58	55	53	51	67	64	61	58	55	51	48	46	46	44	59	56	54	51	49	47	44	42	39	38	50	47	45	43	41	39	36	35	34	29
1.4	85	80	77	73	69	65	62	59	57	52	76	72	68	65	62	59	55	53	50	48	67	63	60	56	53	50	47	45	44	41	59	56	53	49	47	45	41	39	37	36	50	47	44	42	40	38	35	34	32	27
1.6	84	79	75	71	67	63	60	56	53	50	75	71	67	63	59	55	53	50	47	44	66	62	59	55	51	46	45	42	40	38	58	55	52	49	47	44	42	37	35	33	50	47	44	41	38	36	33	32	30	26
1.8	84	78	73	69	64	60	56	53	50	48	75	70	66	62	58	54	50	47	44	41	66	61	58	54	49	45	42	40	38	35	58	55	51	47	45	40	37	35	33	31	50	46	43	40	37	35	32	30	28	25
2.0	83	77	72	67	62	58	53	50	47	43	74	69	64	60	56	52	47	45	42	38	66	60	56	52	49	42	40	38	36	33	58	54	50	46	43	40	35	33	31	29	50	46	43	40	37	34	30	28	26	24
2.2	82	76	70	65	59	54	50	47	44	40	74	68	63	58	54	49	45	42	40	35	66	60	55	51	48	43	38	36	34	32	58	53	49	45	42	37	34	31	29	28	50	46	42	38	35	33	29	27	25	22
2.4	82	75	69	64	58	53	48	45	42	37	73	67	61	56	52	47	43	40	37	33	65	59	54	49	46	41	37	34	31	30	58	53	48	44	41	36	32	30	27	26	50	46	41	37	35	31	27	25	23	21
2.6	81	74	67	62	56	51	46	42	39	35	73	66	60	55	50	45	41	37	35	31	65	58	53	48	45	40	35	32	30	28	58	53	47	43	39	35	31	28	26	24	50	46	41	37	34	30	26	23	21	20
2.8	81	73	66	60	54	49	44	40	36	34	73	65	59	53	48	43	39	36	33	29	65	58	52	47	43	38	33	30	28	26	58	52	47	42	38	34	30	27	25	22	50	45	40	36	33	28	24	22	20	19
3.0	80	72	65	58	52	47	42	38	34	30	72	65	58	52	47	42	37	34	30	27	64	58	51	46	42	37	32	29	27	24	57	52	46	41	37	32	28	25	23	20	50	45	40	35	32	27	23	21	19	17
3.2	79	71	63	56	50	45	40	36	32	28	72	65	57	51	46	40	36	32	29	25	64	58	50	46	41	36	31	28	25	23	57	51	45	41	36	31	27	23	22	18	50	44	39	35	31	27	23	20	18	16
3.4	79	70	62	54	48	43	38	34	30	27	71	63	56	49	44	39	34	32	28	24	64	57	50	45	40	35	29	27	24	22	57	51	44	39	35	30	26	23	19	17	50	44	39	35	31	26	22	19	16	15
3.6	78	69	61	53	47	42	36	32	28	25	71	63	55	48	43	38	33	29	26	23	64	56	49	44	39	34	29	25	23	21	56	50	43	38	34	29	25	21	18	16	50	44	37	34	30	25	21	17	15	14
3.8	78	69	60	51	45	40	35	31	27	23	70	62	54	47	41	36	30	28	25	22	63	56	48	43	38	32	27	24	21	19	56	50	42	37	33	28	24	20	17	15	50	43	37	33	30	24	20	17	13	12
4.0	77	69	58	50	44	39	34	29	25	22	70	61	53	46	40	35	30	27	22	20	63	55	48	42	37	31	26	23	20	17	56	49	42	37	32	28	23	20	15	14	50	43	37	33	28	24	20	17	12	12
4.2	77	62	57	50	43	37	32	28	24	21	69	60	52	45	39	34	29	25	21	18	62	55	47	41	35	30	25	21	19	16	56	49	42	37	32	27	22	19	17	14	50	43	37	32	28	24	20	17	14	12
4.4	76	61	56	49	42	36	31	27	23	20	69	60	51	44	38	32	28	24	19	17	62	54	46	40	34	29	24	21	18	15	55	49	42	36	31	27	22	19	16	13	50	42	37	32	27	23	19	16	13	11
4.6	76	60	55	47	41	35	30	25	21	18	68	60	50	43	37	31	26	23	18	15	62	53	45	39	33	27	22	19	15	14	55	48	42	35	30	26	21	18	14	12	50	42	36	31	26	23	18	15	13	10
4.8	75	59	54	46	40	35	28	24	21	18	68	58	49	42	36	30	25	21	17	14	61	53	44	38	32	27	21	17	14	12	55	48	41	35	29	25	20	17	13	11	49	41	35	30	25	22	17	14	12	09
5.0	75	59	53	45	39	34	28	24	21	16	68	58	48	41	35	30	24	20	16	12	61	52	44	37	31	26	20	16	13	12	57	48	41	34	28	24	20	17	14	11	47	40	34	30	25	21	17	14	12	09
6.0	73	61	50	44	34	29	24	20	16	11	66	55	46	38	30	25	21	18	14	10	60	51	41	35	28	23	19	16	13	09	55	45	37	31	25	21	17	14	11	07	50	44	32	29	23	20	15	13	10	06
7.0	70	58	49	41	30	27	21	18	14	11	64	53	43	35	27	23	19	16	13	09	58	50	40	34	27	22	17	15	12	08	53	42	33	28	22	19	14	12	09	06	49	41	31	27	21	19	14	11	08	05
8.0	68	55	42	38	27	23	18	15	12	08	62	50	38	32	25	21	17	14	11	05	57	46	35	30	23	19	15	13	10	05	50	38	30	26	20	16	12	09	07	04	49	40	30	25	19	16	12	10	07	03
9.0	66	52	38	35	25	21	16	14	11	06	61	49	36	30	23	18	14	11	07	04	56	43	33	27	21	16	12	10	08	05	42	33	28	23	18	15	12	10	07	03	48	39	29	24	18	15	12	09	07	03
10.0	65	51	36	31	29	19	15	11	09	04	59	46	33	27	18	15	13	11	08	03	55	43	31	25	19	12	10	08	06	02	39	29	24	18	15	11	09	08	07	02	47	37	27	21	17	14	10	08	06	02

* Values in this table are based on a length to width ratio of 1.6.

† Ceiling, floor or floor of cavity.

‡ In IES Lighting Handbook, *1984 Reference Volume*

Fig. 59 *Continued**

Per Cent Base† Reflectance	0										10										20										30										40									
Per Cent Wall Reflectance / Cavity Ratio	90	80	70	60	50	40	30	20	10	0	90	80	70	60	50	40	30	20	10	0	90	80	70	60	50	40	30	20	10	0	90	80	70	60	50	40	30	20	10	0	90	80	70	60	50	40	30	20	10	0
0.2	02	02	02	01	01	01	01	00	00	00	09	09	09	08	08	08	07	07	06	06	21	20	20	19	19	19	19	19	18	17	31	31	30	30	29	29	28	28	27	27	40	40	39	39	38	38	37	37	36	36
0.4	04	04	03	03	02	02	01	01	01	00	09	09	08	08	08	07	07	06	06	05	22	21	20	20	19	18	18	17	17	16	31	31	30	29	28	28	27	26	25	25	41	40	39	38	37	36	35	34	34	34
0.6	05	05	04	04	03	02	02	01	01	00	10	09	09	09	08	08	07	06	06	05	23	22	21	20	19	18	17	16	16	15	32	31	30	29	28	27	26	25	24	23	41	40	39	38	37	35	34	33	33	32
0.8	07	06	05	05	04	03	02	02	01	00	11	10	10	09	09	08	07	06	06	05	24	22	21	20	19	18	17	16	15	14	32	31	30	29	28	26	25	24	23	22	41	40	38	37	36	35	33	32	31	31
1.0	08	07	06	05	04	03	03	02	02	00	12	11	10	10	09	08	07	07	06	05	25	23	22	20	19	18	17	16	15	13	33	32	30	29	28	26	25	24	22	21	42	40	38	37	35	33	32	31	30	29
1.2	10	09	08	07	06	05	04	04	03	01	12	11	11	10	09	09	08	07	07	06	25	23	22	20	19	18	16	15	14	12	34	32	31	29	27	26	24	22	21	19	42	40	38	37	36	34	32	31	29	27
1.4	11	10	09	08	07	06	05	04	04	01	13	12	11	11	10	09	08	08	07	06	26	24	22	21	20	18	17	16	14	11	34	32	31	29	27	25	24	22	21	18	42	40	39	37	36	33	31	30	29	27
1.6	12	10	10	09	08	07	06	05	04	01	13	13	12	11	11	10	09	08	07	06	27	24	23	21	20	18	16	15	14	11	34	33	31	29	27	25	23	21	19	17	42	40	39	37	35	33	31	30	28	25
1.8	13	11	11	10	09	08	07	06	05	02	14	13	12	12	11	10	09	08	07	06	27	25	23	21	20	18	17	15	13	10	35	33	31	29	27	25	23	21	19	16	42	40	39	37	35	33	31	29	28	24
2.0	14	12	11	11	10	09	07	06	05	02	14	13	13	12	11	11	09	09	08	07	28	25	23	21	20	18	16	15	13	10	35	33	31	29	28	25	23	20	18	14	42	40	39	37	36	33	31	29	28	23
2.2	15	13	12	11	11	10	09	07	06	02	16	14	14	13	13	12	11	10	09	07	28	25	23	22	20	18	16	14	13	09	36	33	32	29	27	25	23	21	18	13	42	40	39	37	35	33	31	30	27	22
2.4	16	13	13	12	11	10	09	08	07	02	16	14	14	13	13	12	11	10	09	08	29	26	23	22	20	18	16	14	13	09	36	33	32	29	27	25	22	20	17	12	43	40	39	37	35	33	30	29	27	21
2.6	16	14	13	12	12	11	10	09	07	02	17	15	14	14	13	12	11	10	09	08	29	26	24	22	20	18	16	14	12	08	36	34	32	29	27	24	22	19	17	12	43	40	39	37	35	33	29	28	26	20
2.8	17	14	14	13	12	11	10	10	08	02	17	16	15	14	13	13	11	10	09	08	30	27	24	23	20	18	15	13	11	07	37	34	32	29	27	24	21	19	16	11	43	40	38	37	35	32	29	27	25	19
3.0	18	16	14	13	13	11	11	10	08	02	18	16	15	14	13	13	11	11	09	08	30	27	24	23	21	18	15	13	11	07	37	34	33	29	27	24	20	18	15	10	43	40	38	37	35	32	29	27	24	18
3.2	19	16	15	14	13	12	11	11	09	03	19	17	16	15	14	13	12	11	10	09	31	27	25	23	20	18	16	14	12	07	37	35	33	29	28	24	22	19	16	10	43	41	71	37	35	33	30	27	25	18
3.4	19	16	15	14	13	12	11	11	09	03	19	17	16	15	14	13	12	11	10	08	32	27	25	23	20	17	15	13	11	06	38	35	33	29	27	24	21	18	15	09	43	41	39	37	35	33	29	27	24	17
3.6	20	17	16	15	14	13	12	11	10	04	20	18	16	16	14	13	12	11	10	09	32	28	25	23	20	17	15	13	11	06	38	35	33	29	27	23	21	18	15	09	44	41	39	37	35	33	29	26	24	16
3.8	20	18	16	15	14	13	13	12	11	04	21	19	17	16	15	14	13	12	11	09	33	28	25	23	20	17	14	12	10	05	38	35	33	29	26	23	20	17	14	08	44	41	39	37	35	32	29	26	23	15
4.0	21	18	17	16	15	14	13	12	11	04	22	19	18	16	15	14	13	12	11	09	33	28	25	23	20	17	14	12	09	05	38	35	33	28	26	23	20	17	13	07	44	41	38	37	35	32	28	26	23	14
4.2	22	19	18	16	15	14	13	13	12	04	24	21	19	18	17	15	14	13	11	09	33	28	26	23	20	17	14	12	09	04	38	35	33	28	26	23	20	17	13	07	44	41	38	37	35	32	28	25	22	14
4.4	23	19	18	17	16	14	14	13	12	04	24	21	20	18	17	15	14	13	11	09	34	29	26	24	20	17	14	11	09	04	39	33	33	28	26	23	19	16	13	06	44	41	38	37	35	32	28	25	21	13
4.6	23	20	19	17	16	15	14	13	12	04	20	20	19	18	17	15	14	13	12	10	35	30	26	24	20	17	13	11	08	03	39	33	33	28	26	23	19	16	12	06	44	41	38	37	34	32	27	25	21	13
4.8	24	21	19	18	17	15	14	14	13	04	25	21	20	19	17	16	14	13	12	08	35	29	25	23	20	16	13	10	08	03	40	33	33	28	26	23	19	15	12	05	44	41	38	36	34	31	27	24	21	12
5.0	25	22	20	18	17	16	15	14	13	04	27	23	21	19	17	16	14	13	12	08	35	29	25	23	20	16	12	10	07	03	40	33	32	27	25	22	18	15	11	05	45	41	38	36	34	31	26	24	20	10
6.0	27	23	21	19	18	17	16	15	14	06	31	26	24	21	20	18	16	15	13	08	36	30	24	22	19	14	11	08	06	01	39	33	27	23	20	18	15	11	08	04	44	41	37	35	33	31	25	21	18	08
7.0	28	24	23	20	19	17	16	15	15	07	33	27	24	21	20	18	16	15	13	08	37	30	24	22	19	14	11	08	06	01	40	33	26	22	19	17	14	10	07	03	44	41	37	34	32	30	24	20	17	07
8.0	29	25	23	21	20	18	17	15	15	08	33	27	25	22	20	18	16	14	13	07	37	30	23	21	18	13	10	07	05	01	40	33	26	21	18	16	13	09	07	02	44	41	36	34	31	29	23	19	16	06
9.0	30	25	24	21	20	19	17	16	15	08	34	28	25	22	20	18	16	14	13	07	37	29	23	21	18	13	09	07	04	01	40	33	25	20	17	15	12	09	06	02	44	41	35	33	30	28	22	18	16	05
10.0	31	25	24	21	20	19	17	16	15	08	34	28	25	22	21	18	16	14	13	07	37	29	23	20	17	12	09	06	04	01	40	34	25	20	16	14	11	08	05	01	43	41	34	32	29	27	21	17	15	05

* Values in this table are based on a length to width ratio of 1.6.
† Ceiling, floor or floor of cavity.

Fig. 60   Multiplying Factors for Other than 20 Per Cent Effective Floor Cavity Reflectance

% Effective Ceiling Cavity Reflectance, ρcc	80				70				50			30			10		
% Wall Reflectance, ρw	70	50	30	10	70	50	30	10	50	30	10	50	30	10	50	30	10

For 30 Per Cent Effective Floor Cavity Reflectance (20 Per Cent = 1.00)

Room Cavity Ratio																	
1	1.092	1.082	1.075	1.068	1.077	1.070	1.064	1.059	1.049	1.044	1.040	1.028	1.026	1.023	1.012	1.010	1.008
2	1.079	1.066	1.055	1.047	1.068	1.057	1.048	1.039	1.041	1.033	1.027	1.026	1.021	1.017	1.013	1.010	1.006
3	1.070	1.054	1.042	1.033	1.061	1.048	1.037	1.028	1.034	1.027	1.020	1.024	1.017	1.012	1.014	1.009	1.005
4	1.062	1.045	1.033	1.024	1.055	1.040	1.029	1.021	1.030	1.022	1.015	1.022	1.015	1.010	1.014	1.009	1.005
5	1.056	1.038	1.026	1.018	1.050	1.034	1.024	1.015	1.027	1.018	1.012	1.020	1.013	1.008	1.014	1.009	1.004
6	1.052	1.033	1.021	1.014	1.047	1.030	1.020	1.012	1.024	1.015	1.009	1.019	1.012	1.006	1.014	1.008	1.003
7	1.047	1.029	1.018	1.011	1.043	1.026	1.017	1.009	1.022	1.013	1.007	1.018	1.010	1.005	1.014	1.008	1.003
8	1.044	1.026	1.015	1.009	1.040	1.024	1.015	1.007	1.020	1.012	1.006	1.017	1.009	1.004	1.013	1.007	1.003
9	1.040	1.024	1.014	1.007	1.037	1.022	1.014	1.006	1.019	1.011	1.005	1.016	1.009	1.004	1.013	1.007	1.002
10	1.037	1.022	1.012	1.006	1.034	1.020	1.012	1.005	1.017	1.010	1.004	1.015	1.009	1.003	1.013	1.007	1.002

For 10 Per Cent Effective Floor Cavity Reflectance (20 Per Cent = 1.00)

Room Cavity Ratio																	
1	.923	.929	.935	.940	.933	.939	.943	.948	.956	.960	.963	.973	.976	.979	.989	.991	.993
2	.931	.942	.950	.958	.940	.949	.957	.963	.962	.968	.974	.976	.980	.985	.988	.991	.995
3	.939	.951	.961	.969	.945	.957	.966	.973	.967	.975	.981	.978	.983	.988	.988	.991	.995
4	.944	.958	.969	.978	.950	.963	.973	.980	.972	.980	.986	.980	.986	.991	.988	.992	.996
5	.949	.964	.976	.983	.954	.968	.978	.985	.975	.983	.989	.981	.988	.993	.987	.992	.996
6	.953	.969	.980	.986	.958	.972	.982	.989	.977	.985	.992	.982	.989	.995	.987	.993	.997
7	.957	.973	.983	.991	.961	.975	.985	.991	.979	.987	.994	.983	.990	.996	.987	.993	.997
8	.960	.976	.986	.993	.963	.977	.987	.993	.981	.988	.995	.984	.991	.997	.987	.993	.998
9	.963	.978	.987	.994	.965	.979	.989	.994	.983	.990	.996	.985	.992	.998	.987	.994	.998
10	.965	.980	.989	.995	.967	.981	.990	.995	.984	.991	.997	.986	.993	.998	.988	.994	.999

For 0 Per Cent Effective Floor Cavity Reflectance (20 Per Cent = 1.00)

Room Cavity Ratio																	
1	.859	.870	.879	.886	.873	.884	.893	.901	.916	.923	.929	.948	.954	.960	.979	.983	.987
2	.871	.887	.903	.919	.886	.902	.916	.928	.926	.938	.949	.954	.963	.971	.978	.983	.991
3	.882	.904	.915	.942	.898	.918	.934	.947	.936	.950	.964	.958	.969	.979	.976	.984	.993
4	.893	.919	.941	.958	.908	.930	.948	.961	.945	.961	.974	.961	.974	.984	.975	.985	.994
5	.903	.931	.953	.969	.914	.939	.958	.970	.951	.967	.980	.964	.977	.988	.975	.985	.995
6	.911	.940	.961	.976	.920	.945	.965	.977	.955	.972	.985	.966	.979	.991	.975	.986	.996
7	.917	.947	.967	.981	.924	.950	.970	.982	.959	.975	.988	.968	.981	.993	.975	.987	.997
8	.922	.953	.971	.985	.929	.955	.975	.986	.963	.978	.991	.970	.983	.995	.976	.988	.998
9	.928	.958	.975	.988	.933	.959	.980	.989	.966	.980	.993	.971	.985	.996	.976	.988	.998
10	.933	.962	.979	.991	.937	.963	.983	.992	.969	.982	.995	.973	.987	.997	.977	.989	.999

Instructions and Notes for Use of Fig. 61

(The following notes and instructions have been prepared to guide the user of Fig. 61.)

1. The luminaires in this table are organized by source type and luminaire form rather than by applications for convenience in locating luminaires.

    In some cases, luminaire data in this table are based on an actual typical luminaire; in other cases, the data represent a composite of generic luminaire types. Therefore, whenever possible, specific luminaire data should be used in preference to this table of typical luminaires.

2. The polar intensity sketch (candlepower distribution curve) and the corresponding luminaire spacing criterion are representative of many luminaires of each type shown. A specific luminaire may differ in perpendicular plane (crosswise) and parallel plane (lengthwise) intensity distributions and in spacing criterion from the values shown. However, the various coefficients depend only on the average intensity at each polar angle from nadir. The tabulated coefficients can be applied to any luminaire whose average intensity distribution matches the values used to generate the coefficients. The average intensity values used to generate the coefficients are given at the end of the table, normalized to a per thousand lamp lumen basis.

3. The various coefficients depend only on the average intensity distribution curve and are linearly related to the total luminaire efficiency. Consequently, the tabulated coefficients can be applied to luminaires with similarly shaped average intensity distributions using a correcting multiplier equal to the new luminaire total efficiency divided by the tabulated luminaire total efficiency. The use of polarizing diffusers or lenses on fluorescent luminaires has no effect on the coefficients given in this table except as they affect total luminaire efficiency.

4. Satisfactory installations depend on many factors including the environment, space utilization, luminous criteria, etc. as well as the luminaire itself. Consequently, a definitive spacing recommendation cannot be assigned independently to the luminaire. The spacing criterion (SC) values given are only general guides (see page 9-49).* SC values are not assigned to semi-indirect and indirect luminaires since the basis of this technique does not apply to such situations. Also, SC values are not given for those bat-wing luminaires which must be located by criteria other than that of horizontal illuminance.

5. Key: ρ_{CC} = ceiling cavity reflectance (per cent)
    ρ_W = wall reflectance (per cent)
    ρ_{FC} = floor cavity reflectance (per cent)
    RCR = room cavity ratio
    WDRC = wall direct radiation coefficient

SC = luminaire spacing criterion
NA = not applicable

6. Many of the luminaires in this figure appeared in earlier editions of the *IES Lighting Handbook*. The identifying number may be different due to a reordering of the luminaires. In some cases, the data have been modified in terms of more recent or more extensive information. The user should specifically refer to this handbook when referencing luminaires from this handbook.

7. Efficiency, and consequently the coefficients, of fluorescent luminaires is a function of the number of lamps in relation to the size of the luminaire. This is due to temperature changes and due to changes in the blocking of light. In this figure, fluorescent luminaires have been chosen for typical luminaire sizes and numbers of lamps; these are identified under the typical luminaire drawings. Variations of the coefficients with size and number of lamps depend on the many details of luminaire construction. The following correction factors are average values:

    4 lamp, 610 mm (2 ft) wide:
    ×1.05 for 8 lamp,
    1220 mm (4 ft) wide
    ×1.05 for 3 lamp,
    610 mm (2 ft) wide
    ×1.1 for 2 lamp, 610
    mm (2 ft) wide
    ×0.9 for 2 lamp, 300
    mm (1 ft) wide

    2 lamp wraparound:
    ×0.95 for 4 lamp

8. Photometric data for fluorescent luminaires in this table are based upon tests using standard wattage fluorescent lamps. Reduced wattage fluorescent lamps cause lower lamp operating temperatures with some luminaires. Consequently, the efficiency and coefficients may be slightly increased. It is desirable to obtain specific correction factors from the manufacturers. Typical factors for reduced wattage fluorescent lamps (approximately 10 per cent below standard lamp watts) are:

2-lamp strip, surface mounted	× 1.03
4-lamp troffer, enclosed, non air handling	× 1.07
4-lamp wrap around, surface mounted	× 1.07
2-lamp industrial, vented	× 1.00

Electronic ballasts can be designed for any arbitrary operating condition. The manufacturer must be consulted for specific data.

---

*IES Lighting Handbook, *1984 Reference Volume*

Fig. 61 Coefficients of Utilization, Wall Exitance Coefficients, Ceiling Cavity Exitance
(See page 119 for

Typical Luminaire	Typical Intensity Distribution and Per Cent Lamp Lumens	Maint. Cat.	SC	RCR ↓	ρcc 80 pw 50	30	10	ρcc 70 pw 50	30	10	ρcc 50 pw 50	30	10	ρcc 30 pw 50	30	10	ρcc 10 pw 50	30	10	ρcc 0	WDRC	RCR ↓
1 Pendant diffusing sphere with incandescent lamp	35½%↑ 45%↓	V	1.5	0	.87	.87	.87	.81	.81	.81	.70	.70	.70	.59	.59	.59	.49	.49	.49	.45		0
				1	.71	.66	.62	.65	.61	.58	.55	.52	.49	.46	.44	.42	.38	.36	.34	.30	.368	1
				2	.60	.53	.48	.55	.50	.45	.47	.42	.38	.39	.35	.32	.31	.29	.26	.23	.279	2
				3	.52	.44	.38	.48	.41	.36	.40	.35	.31	.33	.29	.26	.27	.24	.21	.18	.227	3
				4	.45	.37	.32	.42	.35	.29	.35	.30	.25	.29	.25	.21	.23	.20	.17	.14	.192	4
				5	.40	.32	.27	.37	.30	.25	.31	.25	.21	.26	.21	.18	.21	.17	.14	.12	.166	5
				6	.35	.28	.23	.33	.26	.21	.28	.22	.18	.23	.19	.15	.19	.15	.12	.10	.146	6
				7	.32	.25	.19	.29	.23	.18	.25	.20	.16	.21	.16	.13	.17	.13	.11	.09	.130	7
				8	.29	.22	.17	.27	.20	.16	.23	.17	.14	.19	.15	.12	.15	.12	.09	.07	.117	8
				9	.26	.19	.15	.24	.18	.14	.21	.16	.12	.17	.13	.10	.14	.11	.08	.07	.107	9
				10	.24	.17	.13	.22	.16	.12	.19	.14	.11	.16	.12	.09	.13	.10	.08	.06	.098	10
2 Concentric ring unit with incandescent silvered-bowl lamp	83%↓ 3½%↓	II	N.A.	0	.83	.83	.83	.72	.72	.72	.50	.50	.50	.30	.30	.30	.12	.12	.12	.03		0
				1	.72	.69	.66	.62	.60	.57	.43	.42	.40	.26	.25	.25	.10	.10	.10	.03	.018	1
				2	.63	.58	.54	.54	.50	.47	.38	.35	.33	.23	.22	.20	.09	.09	.08	.02	.015	2
				3	.55	.49	.45	.47	.43	.39	.33	.30	.28	.20	.19	.17	.08	.07	.07	.02	.013	3
				4	.48	.42	.37	.42	.37	.33	.29	.26	.23	.18	.16	.15	.07	.06	.06	.02	.012	4
				5	.43	.36	.32	.37	.32	.28	.26	.23	.20	.16	.14	.12	.06	.06	.05	.01	.011	5
				6	.38	.32	.27	.33	.28	.24	.23	.20	.17	.14	.12	.11	.06	.05	.04	.01	.010	6
				7	.34	.28	.23	.30	.24	.21	.21	.17	.15	.13	.11	.09	.05	.04	.04	.01	.009	7
				8	.31	.25	.20	.27	.21	.18	.19	.15	.13	.12	.10	.08	.05	.04	.03	.01	.008	8
				9	.28	.22	.18	.24	.19	.16	.17	.14	.11	.10	.09	.07	.04	.03	.03	.01	.008	9
				10	.25	.20	.16	.22	.17	.14	.16	.12	.10	.10	.08	.06	.04	.03	.03	.01	.007	10
3 Porcelain-enameled ventilated standard dome with incandescent lamp	0%↑ 83½%↓	IV	1.3	0	.99	.99	.99	.97	.97	.97	.93	.93	.93	.89	.89	.89	.85	.85	.85	.83		0
				1	.87	.84	.81	.85	.82	.79	.82	.79	.77	.79	.76	.74	.76	.74	.72	.71	.323	1
				2	.76	.70	.65	.74	.69	.65	.71	.67	.63	.69	.65	.62	.66	.63	.60	.59	.311	2
				3	.66	.59	.54	.65	.59	.53	.62	.57	.53	.60	.56	.52	.58	.54	.51	.49	.288	3
				4	.58	.51	.45	.57	.50	.45	.55	.49	.44	.53	.48	.44	.51	.47	.43	.41	.264	4
				5	.52	.44	.39	.51	.44	.38	.49	.43	.38	.47	.42	.37	.46	.41	.37	.35	.241	5
				6	.46	.39	.33	.46	.38	.33	.44	.38	.33	.43	.37	.33	.41	.36	.32	.31	.221	6
				7	.42	.34	.29	.41	.34	.29	.40	.33	.29	.39	.33	.29	.38	.32	.28	.27	.203	7
				8	.38	.31	.26	.37	.31	.26	.36	.30	.26	.35	.30	.26	.35	.29	.25	.24	.187	8
				9	.35	.28	.23	.34	.28	.23	.33	.27	.23	.32	.27	.23	.32	.26	.23	.21	.173	9
				10	.32	.25	.21	.32	.25	.21	.31	.25	.21	.30	.24	.21	.29	.24	.20	.19	.161	10
4 Prismatic square surface drum	18½%↑ 60½%↓	V	1.3	0	.89	.89	.89	.85	.85	.85	.77	.77	.77	.70	.70	.70	.63	.63	.63	.60		0
				1	.77	.74	.71	.74	.71	.68	.67	.65	.63	.61	.59	.57	.55	.54	.53	.50	.264	1
				2	.68	.63	.59	.65	.61	.57	.59	.56	.53	.54	.51	.49	.49	.47	.45	.42	.224	2
				3	.61	.55	.50	.58	.53	.48	.53	.49	.45	.49	.45	.42	.44	.42	.39	.37	.197	3
				4	.54	.48	.43	.52	.46	.42	.48	.43	.39	.44	.40	.37	.40	.37	.34	.32	.176	4
				5	.49	.42	.38	.47	.41	.37	.43	.38	.35	.40	.36	.33	.37	.33	.31	.29	.159	5
				6	.44	.38	.33	.43	.37	.32	.39	.34	.31	.36	.32	.29	.34	.30	.27	.26	.145	6
				7	.40	.34	.30	.39	.33	.29	.36	.31	.27	.33	.29	.26	.31	.27	.25	.23	.133	7
				8	.37	.31	.27	.36	.30	.26	.33	.28	.25	.31	.27	.24	.29	.25	.22	.21	.124	8
				9	.34	.28	.24	.33	.27	.24	.31	.26	.22	.29	.24	.21	.27	.23	.20	.19	.115	9
				10	.32	.26	.22	.30	.25	.21	.28	.24	.21	.27	.23	.20	.25	.21	.19	.17	.108	10
5 R-40 flood without shielding	0%↑ 100%↓	IV	0.8	0	1.19	1.19	1.19	1.16	1.16	1.16	1.11	1.11	1.11	1.06	1.06	1.06	1.02	1.02	1.02	1.00		0
				1	1.08	1.05	1.03	1.06	1.03	1.01	1.02	1.00	.98	.98	.97	.95	.95	.93	.92	.90	.241	1
				2	.99	.94	.89	.97	.92	.88	.93	.90	.86	.90	.87	.84	.88	.85	.83	.81	.238	2
				3	.90	.84	.79	.88	.83	.78	.86	.81	.77	.83	.79	.76	.81	.77	.74	.73	.227	3
				4	.82	.75	.70	.81	.75	.70	.79	.73	.69	.77	.72	.68	.75	.71	.67	.66	.215	4
				5	.76	.68	.63	.75	.68	.63	.73	.67	.62	.71	.66	.62	.69	.65	.61	.59	.202	5
				6	.70	.62	.57	.69	.62	.57	.67	.61	.57	.66	.60	.56	.64	.60	.56	.54	.191	6
				7	.65	.57	.52	.64	.57	.52	.62	.56	.52	.61	.56	.52	.60	.55	.51	.50	.180	7
				8	.60	.53	.48	.59	.53	.48	.58	.52	.48	.57	.52	.47	.56	.51	.47	.46	.169	8
				9	.56	.49	.44	.55	.49	.44	.54	.48	.44	.53	.48	.44	.52	.47	.44	.42	.160	9
				10	.52	.46	.41	.52	.45	.41	.51	.45	.41	.50	.45	.41	.49	.44	.41	.39	.152	10
6 R-40 flood with specular anodized reflector skirt; 45° cutoff	0%↑ 85%↓	IV	0.7	0	1.01	1.01	1.01	.99	.99	.99	.94	.94	.94	.90	.90	.90	.87	.87	.87	.85		0
				1	.95	.93	.91	.93	.91	.89	.89	.88	.87	.86	.85	.84	.83	.82	.82	.80	.115	1
				2	.89	.86	.83	.87	.84	.82	.85	.82	.80	.82	.80	.79	.80	.78	.77	.76	.115	2
				3	.83	.80	.77	.82	.79	.76	.80	.77	.75	.78	.76	.74	.76	.74	.72	.71	.113	3
				4	.79	.74	.71	.78	.74	.71	.76	.73	.70	.74	.71	.69	.73	.70	.68	.67	.110	4
				5	.74	.70	.67	.74	.69	.66	.72	.68	.66	.71	.68	.65	.69	.67	.65	.63	.107	5
				6	.70	.66	.62	.70	.65	.62	.68	.65	.62	.67	.64	.61	.66	.63	.61	.60	.104	6
				7	.67	.62	.59	.66	.62	.59	.65	.61	.58	.64	.61	.58	.63	.60	.58	.57	.100	7
				8	.63	.59	.56	.63	.58	.55	.62	.58	.55	.61	.58	.55	.60	.57	.55	.54	.097	8
				9	.60	.56	.53	.60	.56	.53	.59	.55	.52	.58	.55	.52	.58	.54	.52	.51	.094	9
				10	.57	.53	.50	.57	.53	.50	.56	.52	.50	.56	.52	.50	.55	.52	.49	.48	.091	10

Maint. Cat. / SC — Coefficients of Utilization for 20 Per Cent Effective Floor Cavity Reflectance (ρFC = 20)

Coefficients, Luminaire Spacing Criterion and Maintenance Categories of Typical Luminaires.
instructions and notes)

80			70			50			30			10			80			70			50			30			10		
50	30	10	50	30	10	50	30	10	50	30	10	50	30	10	50	30	10	50	30	10	50	30	10	50	30	10	50	30	10
Wall Exitance Coefficients for 20 Per Cent Effective Floor Cavity Reflectance ($\rho_{FC} = 20$)															Ceiling Cavity Exitance Coefficients for 20 Per Cent Floor Cavity Reflectance ($\rho_{FC} = 20$)														
															.423	.423	.423	.361	.361	.361	.246	.246	.246	.142	.142	.142	.045	.045	.045
.328	.187	.059	.311	.178	.056	.280	.161	.051	.252	.145	.047	.226	.131	.042	.422	.396	.373	.361	.340	.321	.247	.234	.222	.142	.135	.129	.046	.044	.042
.275	.150	.046	.259	.143	.044	.231	.129	.040	.205	.115	.036	.181	.102	.032	.417	.379	.347	.357	.327	.300	.245	.226	.209	.141	.131	.123	.045	.043	.040
.240	.128	.038	.226	.121	.036	.200	.108	.033	.176	.097	.030	.154	.085	.026	.412	.367	.332	.353	.317	.287	.242	.220	.202	.140	.128	.119	.045	.042	.039
.214	.111	.033	.201	.105	.031	.177	.094	.028	.155	.083	.025	.135	.073	.022	.406	.358	.321	.348	.309	.279	.239	.215	.196	.138	.126	.116	.045	.041	.038
.193	.098	.028	.181	.093	.027	.160	.083	.024	.139	.073	.022	.120	.064	.019	.400	.350	.314	.343	.303	.273	.236	.212	.193	.137	.124	.114	.044	.041	.038
.176	.088	.025	.165	.084	.024	.145	.074	.022	.126	.066	.019	.109	.057	.017	.394	.344	.309	.338	.298	.269	.234	.209	.190	.135	.123	.113	.044	.040	.037
.162	.080	.023	.152	.076	.022	.133	.067	.019	.116	.059	.017	.100	.052	.015	.388	.339	.305	.334	.294	.266	.231	.206	.188	.134	.122	.112	.043	.040	.037
.150	.073	.021	.140	.069	.020	.123	.062	.018	.107	.054	.016	.092	.047	.014	.383	.335	.302	.330	.291	.264	.228	.204	.187	.133	.120	.111	.043	.039	.037
.139	.067	.019	.131	.064	.018	.115	.057	.016	.099	.050	.014	.085	.043	.013	.378	.332	.300	.326	.288	.262	.226	.202	.186	.131	.119	.111	.043	.039	.037
.130	.062	.017	.122	.059	.016	.107	.052	.015	.093	.046	.013	.080	.040	.011	.374	.328	.298	.322	.285	.260	.223	.201	.185	.130	.119	.110	.042	.039	.037
															.796	.796	.796	.680	.680	.680	.464	.464	.464	.267	.267	.267	.085	.085	.085
.226	.128	.041	.195	.111	.035	.137	.078	.025	.083	.048	.015	.034	.020	.006	.790	.772	.756	.676	.663	.651	.462	.456	.450	.266	.264	.262	.085	.085	.085
.207	.114	.035	.179	.099	.030	.126	.070	.022	.077	.043	.013	.031	.018	.006	.784	.755	.731	.671	.650	.632	.460	.450	.441	.265	.262	.258	.085	.085	.084
.191	.102	.030	.165	.088	.027	.116	.063	.019	.071	.039	.012	.029	.016	.005	.778	.743	.715	.667	.641	.620	.458	.445	.435	.265	.260	.256	.085	.084	.084
.177	.092	.027	.153	.080	.024	.108	.057	.017	.066	.035	.011	.027	.014	.004	.773	.734	.703	.664	.634	.611	.456	.442	.430	.264	.259	.255	.085	.084	.084
.164	.084	.024	.142	.073	.021	.100	.052	.015	.061	.032	.010	.025	.013	.004	.768	.726	.696	.660	.629	.605	.455	.439	.427	.263	.258	.253	.085	.084	.084
.153	.077	.022	.133	.067	.019	.094	.048	.014	.057	.030	.009	.023	.012	.004	.764	.721	.690	.656	.624	.601	.453	.437	.425	.263	.257	.253	.085	.084	.083
.143	.071	.020	.124	.062	.018	.088	.044	.013	.054	.027	.008	.022	.011	.003	.759	.716	.686	.653	.621	.598	.451	.435	.423	.262	.256	.252	.085	.084	.083
.134	.066	.018	.116	.057	.016	.082	.041	.012	.050	.026	.007	.020	.010	.003	.755	.712	.683	.650	.618	.595	.450	.434	.422	.262	.256	.252	.085	.084	.083
.126	.061	.017	.109	.053	.015	.077	.038	.011	.047	.024	.007	.019	.010	.003	.751	.709	.680	.647	.615	.593	.448	.432	.421	.261	.255	.251	.085	.084	.083
.119	.057	.016	.103	.050	.014	.073	.036	.010	.045	.022	.006	.018	.009	.003	.747	.706	.678	.644	.613	.592	.447	.431	.421	.261	.255	.251	.084	.084	.083
															.159	.159	.159	.136	.136	.136	.093	.093	.093	.053	.053	.053	.017	.017	.017
.248	.141	.045	.242	.138	.044	.231	.133	.042	.221	.128	.041	.212	.123	.040	.150	.130	.113	.128	.112	.097	.088	.077	.067	.050	.045	.039	.016	.014	.013
.240	.131	.040	.235	.129	.040	.225	.125	.039	.216	.121	.038	.208	.117	.037	.143	.110	.082	.123	.095	.071	.084	.066	.050	.048	.038	.029	.016	.012	.009
.225	.120	.036	.220	.118	.036	.212	.115	.035	.204	.112	.034	.196	.109	.034	.137	.095	.062	.118	.082	.054	.081	.057	.038	.047	.033	.022	.015	.011	.007
.209	.109	.032	.205	.107	.032	.197	.105	.031	.190	.102	.031	.184	.100	.030	.131	.084	.048	.113	.073	.042	.077	.051	.030	.045	.030	.018	.014	.010	.006
.194	.099	.029	.191	.098	.029	.184	.096	.028	.177	.094	.028	.171	.092	.028	.125	.076	.039	.108	.065	.034	.074	.046	.024	.043	.027	.014	.014	.009	.005
.181	.091	.026	.177	.090	.026	.171	.088	.026	.166	.086	.025	.160	.084	.025	.119	.069	.032	.103	.060	.028	.071	.042	.020	.041	.025	.012	.013	.008	.004
.168	.083	.024	.165	.082	.023	.160	.081	.023	.155	.079	.023	.150	.078	.023	.114	.063	.027	.098	.055	.024	.068	.038	.017	.039	.023	.010	.013	.007	.003
.157	.077	.022	.155	.076	.022	.150	.075	.021	.145	.074	.021	.141	.072	.021	.109	.058	.024	.093	.050	.021	.065	.035	.015	.038	.021	.009	.012	.007	.003
.147	.071	.020	.145	.071	.020	.141	.070	.020	.136	.068	.020	.133	.067	.019	.103	.054	.021	.089	.047	.018	.062	.033	.013	.036	.019	.008	.012	.006	.003
.138	.066	.018	.136	.066	.018	.132	.065	.018	.129	.064	.018	.125	.063	.018	.099	.051	.018	.085	.044	.016	.059	.031	.011	.034	.018	.007	.011	.006	.002
															.290	.290	.290	.248	.248	.248	.169	.169	.169	.097	.097	.097	.031	.031	.031
.243	.138	.044	.232	.132	.042	.211	.121	.039	.192	.111	.036	.175	.101	.033	.283	.264	.247	.242	.227	.213	.166	.156	.147	.095	.090	.085	.031	.029	.028
.216	.118	.036	.206	.114	.035	.187	.104	.032	.170	.095	.030	.154	.087	.027	.276	.246	.221	.236	.212	.191	.162	.147	.133	.093	.085	.078	.030	.028	.025
.196	.104	.031	.187	.100	.030	.170	.092	.028	.154	.085	.026	.140	.077	.024	.269	.233	.204	.231	.201	.177	.158	.139	.124	.092	.081	.073	.029	.026	.024
.180	.093	.027	.171	.090	.027	.156	.083	.025	.142	.076	.023	.128	.070	.021	.263	.223	.192	.226	.192	.167	.155	.134	.117	090	.079	.069	.029	.026	.023
.166	.084	.024	.158	.081	.024	.144	.075	.022	.131	.069	.021	.119	.064	.019	.257	.215	.183	.221	.186	.160	.152	.130	.113	.088	.076	.067	.028	.025	.022
.154	.077	.022	.147	.074	.021	.134	.069	.020	.122	.064	.019	.111	.058	.017	.252	.208	.177	.216	.180	.154	.149	.126	.109	.087	.074	.065	.028	.024	.021
.143	.071	.020	.137	.068	.019	.125	.063	.018	.114	.059	.017	.104	.054	.016	.246	.203	.173	.212	.176	.151	.146	.123	.107	.085	.073	.063	.027	.024	.021
.134	.066	.018	.129	.063	.018	.118	.059	.017	.108	.055	.016	.098	.050	.015	.242	.199	.169	.208	.172	.148	.144	.121	.105	.084	.071	.062	.027	.023	.021
.126	.061	.017	.121	.059	.017	.111	.055	.016	.102	.051	.015	.093	.047	.014	.237	.195	.166	.204	.169	.145	.142	.119	.103	.083	.070	.061	.027	.023	.020
.119	.057	.016	.114	.055	.015	.105	.051	.014	.096	.048	.014	.088	.044	.013	.233	.192	.164	.201	.167	.143	.139	.117	.102	.081	.069	.061	.026	.023	.020
															.190	.190	.190	.163	.163	.163	.111	.111	.111	.064	.064	.064	.020	.020	.020
.220	.125	.040	.213	.122	.039	.200	.115	.037	.189	.109	.035	.178	.103	.033	.174	.157	.141	.149	.135	.122	.102	.093	.084	.059	.054	.049	.019	.017	.016
.212	.116	.036	.206	.114	.035	.195	.109	.034	.185	.104	.033	.176	.099	.031	.161	.132	.107	.138	.114	.093	.095	.079	.065	.055	.046	.038	.018	.015	.012
.202	.107	.032	.197	.105	.032	.187	.101	.031	.178	.098	.030	.170	.094	.029	.151	.114	.084	.130	.098	.073	.089	.068	.051	.051	.040	.030	.017	.013	.010
.191	.099	.029	.186	.098	.029	.178	.094	.028	.170	.091	.028	.163	.089	.027	.142	.100	.067	.122	.086	.058	.084	.060	.041	.049	.035	.024	.016	.011	.008
.180	.092	.027	.176	.091	.026	.169	.088	.026	.162	.086	.025	.156	.083	.025	.135	.088	.055	.116	.077	.047	.080	.053	.033	.046	.031	.020	.015	.010	.007
.171	.086	.024	.167	.084	.024	.161	.082	.024	.155	.080	.024	.149	.078	.023	.128	.080	.045	.110	.069	.039	.076	.048	.028	.044	.028	.017	.014	.009	.005
.162	.080	.023	.158	.079	.022	.153	.077	.022	.147	.075	.022	.142	.074	.022	.121	.072	.038	.104	.063	.033	.072	.044	.024	.042	.026	.014	.014	.008	.005
.153	.075	.021	.150	.074	.021	.145	.073	.021	.140	.071	.021	.136	.070	.020	.115	.066	.033	.099	.058	.029	.069	.040	.020	.040	.024	.012	.013	.008	.004
.145	.070	.020	.143	.070	.020	.138	.068	.019	.134	.067	.019	.130	.066	.019	.110	.061	.029	.095	.053	.025	.066	.037	.018	.038	.022	.011	.012	.007	.003
.138	.066	.018	.136	.066	.018	.132	.065	.018	.128	.064	.018	.124	.062	.018	.105	.057	.025	.091	.050	.022	.063	.035	.016	.037	.021	.009	.012	.007	.003
															.162	.162	.162	.138	.138	.138	.094	.094	.094	.054	.054	.054	.017	.017	.017
.139	.079	.025	.133	.076	.024	.123	.070	.022	.113	.065	.021	.104	.060	.019	.144	.133	.124	.123	.115	.106	.084	.079	.074	.049	.046	.043	.016	.015	.014
.132	.072	.022	.127	.070	.022	.119	.066	.020	.110	.062	.019	.103	.058	.018	.131	.112	.097	.112	.097	.084	.077	.067	.059	.044	.039	.034	.014	.013	.011
.126	.067	.020	.122	.065	.020	.114	.062	.019	.107	.059	.018	.101	.056	.017	.120	.096	.078	.102	.083	.067	.070	.058	.047	.041	.034	.028	.013	.011	.009
.119	.062	.018	.116	.061	.018	.110	.058	.017	.104	.056	.017	.098	.053	.016	.110	.084	.063	.095	.072	.055	.065	.050	.039	.038	.029	.023	.012	.010	.008
.114	.058	.017	.111	.057	.017	.105	.055	.016	.100	.053	.016	.095	.051	.015	.103	.074	.052	.088	.064	.045	.061	.044	.032	.035	.026	.019	.011	.009	.006
.109	.055	.016	.106	.054	.015	.101	.052	.015	.097	.050	.015	.093	.049	.014	.096	.066	.044	.083	.057	.038	.057	.040	.027	.033	.023	.016	.011	.008	.005
.104	.052	.015	.102	.051	.014	.098	.049	.014	.094	.048	.014	.090	.047	.014	.091	.059	.037	.078	.051	.032	.054	.036	.023	.031	.021	.014	.010	.007	.004
.100	.049	.014	.098	.048	.014	.094	.047	.013	.090	.046	.013	.087	.045	.013	.086	.054	.032	.074	.047	.028	.051	.033	.020	.030	.019	.012	.010	.006	.004
.096	.046	.013	.094	.046	.013	.091	.045	.013	.087	.044	.013	.084	.043	.012	.081	.049	.027	.070	.043	.024	.049	.030	.017	.028	.018	.010	.009	.006	.003
.092	.044	.012	.091	.044	.012	.088	.043	.012	.085	.042	.012	.082	.041	.012	.077	.045	.024	.067	.039	.021	.046	.028	.015	.027	.016	.009	.009	.005	.003

Fig. 61 *Continued (See page 119 for instructions and notes)*

Typical Luminaire	Typical Intensity Distribution and Per Cent Lamp Lumens		ρCC →	80			70			50			30			10			0		ρCC →
			ρW →	50	30	10	50	30	10	50	30	10	50	30	10	50	30	10	0	WDRC	ρW →
	Maint. Cat.	SC	RCR ↓	Coefficients of Utilization for 20 Per Cent Effective Floor Cavity Reflectance (ρFC = 20)																	RCR ↓
7 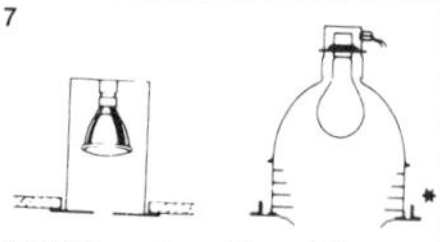	IV	0.7	0	.52	.52	.52	.51	.51	.51	.48	.48	.48	.46	.46	.46	.45	.45	.45	.44		
	0% ↑		1	.49	.48	.47	.48	.47	.46	.46	.45	.45	.44	.44	.43	.43	.43	.42	.41	.055	1
	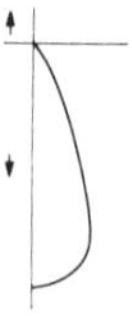		2	.46	.44	.43	.45	.44	.43	.44	.43	.42	.43	.42	.41	.41	.41	.40	.39	.054	2
			3	.43	.41	.40	.43	.41	.40	.42	.40	.39	.41	.39	.38	.40	.39	.38	.37	.053	3
			4	.41	.39	.37	.41	.39	.37	.40	.38	.37	.39	.37	.36	.38	.37	.36	.35	.052	4
	43½% ↓		5	.39	.37	.35	.39	.37	.35	.38	.36	.35	.37	.36	.34	.36	.35	.34	.34	.051	5
EAR-38 lamp above 51 mm (2'') diameter aperture (increase efficiency to 54½% for 76 mm (3'') diameter aperture)*			6	.37	.35	.33	.37	.35	.33	.36	.34	.33	.35	.34	.33	.35	.34	.32	.32	.049	6
			7	.35	.33	.31	.35	.33	.31	.34	.33	.31	.34	.32	.31	.33	.32	.31	.30	.048	7
			8	.34	.31	.30	.33	.31	.30	.33	.31	.30	.32	.31	.29	.32	.31	.29	.29	.046	8
			9	.32	.30	.28	.32	.30	.28	.31	.30	.28	.31	.29	.28	.31	.29	.28	.28	.045	9
			10	.31	.28	.27	.31	.28	.27	.30	.28	.27	.30	.28	.27	.30	.28	.27	.26	.043	10
8 	V	1.0	0	.65	.65	.65	.63	.63	.63	.60	.60	.60	.58	.58	.58	.55	.55	.55	.54		
	0% ↑		1	.59	.57	.56	.58	.56	.55	.56	.54	.53	.53	.52	.52	.52	.51	.50	.49	.133	1
			2	.54	.51	.49	.53	.50	.48	.51	.49	.47	.49	.47	.46	.48	.46	.45	.44	.130	2
			3	.49	.46	.43	.48	.45	.43	.47	.44	.42	.45	.43	.41	.44	.42	.41	.40	.123	3
			4	.45	.41	.38	.44	.41	.38	.43	.40	.38	.42	.39	.37	.41	.39	.37	.36	.116	4
	54½% ↓		5	.41	.37	.35	.41	.37	.34	.40	.36	.34	.39	.36	.34	.38	.35	.33	.32	.109	5
			6	.38	.34	.31	.38	.34	.31	.37	.33	.31	.36	.33	.31	.35	.33	.31	.30	.103	6
			7	.35	.31	.29	.35	.31	.29	.34	.31	.28	.33	.30	.28	.33	.30	.28	.27	.097	7
Medium distribution unit with lens plate and inside frost lamp			8	.33	.29	.26	.32	.29	.26	.32	.28	.26	.31	.28	.26	.31	.28	.26	.25	.092	8
			9	.31	.27	.24	.30	.27	.24	.30	.26	.24	.29	.26	.24	.29	.26	.24	.23	.087	9
			10	.29	.25	.22	.28	.25	.22	.28	.25	.22	.27	.24	.22	.27	.24	.22	.21	.082	10
9 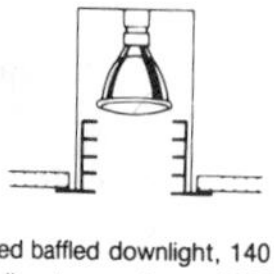	IV	0.5	0	.82	.82	.82	.80	.80	.80	.76	.76	.76	.73	.73	.73	.70	.70	.70	.69		
	0% ↑		1	.78	.77	.75	.76	.75	.74	.74	.73	.72	.71	.70	.70	.69	.68	.68	.67	.051	1
			2	.74	.72	.71	.73	.71	.70	.71	.70	.68	.69	.68	.67	.67	.66	.66	.65	.050	2
			3	.71	.69	.67	.71	.68	.67	.69	.67	.66	.67	.66	.65	.66	.65	.64	.63	.049	3
			4	.69	.66	.64	.68	.66	.64	.67	.65	.63	.66	.64	.63	.64	.63	.62	.61	.048	4
	68½% ↓		5	.67	.64	.62	.66	.63	.62	.65	.63	.61	.64	.62	.61	.63	.61	.60	.59	.047	5
			6	.64	.62	.60	.64	.61	.60	.63	.61	.59	.62	.60	.59	.61	.60	.59	.58	.045	6
Recessed baffled downlight, 140 mm (5 ½'') diameter aperture—150-PAR/FL lamp			7	.63	.60	.58	.62	.60	.58	.61	.59	.57	.61	.59	.57	.60	.58	.57	.56	.044	7
			8	.61	.58	.56	.60	.58	.56	.60	.58	.56	.59	.57	.56	.59	.57	.56	.55	.043	8
			9	.59	.56	.55	.59	.56	.55	.58	.56	.54	.58	.56	.54	.57	.55	.54	.54	.042	9
			10	.58	.55	.53	.57	.55	.53	.57	.55	.53	.56	.54	.53	.56	.54	.53	.52	.041	10
10 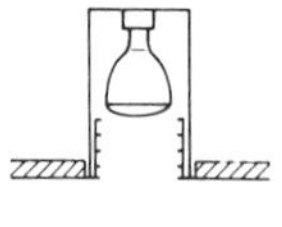	IV	0.5	0	1.01	1.01	1.01	.99	.99	.99	.95	.95	.95	.91	.91	.91	.87	.87	.87	.85		
	0% ↑		1	.96	.94	.93	.94	.93	.91	.91	.89	.88	.88	.87	.86	.85	.84	.83	.82	.085	1
			2	.91	.88	.86	.90	.87	.85	.87	.85	.83	.84	.83	.81	.82	.81	.80	.79	.084	2
			3	.87	.83	.81	.86	.83	.80	.83	.81	.79	.81	.79	.78	.80	.78	.77	.75	.082	3
			4	.83	.79	.76	.82	.79	.76	.80	.77	.75	.79	.76	.74	.77	.75	.73	.72	.080	4
	85% ↓		5	.79	.76	.73	.79	.75	.72	.77	.74	.72	.76	.73	.71	.75	.72	.71	.70	.078	5
			6	.76	.72	.70	.76	.72	.69	.74	.71	.69	.73	.71	.68	.72	.70	.68	.67	.076	6
Recessed baffled downlight, 140 mm (5½'') diameter aperture—75ER30 lamp			7	.73	.69	.67	.73	.69	.67	.72	.69	.66	.71	.68	.66	.70	.68	.66	.65	.073	7
			8	.71	.67	.64	.70	.67	.64	.69	.66	.64	.69	.66	.64	.68	.65	.63	.62	.071	8
			9	.68	.64	.62	.68	.64	.62	.67	.64	.62	.66	.63	.61	.66	.63	.61	.60	.069	9
			10	.66	.62	.60	.66	.62	.60	.65	.62	.59	.64	.61	.59	.64	.61	.59	.58	.067	10
11 	V	1.4	0	.63	.63	.63	.62	.62	.62	.59	.59	.59	.57	.57	.57	.54	.54	.54	.53		
	0% ↑		1	.57	.55	.54	.56	.54	.53	.54	.52	.51	.52	.51	.50	.50	.49	.48	.47	.153	1
			2	.51	.48	.46	.50	.48	.45	.48	.46	.44	.47	.45	.43	.45	.44	.42	.41	.150	2
			3	.46	.42	.40	.45	.42	.39	.44	.41	.39	.42	.40	.38	.41	.39	.37	.36	.142	3
			4	.42	.38	.35	.41	.37	.34	.40	.36	.34	.39	.36	.33	.37	.35	.33	.32	.133	4
	53½% ↓		5	.38	.34	.30	.37	.33	.30	.36	.33	.30	.35	.32	.30	.34	.32	.29	.28	.124	5
			6	.34	.30	.27	.34	.30	.27	.33	.29	.27	.32	.29	.27	.31	.28	.26	.25	.117	6
			7	.31	.27	.24	.31	.27	.24	.30	.27	.24	.29	.26	.24	.29	.26	.24	.23	.109	7
Wide distribution unit with lens plate and inside frost lamp			8	.29	.25	.22	.28	.24	.22	.28	.24	.22	.27	.24	.22	.27	.24	.21	.20	.103	8
			9	.26	.22	.20	.26	.22	.20	.26	.22	.20	.25	.22	.20	.25	.22	.19	.19	.097	9
			10	.24	.21	.18	.24	.20	.18	.24	.20	.18	.23	.20	.18	.23	.20	.18	.17	.091	10
12 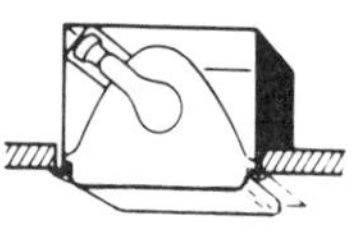	V	1.3	0	.62	.62	.62	.60	.60	.60	.57	.57	.57	.54	.54	.54	.52	.52	.52	.51		
	1½% ↑		1	.52	.50	.48	.51	.49	.47	.49	.47	.45	.46	.45	.43	.44	.43	.42	.40	.256	1
			2	.45	.41	.38	.44	.40	.37	.42	.39	.36	.40	.37	.35	.38	.36	.34	.33	.222	2
			3	.39	.35	.31	.38	.34	.31	.37	.33	.30	.35	.32	.29	.33	.31	.28	.27	.195	3
			4	.35	.30	.26	.34	.29	.26	.32	.28	.25	.31	.27	.25	.29	.27	.24	.23	.173	4
	50½% ↓		5	.31	.26	.22	.30	.25	.22	.29	.25	.22	.28	.24	.21	.26	.23	.21	.20	.154	5
			6	.28	.23	.19	.27	.22	.19	.26	.22	.19	.25	.21	.19	.24	.21	.18	.17	.139	6
			7	.25	.20	.17	.25	.20	.17	.24	.20	.17	.23	.19	.16	.22	.19	.16	.15	.127	7
Recessed unit with dropped diffusing glass			8	.23	.18	.15	.22	.18	.15	.22	.18	.15	.21	.17	.15	.20	.17	.14	.13	.116	8
			9	.21	.16	.14	.21	.16	.13	.20	.16	.13	.19	.16	.13	.18	.15	.13	.12	.107	9
			10	.19	.15	.12	.19	.15	.12	.18	.15	.12	.18	.14	.12	.17	.14	.12	.11	.099	10

* Also, reflector downlight with baffles and inside frosted lamp.

80			70			50			30			10			80			70			50			30			10		
50	30	10	50	30	10	50	30	10	50	30	10	50	30	10	50	30	10	50	30	10	50	30	10	50	30	10	50	30	10
Wall Exitance Coefficients for 20 Per Cent Effective Floor Cavity Reflectance ($\rho_{FC} = 20$)															Ceiling Cavity Exitance Coefficients for 20 Per Cent Floor Cavity Reflectance ($\rho_{FC} = 20$)														
															.083	.083	.083	.071	.071	.071	.048	.048	.048	.028	.028	.028	.009	.009	.009
.069	.039	.012	.066	.038	.012	.061	.035	.011	.056	.032	.010	.051	.030	.010	.074	.069	.064	.063	.059	.055	.043	.041	.038	.025	.023	.022	.008	.008	.007
.065	.036	.011	.063	.035	.011	.058	.032	.010	.054	.030	.010	.050	.028	.009	.067	.058	.050	.057	.050	.043	.039	.034	.030	.023	.020	.018	.007	.006	.006
.062	.033	.010	.060	.032	.010	.056	.030	.009	.052	.029	.009	.049	.027	.008	.061	.050	.040	.052	.043	.035	.036	.030	.025	.021	.017	.014	.007	.006	.005
.059	.031	.009	.057	.030	.009	.054	.028	.009	.051	.027	.008	.048	.026	.008	.056	.043	.033	.048	.037	.029	.033	.026	.020	.019	.015	.012	.006	.005	.004
.056	.029	.008	.054	.028	.008	.052	.027	.008	.049	.026	.008	.046	.025	.007	.052	.038	.027	.045	.033	.024	.031	.023	.017	.018	.013	.010	.006	.004	.003
.053	.027	.008	.052	.026	.008	.050	.025	.007	.047	.025	.007	.045	.024	.007	.049	.034	.023	.042	.029	.020	.029	.020	.014	.017	.012	.008	.005	.004	.003
.051	.025	.007	.050	.025	.007	.048	.024	.007	.046	.023	.007	.044	.023	.007	.046	.030	.019	.039	.026	.017	.027	.018	.012	.016	.011	.007	.005	.004	.002
.049	.024	.007	.048	.024	.007	.046	.023	.007	.044	.022	.006	.043	.022	.006	.043	.027	.017	.037	.024	.014	.026	.017	.010	.015	.010	.006	.005	.003	.002
.047	.023	.006	.046	.023	.006	.044	.022	.006	.043	.021	.006	.041	.021	.006	.041	.025	.014	.035	.022	.013	.024	.015	.009	.014	.009	.005	.005	.003	.002
.045	.022	.006	.045	.022	.006	.043	.021	.006	.041	.021	.006	.040	.020	.006	.039	.023	.013	.034	.020	.011	.023	.014	.008	.014	.008	.005	.004	.003	.002
															.104	.104	.104	.088	.088	.088	.060	.060	.060	.035	.035	.035	.011	.011	.011
.121	.069	.022	.117	.067	.021	.110	.063	.020	.104	.060	.019	.098	.057	.018	.095	.085	.077	.081	.073	.066	.055	.050	.046	.032	.029	.027	.010	.009	.009
.116	.063	.019	.113	.062	.019	.107	.059	.018	.101	.057	.018	.096	.054	.017	.088	.072	.058	.075	.062	.051	.052	.043	.035	.030	.025	.021	.010	.008	.007
.110	.058	.017	.107	.057	.017	.102	.055	.017	.097	.053	.016	.092	.051	.016	.082	.062	.046	.070	.053	.040	.048	.037	.028	.028	.022	.016	.009	.007	.005
.103	.054	.016	.101	.053	.016	.097	.051	.015	.092	.050	.015	.088	.048	.015	.077	.054	.036	.066	.047	.032	.046	.033	.022	.026	.019	.013	.008	.006	.004
.098	.050	.014	.096	.049	.014	.092	.048	.014	.088	.046	.014	.084	.045	.014	.073	.048	.030	.063	.042	.026	.043	.029	.018	.025	.017	.011	.008	.006	.004
.092	.046	.013	.091	.046	.013	.087	.045	.013	.084	.043	.013	.081	.042	.013	.069	.043	.025	.060	.038	.021	.041	.026	.015	.024	.015	.009	.008	.005	.003
.088	.043	.012	.086	.043	.012	.083	.042	.012	.080	.041	.012	.077	.040	.012	.066	.039	.021	.057	.034	.018	.039	.024	.013	.023	.014	.008	.007	.005	.003
.083	.041	.011	.082	.040	.011	.079	.039	.011	.076	.039	.011	.074	.038	.011	.063	.036	.018	.054	.031	.016	.037	.022	.011	.022	.013	.007	.007	.004	.002
.079	.038	.011	.078	.038	.011	.075	.037	.011	.073	.036	.010	.070	.036	.010	.060	.033	.016	.052	.029	.014	.036	.020	.010	.021	.012	.006	.007	.004	.002
.075	.036	.010	.074	.036	.010	.072	.035	.010	.069	.035	.010	.067	.034	.010	.057	.031	.014	.049	.027	.012	.034	.019	.008	.020	.011	.005	.006	.004	.002
															.131	.131	.131	.112	.112	.112	.076	.076	.076	.044	.044	.044	.014	.014	.014
.090	.051	.016	.086	.049	.016	.077	.044	.014	.069	.040	.013	.062	.036	.012	.115	.108	.102	.099	.093	.088	.068	.064	.061	.039	.037	.035	.012	.012	.011
.083	.046	.014	.079	.044	.013	.072	.040	.013	.066	.037	.012	.060	.034	.011	.103	.091	.082	.088	.078	.070	.060	.054	.049	.035	.032	.029	.011	.010	.009
.077	.041	.012	.074	.040	.012	.068	.037	.011	.063	.034	.011	.058	.032	.010	.092	.078	.066	.079	.067	.058	.054	.047	.040	.031	.027	.024	.010	.009	.008
.072	.038	.011	.070	.036	.011	.065	.034	.010	.060	.032	.010	.056	.030	.009	.083	.067	.055	.072	.058	.048	.049	.040	.034	.028	.024	.020	.009	.008	.007
.068	.035	.010	.066	.034	.010	.062	.032	.009	.058	.030	.009	.054	.029	.009	.076	.059	.046	.065	.051	.040	.045	.035	.028	.026	.021	.017	.008	.007	.006
.064	.032	.009	.062	.031	.009	.059	.030	.009	.055	.029	.008	.052	.028	.008	.070	.052	.039	.060	.045	.034	.041	.031	.024	.024	.018	.014	.008	.006	.005
.061	.030	.009	.059	.030	.008	.056	.028	.008	.053	.027	.008	.051	.026	.008	.064	.046	.033	.055	.040	.029	.038	.028	.020	.022	.016	.012	.007	.005	.004
.058	.028	.008	.057	.028	.008	.054	.027	.008	.052	.026	.008	.049	.025	.007	.060	.041	.029	.052	.036	.025	.036	.025	.018	.021	.015	.011	.007	.005	.003
.056	.027	.008	.054	.026	.007	.052	.026	.007	.050	.025	.007	.048	.024	.007	.056	.037	.025	.048	.032	.022	.033	.023	.015	.019	.013	.009	.006	.004	.003
.053	.026	.007	.052	.025	.007	.050	.025	.007	.048	.024	.007	.046	.023	.007	.053	.034	.022	.045	.030	.019	.032	.021	.014	.018	.012	.008	.006	.004	.003
															.162	.162	.162	.139	.139	.139	.095	.095	.095	.054	.054	.054	.017	.017	.017
.123	.070	.022	.117	.067	.021	.107	.061	.020	.097	.056	.018	.088	.051	.016	.144	.134	.126	.123	.115	.108	.084	.079	.075	.048	.046	.043	.016	.015	.014
.115	.063	.019	.110	.061	.019	.102	.057	.018	.094	.053	.016	.086	.049	.015	.129	.113	.100	.110	.097	.086	.076	.067	.060	.044	.039	.035	.014	.013	.011
.108	.058	.017	.104	.056	.017	.097	.052	.016	.090	.049	.015	.084	.046	.014	.117	.097	.081	.100	.083	.070	.069	.058	.049	.040	.034	.029	.013	.011	.009
.102	.053	.016	.099	.052	.015	.092	.049	.015	.087	.046	.014	.081	.044	.013	.106	.084	.066	.091	.072	.057	.063	.050	.040	.036	.029	.024	.012	.010	.008
.096	.049	.014	.094	.048	.014	.088	.046	.014	.083	.044	.013	.079	.042	.013	.098	.073	.055	.084	.063	.048	.058	.044	.034	.034	.026	.020	.011	.008	.007
.092	.046	.013	.089	.045	.013	.085	.043	.013	.080	.042	.012	.076	.040	.012	.091	.065	.046	.078	.056	.040	.054	.039	.029	.031	.023	.017	.010	.008	.006
.087	.043	.012	.085	.042	.012	.081	.041	.012	.077	.040	.012	.074	.038	.011	.084	.058	.039	.073	.050	.034	.050	.035	.024	.029	.021	.015	.009	.007	.005
.084	.041	.011	.082	.040	.011	.078	.039	.011	.075	.038	.011	.072	.037	.011	.079	.052	.034	.068	.045	.030	.047	.032	.021	.027	.019	.013	.009	.006	.004
.080	.039	.011	.078	.038	.011	.075	.037	.011	.072	.036	.010	.069	.035	.010	.074	.048	.030	.064	.041	.026	.044	.029	.018	.026	.017	.011	.008	.006	.004
.077	.037	.010	.075	.036	.010	.072	.036	.010	.070	.035	.010	.067	.034	.010	.070	.044	.026	.061	.038	.023	.042	.027	.016	.025	.016	.010	.008	.005	.003
															.101	.101	.101	.087	.087	.087	.059	.059	.059	.034	.034	.034	.011	.011	.011
.131	.074	.024	.127	.072	.023	.120	.069	.022	.114	.066	.021	.108	.062	.020	.094	.084	.074	.080	.072	.064	.055	.049	.044	.032	.029	.026	.010	.009	.008
.126	.069	.021	.123	.068	.021	.117	.065	.020	.111	.062	.020	.106	.060	.019	.088	.070	.056	.075	.061	.048	.052	.042	.034	.030	.024	.020	.010	.008	.006
.119	.064	.019	.117	.062	.019	.111	.060	.018	.107	.058	.018	.102	.057	.017	.083	.061	.043	.071	.052	.037	.049	.036	.026	.028	.021	.015	.009	.007	.005
.113	.059	.017	.110	.058	.017	.106	.056	.017	.101	.054	.016	.097	.053	.016	.079	.053	.034	.067	.046	.030	.046	.032	.021	.027	.019	.012	.009	.006	.004
.106	.054	.016	.104	.053	.016	.100	.052	.015	.096	.051	.015	.093	.050	.015	.075	.048	.028	.064	.041	.024	.044	.029	.017	.026	.017	.010	.008	.006	.003
.100	.050	.014	.098	.050	.014	.095	.049	.014	.091	.047	.014	.088	.046	.014	.071	.043	.023	.061	.037	.020	.042	.026	.014	.024	.015	.008	.008	.005	.003
.095	.047	.013	.093	.046	.013	.090	.045	.013	.087	.044	.013	.084	.044	.013	.068	.039	.019	.058	.034	.017	.040	.024	.012	.023	.014	.007	.008	.005	.002
.090	.044	.012	.088	.043	.012	.085	.043	.012	.082	.042	.012	.080	.041	.012	.065	.036	.016	.056	.031	.014	.039	.022	.010	.022	.013	.006	.007	.004	.002
.085	.041	.011	.083	.041	.011	.081	.040	.011	.078	.039	.011	.076	.039	.011	.062	.034	.014	.053	.029	.013	.037	.020	.009	.022	.012	.005	.007	.004	.002
.081	.039	.011	.079	.038	.011	.077	.038	.011	.075	.037	.011	.073	.037	.010	.059	.031	.013	.051	.027	.011	.035	.019	.008	.021	.011	.005	.007	.004	.002
															.112	.112	.112	.095	.095	.095	.065	.065	.065	.037	.037	.037	.012	.012	.012
.187	.106	.034	.182	.104	.033	.174	.100	.032	.167	.096	.031	.160	.093	.030	.108	.094	.080	.092	.080	.069	.063	.055	.048	.036	.032	.028	.012	.010	.009
.168	.092	.028	.164	.090	.028	.157	.087	.027	.150	.084	.026	.144	.082	.026	.104	.081	.062	.089	.070	.053	.061	.048	.037	.035	.028	.022	.011	.009	.007
.151	.081	.024	.148	.079	.024	.142	.077	.023	.136	.074	.023	.130	.072	.022	.100	.072	.050	.086	.062	.043	.059	.043	.030	.034	.025	.018	.011	.008	.006
.138	.072	.021	.135	.070	.021	.129	.068	.020	.124	.066	.020	.119	.065	.020	.096	.065	.042	.083	.057	.036	.057	.039	.026	.033	.023	.015	.011	.008	.005
.126	.064	.019	.123	.063	.018	.118	.062	.018	.113	.060	.018	.109	.058	.018	.092	.060	.036	.079	.052	.032	.055	.036	.022	.032	.021	.013	.010	.007	.004
.116	.058	.017	.113	.057	.016	.109	.056	.016	.105	.054	.016	.101	.053	.016	.088	.056	.032	.076	.048	.028	.052	.034	.020	.030	.020	.012	.010	.007	.004
.107	.053	.015	.105	.052	.015	.101	.051	.015	.097	.050	.014	.094	.049	.014	.085	.052	.030	.073	.045	.026	.050	.032	.018	.029	.019	.011	.009	.006	.004
.100	.049	.014	.098	.048	.014	.094	.047	.013	.091	.046	.013	.087	.045	.013	.081	.049	.027	.070	.043	.024	.048	.030	.017	.028	.018	.010	.009	.006	.003
.093	.045	.013	.091	.044	.012	.088	.043	.012	.085	.043	.012	.082	.042	.012	.078	.047	.026	.067	.040	.022	.046	.028	.016	.027	.017	.009	.009	.006	.003
.087	.042	.012	.086	.041	.012	.083	.040	.011	.080	.040	.011	.077	.039	.011	.075	.044	.024	.064	.039	.021	.045	.027	.015	.026	.016	.009	.008	.005	.003

Fig. 61 Continued (See page 119 for instructions and notes)

Coefficients of Utilization for 20 Per Cent Effective Floor Cavity Reflectance (ρ_{FC} = 20)

Typical Luminaire	Maint. Cat.	SC	RCR	ρCC→ 80, ρW 50	80 30	80 10	70 50	70 30	70 10	50 50	50 30	50 10	30 50	30 30	30 10	10 50	10 30	10 10	0 0	WDRC	RCR
13 Bilateral batwing distribution—clear HID with dropped prismatic lens (2½%↑, 71%↓)	V	N.A.	0	.87	.87	87	.85	.85	.85	.80	.80	.80	.76	.76	.76	.73	.73	.73	.71		0
			1	.75	.72	.69	.73	.70	.68	.70	.67	.65	.66	.64	.63	.63	.62	.60	.59	.312	1
			2	.66	.60	.56	.64	.59	.55	.61	.57	.54	.58	.55	.52	.56	.53	.51	.49	.279	2
			3	.58	.51	.47	.56	.51	.46	.54	.49	.45	.51	.47	.44	.49	.46	.43	.41	.251	3
			4	.51	.44	.39	.50	.44	.39	.48	.42	.38	.46	.41	.37	.44	.40	.37	.35	.226	4
			5	.45	.39	.34	.44	.38	.33	.42	.37	.33	.41	.36	.32	.39	.35	.32	.30	.206	5
			6	.41	.34	.29	.40	.33	.29	.38	.33	.28	.37	.32	.28	.35	.31	.28	.26	.188	6
			7	.37	.30	.26	.36	.30	.25	.35	.29	.25	.33	.28	.25	.32	.28	.24	.23	.173	7
			8	.33	.27	.23	.33	.27	.22	.31	.26	.22	.30	.25	.22	.29	.25	.22	.20	.159	8
			9	.30	.24	.20	.30	.24	.20	.29	.23	.20	.28	.23	.19	.27	.22	.19	.18	.148	9
			10	.28	.22	.18	.27	.22	.18	.26	.21	.18	.26	.21	.17	.25	.20	.17	.16	.138	10
14 Clear HID lamp and glass refractor above plastic lens panel (0%↑, 66%↓)	V	1.3	0	.78	.78	.78	.77	.77	.77	.73	.73	.73	.70	.70	.70	.67	.67	.67	.66		0
			1	.71	.69	.67	.69	.67	.65	.67	.65	.63	.64	.62	.61	.62	.61	.60	.58	.188	1
			2	.64	.60	.57	.62	.59	.56	.60	.57	.55	.58	.56	.54	.56	.54	.53	.51	.183	2
			3	.57	.53	.49	.56	.52	.49	.54	.51	.48	.53	.50	.47	.51	.49	.46	.45	.173	3
			4	.52	.47	.43	.51	.46	.43	.49	.46	.42	.48	.45	.42	.47	.44	.41	.40	.161	4
			5	.47	.42	.38	.46	.42	.38	.45	.41	.38	.44	.40	.37	.43	.40	.37	.36	.151	5
			6	.43	.38	.34	.42	.38	.34	.41	.37	.34	.40	.36	.34	.39	.36	.33	.32	.141	6
			7	.39	.34	.31	.39	.34	.31	.38	.34	.30	.37	.33	.30	.36	.33	.30	.29	.132	7
			8	.36	.31	.28	.36	.31	.28	.35	.31	.28	.34	.30	.27	.34	.30	.27	.26	.124	8
			9	.34	.29	.25	.33	.28	.25	.32	.28	.25	.32	.28	.25	.31	.28	.25	.24	.117	9
			10	.31	.26	.23	.31	.26	.23	.30	.26	.23	.30	.26	.23	.29	.25	.23	.22	.110	10
15 Enclosed reflector with an incandescent lamp (0%↑, 71½%↓)	V	1.4	0	.85	.85	.85	.83	.83	.83	.80	.80	.80	.76	.76	.76	.73	.73	.73	.72		0
			1	.77	.75	.73	.76	.74	.72	.73	.71	.69	.70	.69	.67	.67	.66	.65	.64	.189	1
			2	.70	.66	.63	.68	.65	.62	.66	.63	.60	.64	.61	.59	.61	.60	.58	.56	.190	2
			3	.63	.58	.54	.62	.57	.54	.60	.56	.53	.58	.54	.52	.56	.53	.51	.50	.183	3
			4	.56	.51	.47	.56	.51	.47	.54	.50	.46	.52	.49	.46	.51	.48	.45	.44	.174	4
			5	.51	.46	.42	.50	.45	.41	.49	.44	.41	.48	.44	.40	.46	.43	.40	.39	.164	5
			6	.46	.41	.37	.46	.41	.37	.45	.40	.37	.43	.39	.36	.42	.39	.36	.34	.155	6
			7	.42	.37	.33	.42	.37	.33	.41	.36	.33	.40	.36	.32	.39	.35	.32	.31	.146	7
			8	.39	.33	.30	.38	.33	.29	.37	.33	.29	.37	.32	.29	.36	.32	.29	.28	.137	8
			9	.36	.30	.27	.35	.30	.27	.35	.30	.27	.34	.30	.26	.33	.29	.26	.25	.129	9
			10	.33	.28	.24	.33	.28	.24	.32	.27	.24	.31	.27	.24	.31	.27	.24	.23	.122	10
16 "High bay" narrow distribution ventilated reflector with clear HID lamp (1½%↑, 77%↓)	III	0.7	0	.93	.93	.93	.90	.90	.90	.86	.86	.86	.82	.82	.82	.78	.78	.78	.77		0
			1	.86	.84	.82	.84	.82	.80	.80	.79	.77	.77	.76	.75	.74	.74	.73	.71	.138	1
			2	.79	.76	.73	.78	.75	.72	.75	.73	.71	.73	.71	.69	.70	.69	.67	.66	.136	2
			3	.74	.70	.66	.73	.69	.66	.70	.67	.65	.68	.66	.63	.66	.64	.62	.61	.132	3
			4	.69	.64	.61	.68	.64	.60	.66	.62	.60	.64	.61	.59	.63	.60	.58	.57	.126	4
			5	.64	.60	.56	.63	.59	.56	.62	.58	.55	.60	.57	.55	.59	.56	.54	.53	.120	5
			6	.60	.55	.52	.60	.55	.52	.58	.54	.51	.57	.54	.51	.56	.53	.50	.49	.115	6
			7	.57	.52	.49	.56	.52	.48	.55	.51	.48	.54	.50	.48	.53	.50	.47	.46	.109	7
			8	.53	.49	.45	.53	.48	.45	.52	.48	.45	.51	.47	.45	.50	.47	.44	.43	.104	8
			9	.51	.46	.43	.50	.46	.43	.49	.45	.42	.48	.45	.42	.47	.44	.42	.41	.100	9
			10	.48	.43	.40	.48	.43	.40	.47	.43	.40	.46	.42	.40	.45	.42	.40	.39	.095	10
17 "High bay" intermediate distribution ventilated reflector with clear HID lamp (1%↑, 76%↓)	III	1.0	0	.91	.91	.91	.89	.89	.89	.85	.85	.85	.81	.81	.81	.78	.78	.78	.76		0
			1	.83	.81	.79	.81	.79	.77	.78	.76	.75	.75	.74	.72	.72	.71	.70	.68	.187	1
			2	.75	.71	.68	.74	.70	.67	.71	.68	.65	.68	.66	.64	.66	.64`	.62	.61	.189	2
			3	.68	.63	.59	.67	.62	.59	.65	.61	.58	.62	.59	.57	.61	.58	.56	.54	.183	3
			4	.62	.56	.52	.61	.56	.52	.59	.54	.51	.57	.53	.50	.55	.52	.50	.48	.174	4
			5	.56	.50	.46	.55	.50	.46	.54	.49	.45	.52	.48	.45	.51	.47	.44	.43	.165	5
			6	.51	.46	.41	.51	.45	.41	.49	.44	.41	.48	.44	.40	.47	.43	.40	.39	.155	6
			7	.47	.41	.37	.47	.41	.37	.45	.40	.37	.44	.40	.37	.43	.39	.36	.35	.147	7
			8	.43	.38	.34	.43	.37	.34	.42	.37	.33	.41	.36	.33	.40	.36	.33	.32	.138	8
			9	.40	.35	.31	.40	.34	.31	.39	.34	.31	.38	.34	.30	.37	.33	.30	.29	.131	9
			10	.37	.32	.28	.37	.32	.28	.36	.31	.28	.35	.31	.28	.35	.31	.28	.27	.124	10
18 "High bay" wide distribution ventilated reflector with clear HID lamp (½%↑, 77½%↓)	III	1.5	0	.93	.93	.93	.91	.91	.91	.87	.87	.87	.83	.83	.83	.79	.79	.79	.78		0
			1	.84	.81	.79	.82	.80	.78	.79	.77	.75	.76	.74	.73	.73	.72	.70	.69	.217	1
			2	.75	.71	.67	.74	.70	.66	.71	.68	.65	.68	.66	.63	.66	.64	.62	.60	.219	2
			3	.67	.62	.57	.66	.61	.57	.64	.59	.56	.61	.58	.55	.59	.56	.54	.52	.211	3
			4	.60	.54	.50	.59	.54	.49	.57	.52	.48	.55	.51	.48	.54	.50	.47	.46	.200	4
			5	.54	.48	.43	.53	.47	.43	.52	.46	.42	.50	.45	.42	.49	.45	.41	.40	.189	5
			6	.49	.42	.38	.48	.42	.38	.47	.41	.37	.45	.41	.37	.44	.40	.37	.35	.177	6
			7	.44	.38	.34	.44	.38	.33	.42	.37	.33	.41	.36	.33	.40	.36	.33	.31	.166	7
			8	.40	.34	.30	.40	.34	.30	.39	.33	.30	.38	.33	.29	.37	.32	.29	.28	.156	8
			9	.37	.31	.27	.37	.31	.27	.36	.30	.27	.35	.30	.26	.34	.29	.26	.25	.146	9
			10	.34	.28	.24	.34	.28	.24	.33	.28	.24	.32	.27	.24	.31	.27	.24	.22	.138	10

80			70			50			30			10			80			70			50			30			10		
50	30	10	50	30	10	50	30	10	50	30	10	50	30	10	50	30	10	50	30	10	50	30	10	50	30	10	50	30	10
Wall Exitance Coefficients for 20 Per Cent Effective Floor Cavity Reflectance (ρ_{FC} = 20)															Ceiling Cavity Exitance Coefficients for 20 Per Cent Floor Cavity Reflectance (ρ_{FC} = 20)														
															.159	.159	.159	.136	.136	.136	.093	.093	.093	.053	.053	.053	.017	.017	.017
.238	.135	.043	.232	.132	.042	.220	.126	.040	.210	.121	.039	.201	.116	.038	.152	.134	.117	.130	.115	.101	.089	.079	.070	.051	.046	.041	.016	.015	.013
.218	.119	.037	.212	.117	.036	.202	.113	.035	.193	.108	.034	.185	.105	.033	.146	.116	.091	.125	.100	.079	.086	.069	.055	.050	.040	.032	.016	.013	.010
.200	.106	.032	.195	.105	.031	.186	.101	.031	.178	.098	.030	.171	.095	.029	.141	.103	.074	.121	.089	.064	.083	.062	.045	.048	.036	.026	.015	.012	.009
.184	.096	.028	.180	.094	.028	.172	.091	.027	.165	.089	.027	.158	.086	.026	.135	.094	.062	.116	.081	.054	.080	.056	.038	.046	.033	.022	.015	.011	.007
.170	.087	.025	.167	.086	.025	.160	.083	.024	.153	.081	.024	.147	.079	.024	.130	.086	.054	.111	.075	.047	.077	.052	.033	.044	.031	.020	.014	.010	.006
.158	.079	.023	.155	.078	.023	.149	.076	.022	.143	.074	.022	.137	.072	.021	.125	.080	.048	.107	.069	.042	.074	.049	.030	.043	.029	.018	.014	.009	.006
.147	.073	.021	.144	.072	.021	.139	.070	.020	.133	.068	.020	.128	.067	.020	.120	.075	.044	.103	.065	.038	.071	.046	.027	.041	.027	.016	.013	.009	.005
.138	.067	.019	.135	.067	.019	.130	.065	.019	.125	.063	.018	.121	.062	.018	.115	.071	.041	.099	.061	.035	.068	.043	.025	.040	.025	.015	.013	.008	.005
.129	.063	.017	.127	.062	.017	.122	.060	.017	.118	.059	.017	.114	.058	.017	.111	.067	.038	.095	.058	.033	.066	.041	.024	.038	.024	.014	.012	.008	.005
.122	.058	.016	.119	.058	.016	.115	.056	.016	.111	.055	.016	.107	.054	.015	.106	.064	.036	.092	.056	.031	.064	.039	.022	.037	.023	.013	.012	.008	.004
															.126	.126	.126	.107	.107	.107	.073	.073	.073	.042	.042	.042	.013	.013	.013
.161	.091	.029	.156	.089	.028	.148	.085	.027	.140	.081	.026	.133	.077	.025	.116	.103	.092	.099	.089	.079	.068	.061	.055	.039	.035	.032	.013	.011	.010
.154	.085	.026	.151	.083	.026	.143	.080	.025	.136	.077	.024	.130	.074	.023	.109	.087	.069	.093	.075	.060	.064	.052	.042	.037	.030	.024	.012	.010	.008
.146	.078	.023	.143	.076	.023	.136	.074	.022	.130	.071	.022	.125	.069	.021	.102	.075	.054	.088	.065	.046	.060	.045	.033	.035	.026	.019	.011	.009	.006
.138	.072	.021	.135	.070	.021	.129	.068	.020	.124	.066	.020	.119	.065	.020	.097	.066	.042	.083	.057	.037	.057	.040	.026	.033	.023	.015	.011	.008	.005
.130	.066	.019	.127	.065	.019	.122	.063	.019	.117	.062	.018	.113	.060	.018	.092	.059	.034	.079	.051	.030	.054	.036	.021	.032	.021	.013	.010	.007	.004
.122	.061	.018	.120	.061	.017	.115	.059	.017	.111	.058	.017	.107	.056	.017	.087	.053	.028	.075	.046	.025	.052	.032	.018	.030	.019	.010	.010	.006	.003
.115	.057	.016	.113	.056	.016	.109	.055	.016	.106	.054	.016	.102	.053	.016	.083	.048	.024	.072	.042	.021	.050	.029	.015	.029	.017	.009	.009	.006	.003
.109	.053	.015	.107	.053	.015	.104	.052	.015	.100	.051	.015	.097	.050	.015	.080	.045	.021	.068	.039	.018	.047	.027	.013	.028	.016	.008	.009	.005	.003
.103	.050	.014	.102	.050	.014	.098	.049	.014	.095	.048	.014	.093	.047	.014	.076	.041	.018	.065	.036	.016	.045	.025	.011	.026	.015	.007	.009	.005	.002
.098	.047	.013	.097	.047	.013	.094	.046	.013	.091	.045	.013	.088	.044	.013	.073	.039	.016	.063	.033	.014	.043	.024	.010	.025	.014	.006	.008	.005	.002
															.137	.137	.137	.117	.117	.117	.080	.080	.080	.046	.046	.046	.015	.015	.015
.167	.095	.030	.162	.092	.029	.152	.087	.028	.144	.083	.027	.136	.079	.025	.126	.113	.102	.108	.097	.087	.074	.067	.060	.043	.039	.035	.014	.012	.011
.163	.089	.027	.159	.088	.027	.151	.084	.026	.143	.080	.025	.137	.077	.024	.118	.096	.077	.101	.082	.066	.069	.057	.046	.040	.033	.027	.013	.011	.009
.157	.083	.025	.153	.082	.025	.146	.079	.024	.139	.076	.023	.133	.074	.023	.112	.083	.059	.096	.071	.051	.066	.049	.036	.038	.029	.021	.012	.009	.007
.149	.077	.023	.146	.076	.023	.139	.074	.022	.134	.072	.022	.128	.070	.021	.106	.073	.047	.091	.063	.041	.063	.044	.029	.036	.026	.017	.012	.008	.006
.141	.072	.021	.138	.071	.021	.133	.069	.020	.128	.067	.020	.123	.066	.020	.101	.065	.038	.087	.056	.033	.060	.039	.023	.035	.023	.014	.011	.008	.005
.134	.067	.019	.131	.066	.019	.126	.065	.019	.122	.063	.019	.117	.062	.018	.096	.059	.032	.083	.051	.028	.057	.036	.020	.033	.021	.012	.011	.007	.004
.127	.063	.018	.124	.062	.018	.120	.061	.017	.116	.059	.017	.112	.058	.017	.092	.054	.027	.079	.047	.023	.055	.033	.017	.032	.019	.010	.010	.006	.003
.120	.059	.016	.118	.058	.016	.114	.057	.016	.110	.056	.016	.107	.055	.016	.088	.049	.023	.076	.043	.020	.052	.030	.014	.030	.018	.009	.010	.006	.003
.114	.055	.015	.112	.055	.015	.108	.054	.015	.105	.053	.015	.102	.052	.015	.084	.046	.020	.072	.040	.018	.050	.028	.013	.029	.017	.007	.009	.005	.002
.108	.052	.014	.106	.051	.014	.103	.051	.014	.100	.050	.014	.097	.049	.014	.081	.043	.018	.069	.037	.016	.048	.026	.011	.028	.016	.007	.009	.005	.002
															.158	.158	.158	.135	.135	.135	.092	.092	.092	.053	.053	.053	.017	.017	.017
.147	.084	.026	.141	.081	.026	.131	.075	.024	.121	.070	.022	.112	.065	.021	.144	.133	.122	.123	.114	.105	.084	.078	.073	.049	.045	.042	.016	.015	.014
.140	.077	.024	.136	.075	.023	.127	.070	.022	.118	.066	.021	.111	.063	.020	.133	.113	.097	.114	.097	.084	.078	.067	.058	.045	.039	.034	.014	.013	.011
.133	.071	.021	.129	.069	.021	.121	.066	.020	.114	.063	.019	.107	.059	.018	.123	.099	.079	.106	.085	.068	.073	.059	.048	.042	.035	.028	.013	.011	.009
.126	.066	.019	.123	.064	.019	.116	.061	.018	.110	.059	.018	.104	.056	.017	.116	.087	.066	.099	.075	.057	.068	.053	.040	.039	.031	.024	.013	.010	.008
.120	.061	.018	.117	.060	.017	.111	.058	.017	.105	.055	.016	.100	.053	.016	.109	.078	.056	.094	.068	.049	.064	.047	.034	.037	.028	.020	.012	.009	.007
.114	.057	.016	.111	.056	.016	.106	.054	.016	.101	.052	.015	.096	.051	.015	.103	.071	.048	.089	.062	.042	.061	.043	.030	.035	.025	.018	.011	.008	.006
.108	.053	.015	.106	.053	.015	.101	.051	.015	.097	.050	.014	.092	.048	.014	.098	.065	.042	.084	.057	.037	.058	.040	.026	.034	.023	.016	.011	.008	.005
.103	.050	.014	.101	.050	.014	.097	.048	.014	.093	.047	.014	.089	.046	.013	.093	.060	.038	.080	.052	.033	.056	.037	.023	.032	.022	.014	.010	.007	.005
.098	.048	.013	.096	.047	.013	.093	.046	.013	.089	.045	.013	.086	.043	.013	.089	.056	.034	.077	.049	.030	.053	.034	.021	.031	.020	.013	.010	.007	.004
.094	.045	.013	.092	.045	.012	.089	.043	.012	.086	.042	.012	.082	.041	.012	.086	.053	.031	.074	.046	.027	.051	.032	.019	.030	.019	.012	.010	.006	.004
															.153	.153	.153	.131	.131	.131	.089	.089	.089	.051	.051	.051	.016	.016	.016
.171	.098	.031	.166	.095	.030	.156	.089	.029	.146	.084	.027	.137	.080	.026	.140	.127	.115	.120	.109	.099	.082	.075	.068	.047	.043	.040	.015	.014	.013
.168	.092	.028	.163	.090	.028	.154	.086	.027	.146	.082	.026	.138	.078	.025	.131	.108	.089	.113	.093	.077	.077	.064	.053	.044	.037	.031	.014	.012	.010
.161	.086	.026	.157	.084	.025	.149	.081	.025	.142	.078	.024	.135	.075	.023	.124	.094	.070	.106	.081	.061	.073	.056	.043	.042	.033	.025	.014	.011	.008
.153	.080	.023	.149	.078	.023	.143	.076	.023	.136	.073	.022	.130	.071	.022	.118	.083	.057	.101	.072	.050	.069	.050	.035	.040	.029	.021	.013	.010	.007
.145	.074	.021	.142	.073	.021	.136	.071	.021	.130	.069	.020	.125	.067	.020	.112	.075	.048	.096	.065	.041	.066	.045	.029	.038	.027	.017	.012	.009	.006
.138	.069	.020	.135	.068	.020	.129	.066	.019	.124	.065	.019	.120	.063	.019	.107	.068	.041	.092	.059	.035	.063	.041	.025	.037	.024	.015	.012	.008	.005
.131	.065	.018	.128	.064	.018	.123	.062	.018	.119	.061	.018	.114	.059	.017	.102	.063	.035	.088	.055	.031	.061	.038	.022	.035	.023	.013	.011	.007	.004
.124	.061	.017	.122	.060	.017	.117	.059	.017	.113	.057	.017	.109	.056	.016	.098	.058	.031	.084	.051	.027	.058	.036	.019	.034	.021	.012	.011	.007	.004
.118	.057	.016	.116	.056	.016	.112	.055	.016	.108	.054	.015	.104	.053	.015	.094	.055	.028	.081	.047	.024	.056	.033	.017	.033	.020	.010	.011	.006	.003
.112	.054	.015	.110	.053	.015	.106	.052	.015	.103	.051	.015	.100	.050	.014	.090	.051	.026	.078	.045	.022	.054	.031	.016	.031	.019	.009	.010	.006	.003
															.154	.154	.154	.132	.132	.132	.090	.090	.090	.052	.052	.052	.017	.017	.017
.188	.107	.034	.183	.104	.033	.172	.099	.032	.162	.094	.030	.154	.089	.029	.143	.128	.115	.122	.110	.099	.084	.076	.068	.048	.044	.040	.015	.014	.013
.186	.102	.031	.181	.100	.031	.172	.095	.030	.163	.092	.029	.155	.088	.028	.135	.109	.087	.115	.094	.076	.079	.065	.053	.046	.038	.031	.015	.012	.010
.178	.095	.028	.174	.093	.028	.166	.090	.027	.158	.087	.027	.152	.084	.026	.128	.095	.068	.110	.082	.059	.075	.057	.042	.043	.033	.024	.014	.011	.008
.170	.088	.026	.166	.087	.026	.159	.084	.025	.152	.082	.025	.146	.079	.024	.122	.084	.055	.105	.073	.048	.072	.051	.034	.042	.030	.020	.013	.010	.007
.161	.082	.024	.157	.081	.024	.151	.079	.023	.145	.076	.023	.139	.074	.022	.117	.076	.045	.100	.065	.039	.069	.046	.028	.040	.027	.017	.013	.009	.005
.152	.076	.022	.149	.075	.022	.143	.073	.021	.138	.072	.021	.133	.070	.021	.112	.069	.038	.096	.060	.033	.066	.042	.024	.038	.025	.014	.012	.008	.005
.143	.071	.020	.141	.070	.020	.136	.069	.020	.131	.067	.020	.126	.066	.019	.107	.064	.033	.092	.055	.029	.064	.039	.020	.037	.023	.012	.012	.007	.004
.136	.066	.019	.133	.066	.019	.129	.064	.018	.124	.063	.018	.120	.062	.018	.102	.059	.029	.088	.051	.026	.061	.036	.018	.036	.021	.011	.011	.007	.004
.128	.062	.017	.126	.062	.017	.122	.060	.017	.118	.059	.017	[illegible]	.058	.017	.098	.055	.026	.085	.048	.023	.059	.034	.016	.034	.020	.010	.011	.007	.003
.122	.058	.016	.120	.058	.016	.116	.057	.016	.112	.056	.016	.109	.055	.016	.094	.052	.024	.081	.045	.021	.056	.032	.015	.033	.019	.009	.011	.006	.003

Fig. 61 *Continued (See page 119 for instructions and notes)*

Typical Luminaire	Typical Intensity Distribution and Per Cent Lamp Lumens	ρcc →	80			70			50			30			10			0	WDRC	ρcc → RCR
	Maint. Cat. / SC	ρw → / RCR ↓	50	30	10	50	30	10	50	30	10	50	30	10	50	30	10	0		ρw →

Coefficients of Utilization for 20 Per Cent Effective Floor Cavity Reflectance (ρFC = 20)

19 — "High bay" intermediate distribution ventilated reflector with phosphor coated HID lamp. (6¼% ↑, 75½% ↓) — Maint. Cat. III, SC 1.0

RCR	50	30	10	50	30	10	50	30	10	50	30	10	50	30	10	0	WDRC
0	.96	.96	.96	.93	.93	.93	.88	.88	.88	.83	.83	.83	.78	.78	.78	.76	
1	.88	.86	.83	.86	.83	.81	.81	.79	.78	.77	.75	.74	.73	.72	.71	.69	.167
2	.80	.76	.73	.78	.74	.71	.74	.71	.69	.71	.68	.66	.68	.66	.64	.62	.168
3	.73	.68	.64	.71	.67	.63	.68	.64	.61	.65	.62	.60	.63	.60	.58	.56	.162
4	.67	.61	.57	.65	.60	.57	.63	.59	.55	.60	.57	.54	.58	.55	.52	.51	.155
5	.61	.56	.52	.60	.55	.51	.58	.53	.50	.56	.52	.49	.54	.50	.48	.46	.147
6	.57	.51	.47	.56	.50	.46	.54	.49	.45	.52	.48	.45	.50	.46	.44	.42	.139
7	.52	.47	.43	.51	.46	.42	.50	.45	.42	.48	.44	.41	.47	.43	.40	.39	.132
8	.49	.43	.39	.48	.42	.39	.46	.42	.38	.45	.41	.38	.44	.40	.37	.36	.125
9	.45	.40	.36	.45	.39	.36	.43	.39	.35	.42	.38	.35	.41	.37	.34	.33	.118
10	.42	.37	.33	.42	.37	.33	.41	.36	.33	.39	.35	.32	.38	.35	.32	.31	.112

20 — "High bay" wide distribution ventilated reflector with phosphor coated HID lamp. (12% ↑, 69% ↓) — Maint. Cat. III, SC 1.5

RCR	50	30	10	50	30	10	50	30	10	50	30	10	50	30	10	0	WDRC
0	.93	.93	.93	.90	.90	.90	.83	.83	.83	.77	.77	.77	.72	.72	.72	.69	
1	.85	.82	.80	.82	.79	.77	.76	.74	.73	.71	.70	.69	.66	.65	.65	.62	.168
2	.76	.72	.69	.74	.70	.67	.69	.66	.64	.65	.63	.61	.61	.59	.58	.56	.168
3	.69	.64	.60	.67	.62	.59	.63	.59	.56	.59	.56	.54	.56	.54	.51	.49	.163
4	.62	.57	.52	.61	.55	.51	.57	.53	.50	.54	.51	.48	.51	.48	.46	.44	.156
5	.57	.51	.46	.55	.50	.46	.52	.48	.44	.49	.46	.43	.47	.44	.41	.39	.148
6	.52	.45	.41	.50	.45	.40	.48	.43	.39	.45	.41	.38	.43	.40	.37	.35	.141
7	.47	.41	.37	.46	.40	.36	.44	.39	.35	.42	.37	.34	.40	.36	.33	.32	.133
8	.43	.37	.33	.42	.36	.33	.40	.35	.32	.38	.34	.31	.37	.33	.30	.29	.126
9	.40	.34	.30	.39	.33	.29	.37	.32	.29	.35	.31	.28	.34	.30	.27	.26	.120
10	.37	.31	.27	.36	.30	.27	.34	.29	.26	.33	.28	.25	.31	.28	.25	.24	.114

21 — "Low bay" rectangular pattern, lensed bottom reflector unit with clear HID lamp. (0° ↑, 45, 68½° ↓, II, ⊥) — Maint. Cat. V, SC 1.8

RCR	50	30	10	50	30	10	50	30	10	50	30	10	50	30	10	0	WDRC
0	.82	.82	.82	.80	.80	.80	.76	.76	.76	.73	.73	.73	.70	.70	.70	.68	
1	.73	.70	.68	.71	.69	.67	.68	.66	.64	.65	.64	.62	.63	.62	.61	.59	.231
2	.64	.60	.56	.63	.59	.55	.60	.57	.54	.58	.55	.53	.56	.54	.52	.50	.227
3	.56	.51	.47	.55	.51	.47	.53	.49	.46	.52	.48	.45	.50	.47	.44	.43	.213
4	.50	.44	.40	.49	.44	.40	.48	.43	.39	.46	.42	.39	.44	.41	.38	.37	.199
5	.45	.39	.34	.44	.38	.34	.42	.38	.34	.41	.37	.33	.40	.36	.33	.32	.184
6	.40	.34	.30	.39	.34	.30	.38	.33	.29	.37	.33	.29	.36	.32	.29	.28	.171
7	.36	.30	.26	.36	.30	.26	.35	.29	.26	.34	.29	.26	.33	.29	.25	.24	.159
8	.33	.27	.23	.32	.27	.23	.31	.26	.23	.31	.26	.23	.30	.26	.23	.21	.148
9	.30	.24	.20	.29	.24	.20	.29	.24	.20	.28	.23	.20	.27	.23	.20	.19	.138
10	.27	.22	.18	.27	.22	.18	.26	.22	.18	.26	.21	.18	.25	.21	.18	.17	.129

22 — "Low bay" lensed bottom reflector unit with clear HID lamp. (3 ↑, 68 ↓) — Maint. Cat. V, SC 1.9

RCR	50	30	10	50	30	10	50	30	10	50	30	10	50	30	10	0	WDRC
0	.83	.83	.83	.81	.81	.81	.77	.77	.77	.73	.73	.73	.70	.70	.70	.68	
1	.72	.69	.66	.70	.67	.65	.67	.64	.62	.63	.62	.60	.60	.59	.57	.56	.302
2	.62	.57	.53	.61	.56	.52	.58	.54	.50	.55	.52	.50	.52	.50	.47	.46	.279
3	.54	.48	.43	.53	.47	.43	.50	.45	.41	.48	.44	.40	.46	.42	.39	.38	.253
4	.47	.41	.36	.46	.40	.35	.44	.39	.35	.42	.37	.34	.40	.36	.33	.31	.229
5	.42	.35	.30	.41	.34	.30	.39	.33	.29	.37	.32	.29	.36	.31	.28	.26	.208
6	.37	.30	.26	.36	.30	.25	.35	.29	.25	.33	.28	.25	.32	.27	.24	.23	.189
7	.33	.27	.22	.33	.26	.22	.31	.26	.22	.30	.25	.21	.29	.24	.21	.19	.173
8	.30	.24	.19	.29	.23	.19	.28	.23	.19	.27	.22	.19	.26	.22	.18	.17	.159
9	.27	.21	.17	.27	.21	.17	.26	.20	.17	.25	.20	.17	.24	.19	.16	.15	.147
10	.25	.19	.15	.24	.19	.15	.24	.18	.15	.22	.18	.15	.21	.17	.15	.13	.137

23 — Wide spread, recessed, small open bottom reflector with low wattage diffuse HID lamp. (0° ↑, 56° ↓) — Maint. Cat. IV, SC 1.7

RCR	50	30	10	50	30	10	50	30	10	50	30	10	50	30	10	0	WDRC
0	.67	.67	.67	.65	.65	.65	.62	.62	.62	.60	.60	.60	.57	.57	.57	.56	
1	.60	.58	.56	.58	.57	.55	.56	.55	.53	.54	.53	.52	.52	.51	.50	.49	.177
2	.53	.49	.46	.52	.48	.46	.50	.47	.44	.48	.45	.43	.46	.44	.43	.42	.179
3	.46	.42	.39	.46	.42	.38	.44	.41	.38	.42	.40	.37	.41	.39	.37	.35	.172
4	.41	.36	.33	.40	.36	.33	.39	.35	.32	.38	.34	.32	.37	.34	.31	.30	.161
5	.37	.32	.28	.36	.31	.28	.35	.31	.28	.34	.30	.27	.33	.30	.27	.26	.150
6	.33	.28	.24	.32	.28	.24	.31	.27	.24	.30	.27	.24	.30	.26	.24	.23	.139
7	.30	.25	.21	.29	.25	.21	.29	.25	.21	.28	.24	.21	.27	.23	.21	.20	.129
8	.27	.22	.19	.26	.22	.19	.26	.22	.19	.25	.21	.19	.24	.21	.19	.17	.120
9	.25	.20	.17	.24	.20	.17	.24	.20	.17	.23	.19	.17	.22	.19	.17	.16	.112
10	.22	.18	.15	.22	.18	.15	.22	.18	.15	.21	.18	.15	.21	.17	.15	.14	.105

24 — Open top, indirect, reflector type unit with HID lamp (mult. by 0.9 for lens top). (78 ↑, 0 ↓) — Maint. Cat. VI, SC N.A.

RCR	50	30	10	50	30	10	50	30	10	50	30	10	50	30	10	0	WDRC
0	.74	.74	.74	.63	.63	.63	.43	.43	.43	.25	.25	.25	.08	.08	.08	.00	
1	.64	.62	.59	.55	.53	.51	.38	.36	.35	.22	.21	.20	.07	.07	.07	.00	.000
2	.56	.52	.48	.48	.45	.42	.33	.31	.29	.19	.18	.17	.06	.06	.06	.00	.000
3	.49	.44	.40	.42	.38	.35	.29	.26	.24	.17	.15	.14	.05	.05	.05	.00	.000
4	.43	.38	.34	.37	.33	.29	.26	.23	.20	.15	.13	.12	.05	.04	.04	.00	.000
5	.38	.33	.28	.33	.28	.25	.23	.20	.17	.13	.12	.10	.04	.04	.03	.00	.000
6	.34	.28	.24	.29	.25	.21	.20	.17	.15	.12	.10	.09	.04	.03	.03	.00	.000
7	.31	.25	.21	.26	.22	.18	.18	.15	.13	.11	.09	.08	.03	.03	.03	.00	.000
8	.28	.22	.18	.24	.19	.16	.16	.13	.11	.10	.08	.07	.03	.03	.02	.00	.000
9	.25	.20	.16	.21	.17	.14	.15	.12	.10	.09	.07	.06	.03	.02	.02	.00	.000
10	.23	.17	.14	.20	.15	.12	.14	.11	.09	.08	.06	.05	.03	.02	.02	.00	.000

80			70			50			30			10			80			70			50			30			10		
50	30	10	50	30	10	50	30	10	50	30	10	50	30	10	50	30	10	50	30	10	50	30	10	50	30	10	50	30	10
Wall Exitance Coefficients for 20 Per Cent Effective Floor Cavity Reflectance (ρ_{FC} = 20)															Ceiling Cavity Exitance Coefficients for 20 Per Cent Floor Cavity Reflectance (ρ_{FC} = 20)														
															.207	.207	.207	.177	.177	.177	.121	.121	.121	.069	.069	.069	.022	.022	.022
.176	.100	.032	.168	.096	.030	.154	.088	.028	.141	.081	.026	.128	.074	.024	.194	.180	.168	.166	.155	.144	.113	.106	.100	.065	.062	.058	.021	.020	.019
.170	.093	.029	.163	.090	.028	.150	.084	.026	.139	.078	.024	.128	.073	.023	.184	.160	.140	.157	.138	.121	.108	.095	.085	.062	.055	.050	.020	.018	.016
.162	.086	.026	.156	.083	.025	.145	.078	.024	.135	.074	.023	.125	.069	.021	.175	.145	.121	.150	.125	.105	.103	.087	.074	.060	.051	.043	.019	.017	.014
.153	.080	.023	.148	.078	.023	.138	.073	.022	.129	.069	.021	.121	.066	.020	.168	.134	.107	.144	.116	.093	.099	.081	.066	.057	.047	.039	.018	.015	.013
.145	.074	.021	.141	.072	.021	.132	.069	.020	.124	.065	.019	.116	.062	.019	.162	.125	.097	.139	.108	.085	.096	.075	.060	.055	.044	.035	.018	.014	.012
.138	.069	.020	.133	.067	.019	.126	.064	.019	.118	.061	.018	.111	.058	.017	.156	.118	.090	.134	.102	.078	.093	.071	.055	.054	.042	.033	.017	.014	.011
.131	.065	.018	.127	.063	.018	.120	.060	.017	.113	.058	.017	.106	.055	.016	.151	.112	.084	.130	.097	.073	.090	.068	.052	.052	.040	.031	.017	.013	.010
.124	.061	.017	.120	.059	.017	.114	.057	.016	.108	.055	.016	.102	.052	.015	.147	.107	.080	.126	.093	.069	.087	.065	.049	.051	.038	.029	.016	.013	.010
.118	.057	.016	.115	.056	.016	.109	.054	.015	.103	.052	.015	.098	.050	.014	.142	.103	.076	.123	.089	.066	.085	.063	.047	.049	.037	.028	.016	.012	.009
.112	.054	.015	.109	.053	.015	.104	.051	.014	.099	.049	.014	.094	.047	.014	.138	.099	.073	.119	.086	.064	.083	.061	.046	.048	.036	.027	.016	.012	.009
															.244	.244	.244	.209	.209	.209	.143	.143	.143	.082	.082	.082	.026	.026	.026
.183	.104	.033	.174	.099	.032	.157	.090	.029	.141	.081	.026	.127	.073	.024	.232	.218	.205	.199	.187	.177	.136	.129	.122	.078	.075	.071	.025	.024	.023
.177	.097	.030	.168	.093	.029	.153	.085	.026	.139	.078	.024	.126	.071	.023	.223	.199	.178	.191	.171	.154	.131	.118	.107	.076	.069	.063	.024	.022	.021
.168	.090	.027	.161	.086	.026	.148	.080	.024	.135	.074	.023	.123	.068	.021	.216	.184	.159	.185	.159	.138	.127	.111	.097	.073	.065	.057	.024	.021	.019
.160	.083	.024	.154	.080	.024	.141	.075	.022	.130	.070	.021	.119	.065	.020	.209	.173	.146	.180	.150	.127	.124	.104	.089	.071	.061	.053	.023	.020	.017
.152	.077	.022	.146	.075	.022	.135	.070	.021	.125	.066	.020	.115	.061	.018	.204	.165	.136	.175	.143	.119	.121	.100	.084	.070	.059	.050	.023	.019	.016
.144	.072	.021	.139	.070	.020	.129	.066	.019	.119	.062	.018	.110	.058	.017	.199	.158	.129	.171	.137	.112	.118	.096	.080	.068	.056	.047	.022	.018	.016
.137	.068	.019	.132	.066	.019	.123	.062	.018	.114	.058	.017	.106	.055	.016	.194	.153	.124	.167	.132	.108	.115	.093	.076	.067	.055	.045	.022	.018	.015
.130	.063	.018	.125	.062	.017	.117	.058	.017	.109	.055	.016	.101	.052	.015	.190	.148	.120	.163	.128	.104	.113	.090	.074	.066	.053	.044	.021	.017	.015
.124	.060	.017	.119	.058	.016	.112	.055	.016	.104	.052	.015	.097	.049	.014	.186	.144	.116	.160	.125	.101	.111	.088	.072	.065	.052	.043	.021	.017	.014
.118	.056	.016	.114	.055	.015	.106	.052	.015	.100	.049	.014	.093	.047	.013	.182	.141	.114	.157	.122	.099	.109	.086	.071	.063	.051	.042	.021	.017	.014
															.130	.130	.130	.112	.112	.112	.076	.076	.076	.044	.044	.044	.014	.014	.014
.186	.106	.033	.181	.103	.033	.172	.099	.032	.164	.094	.030	.156	.091	.029	.122	.107	.094	.104	.092	.081	.071	.063	.056	.041	.037	.033	.013	.012	.011
.181	.099	.030	.177	.097	.030	.169	.094	.029	.162	.091	.028	.155	.088	.028	.115	.090	.069	.099	.078	.060	.068	.054	.042	.039	.031	.025	.013	.010	.008
.171	.091	.027	.168	.090	.027	.161	.087	.026	.154	.085	.026	.148	.082	.025	.110	.078	.053	.094	.067	.046	.065	.047	.032	.037	.027	.019	.012	.009	.006
.161	.084	.025	.158	.083	.024	.152	.081	.024	.146	.078	.024	.141	.076	.023	.105	.069	.041	.090	.060	.036	.062	.042	.025	.036	.024	.015	.012	.008	.005
.151	.077	.022	.148	.076	.022	.143	.074	.022	.138	.073	.022	.133	.071	.021	.100	.062	.033	.086	.053	.029	.059	.037	.020	.034	.022	.012	.011	.007	.004
.142	.071	.020	.139	.070	.020	.134	.069	.020	.130	.067	.020	.126	.066	.020	.096	.056	.027	.082	.049	.024	.057	.034	.017	.033	.020	.010	.011	.007	.003
.133	.066	.019	.131	.065	.019	.127	.064	.018	.122	.063	.018	.119	.062	.018	.092	.051	.023	.079	.045	.020	.054	.031	.014	.032	.018	.009	.010	.006	.003
.125	.061	.017	.123	.061	.017	.119	.060	.017	.116	.059	.017	.112	.058	.017	.088	.047	.020	.075	.041	.017	.052	.029	.012	.030	.017	.007	.010	.006	.002
.118	.057	.016	.116	.057	.016	.113	.056	.016	.109	.055	.016	.106	.054	.016	.084	.044	.017	.072	.038	.015	.050	.027	.011	.029	.016	.006	.009	.005	.002
.112	.053	.015	.110	.053	.015	.107	.052	.015	.104	.051	.015	.101	.051	.015	.080	.041	.015	.069	.036	.013	.048	.025	.010	.028	.015	.006	.009	.005	.002
															.156	.156	.156	.133	.133	.133	.091	.091	.091	.052	.052	.052	.017	.017	.017
.230	.131	.041	.224	.128	.041	.213	.122	.039	.203	.117	.038	.194	.112	.036	.149	.131	.115	.128	.113	.099	.087	.078	.069	.050	.045	.040	.016	.014	.013
.216	.118	.036	.210	.116	.036	.201	.112	.035	.192	.108	.034	.183	.104	.033	.144	.114	.089	.124	.099	.077	.085	.068	.054	.049	.040	.032	.016	.013	.010
.200	.106	.032	.195	.104	.031	.186	.101	.031	.178	.098	.030	.171	.094	.029	.139	.102	.073	.119	.088	.063	.082	.061	.044	.047	.036	.026	.015	.012	.009
.184	.096	.028	.180	.094	.028	.172	.091	.027	.165	.089	.027	.158	.086	.026	.134	.093	.061	.115	.080	.053	.079	.056	.038	.046	.033	.022	.015	.011	.007
.170	.087	.025	.166	.085	.025	.159	.083	.024	.153	.081	.024	.147	.078	.024	.129	.086	.054	.111	.074	.047	.077	.052	.033	.044	.030	.020	.014	.010	.006
.158	.079	.023	.154	.078	.022	.148	.076	.022	.142	.074	.022	.137	.072	.021	.124	.080	.048	.107	.069	.042	.074	.049	.030	.043	.029	.018	.014	.009	.006
.147	.072	.021	.144	.072	.020	.138	.070	.020	.132	.068	.020	.127	.066	.019	.120	.075	.044	.103	.065	.039	.071	.046	.027	.041	.027	.016	.013	.009	.005
.137	.067	.019	.134	.066	.019	.129	.064	.018	.124	.063	.018	.119	.061	.018	.115	.071	.041	.099	.062	.036	.069	.043	.026	.040	.026	.015	.013	.008	.005
.128	.062	.017	.125	.061	.017	.121	.060	.017	.116	.058	.017	.112	.057	.016	.111	.068	.039	.095	.059	.034	.066	.041	.024	.039	.024	.014	.012	.008	.005
.120	.058	.016	.118	.057	.016	.113	.056	.016	.109	.054	.015	.106	.053	.015	.107	.065	.037	.092	.056	.032	.064	.040	.023	.037	.023	.014	.012	.008	.005
															.107	.107	.107	.091	.091	.091	.062	.062	.062	.036	.036	.036	.011	.011	.011
.146	.083	.026	.142	.081	.026	.135	.077	.025	.128	.074	.024	.122	.071	.023	.099	.088	.077	.085	.075	.067	.058	.052	.046	.033	.030	.027	.011	.010	.009
.145	.079	.024	.141	.078	.024	.135	.075	.023	.129	.072	.023	.123	.070	.022	.094	.074	.057	.080	.064	.049	.055	.044	.034	.032	.026	.020	.010	.008	.007
.138	.074	.022	.136	.073	.022	.130	.070	.021	.125	.068	.021	.120	.066	.021	.090	.064	.043	.077	.055	.038	.053	.038	.026	.030	.022	.016	.010	.007	.005
.131	.068	.020	.128	.067	.020	.123	.065	.020	.119	.064	.019	.114	.062	.019	.086	.056	.034	.074	.049	.029	.051	.034	.021	.029	.020	.012	.009	.006	.004
.123	.063	.018	.121	.062	.018	.116	.061	.018	.112	.059	.018	.108	.058	.017	.082	.050	.027	.070	.044	.024	.048	.031	.017	.028	.018	.010	.009	.006	.003
.116	.058	.017	.114	.057	.017	.110	.056	.016	.106	.055	.016	.102	.054	.016	.078	.046	.022	.067	.040	.020	.046	.028	.014	.027	016	.008	.009	.005	.003
.109	.054	.015	.107	.053	.015	.103	.052	.015	.100	.051	.015	.097	.050	.015	.075	.042	.019	.064	.036	.017	.044	.026	.012	.026	.015	.007	.008	.005	.002
.102	.050	.014	.101	.050	.014	.097	.049	.014	.094	.048	.014	.091	.047	.014	.071	.039	.016	.062	.034	.014	.043	.024	.010	.025	.014	.006	.008	.005	.002
.096	.047	.013	.095	.046	.013	.092	.045	.013	.089	.045	.013	.087	.044	.013	.068	.036	.014	.059	.031	.012	.041	.022	.009	.024	.013	.005	.008	.004	.002
.091	.044	.012	.090	.043	.012	.087	.043	.012	.084	.042	.012	.082	.041	.012	.065	.034	.013	.056	.029	.011	.039	.021	.008	.023	.012	.005	.007	.004	.002
															.743	.743	.743	.635	.635	.635	.433	.433	.433	.249	.249	.249	.080	.080	.080
.201	.114	.036	.172	.098	.031	.117	.067	.022	.068	.039	.013	.022	.013	.004	.737	.721	.707	.631	.619	.609	.431	.426	.421	.248	.247	.245	.080	.079	.079
.184	.101	.031	.158	.087	.027	.108	.060	.019	.062	.035	.011	.020	.011	.004	.732	.706	.685	.627	.608	.592	.430	.421	.413	.248	.245	.242	.079	.079	.079
.170	.091	.027	.146	.078	.024	.100	.054	.017	.058	.032	.010	.019	.010	.003	.727	.695	.670	.623	.600	.581	.428	.417	.408	.247	.243	.240	.079	.079	.079
.158	.082	.024	.135	.071	.021	.093	.049	.015	.054	.029	.009	.017	.009	.003	.723	.687	.660	.620	.594	.573	.426	.414	.404	.247	.242	.239	.079	.079	.079
.147	.075	.022	.126	.065	.019	.087	.045	.013	.050	.027	.008	.016	.009	.003	.718	.681	.653	.617	.589	.568	.425	.411	.401	.246	.242	.238	.079	.079	.078
.137	.069	.020	.117	.059	.017	.081	.042	.012	.047	.024	.007	.015	.008	.002	.714	.676	.648	.614	.585	.564	.423	.410	.399	.246	.241	.237	.079	.079	.078
.128	.063	.018	.110	.055	.016	.076	.038	.011	.044	.023	.007	.014	.007	.002	.710	.671	.644	.611	.582	.562	.422	.408	.398	.245	.240	.237	.079	.079	.078
.120	.059	.016	.103	.051	.014	.071	.036	.010	.042	.021	.006	.013	.007	.002	.706	.668	.642	.608	.579	.559	.421	.407	.397	.245	.240	.236	.079	.079	.078
.113	.054	.015	.097	.047	.013	.067	.033	.009	.039	.020	.006	.013	.006	.002	.703	.665	.639	.605	.577	.558	.419	.406	.396	.244	.240	.236	.079	.079	.078
.106	.051	.014	.091	.044	.012	.063	.031	.009	.037	.018	.005	.012	.006	.002	.699	.662	.638	.603	.575	.556	.418	.405	.395	.244	.239	.236	.079	.079	.078

Fig. 61 Continued (See page 119 for instructions and notes)

Coefficients of Utilization for 20 Per Cent Effective Floor Cavity Reflectance ($\rho_{FC} = 20$)

25 — Porcelain-enameled reflector with 35°CW shielding

Maint. Cat. II — SC 1.3 — 22½%↑, 65%↓

ρ_{CC}	80			70			50			30			10			0	WDRC	RCR
ρ_W	50	30	10	50	30	10	50	30	10	50	30	10	50	30	10	0		
0	.99	.99	.99	.94	.94	.94	.85	.85	.85	.77	.77	.77	.69	.69	.69	.65		0
1	.87	.84	.81	.83	.80	.77	.75	.73	.71	.68	.66	.65	.62	.60	.59	.56	.236	1
2	.77	.71	.67	.73	.68	.64	.67	.63	.60	.60	.58	.55	.55	.53	.51	.48	.220	2
3	.68	.62	.56	.65	.59	.54	.59	.55	.51	.54	.50	.47	.49	.46	.44	.41	.203	3
4	.61	.54	.48	.58	.52	.47	.53	.48	.44	.48	.44	.41	.44	.41	.38	.35	.186	4
5	.54	.47	.42	.52	.46	.41	.48	.42	.38	.44	.39	.36	.40	.36	.33	.31	.170	5
6	.49	.42	.37	.47	.40	.36	.43	.38	.34	.40	.35	.32	.36	.33	.30	.27	.157	6
7	.45	.37	.32	.43	.36	.32	.39	.34	.30	.36	.32	.28	.33	.29	.26	.24	.145	7
8	.41	.34	.29	.39	.33	.28	.36	.31	.27	.33	.29	.25	.31	.27	.24	.22	.135	8
9	.37	.31	.26	.36	.30	.25	.33	.28	.24	.31	.26	.23	.28	.24	.22	.20	.126	9
10	.34	.28	.24	.33	.27	.23	.31	.25	.22	.28	.24	.21	.26	.22	.20	.18	.118	10

26 — Diffuse aluminum reflector with 35°CW shielding

Maint. Cat. II — SC 1.5/1.3 — 17%↑, 66%↓

ρ_{CC}	80			70			50			30			10			0	WDRC	RCR
ρ_W	50	30	10	50	30	10	50	30	10	50	30	10	50	30	10	0		
0	.95	.95	.95	.91	.91	.91	.83	.83	.83	.76	.76	.76	.69	.69	.69	.66		0
1	.85	.82	.79	.81	.79	.76	.75	.73	.71	.69	.67	.66	.63	.62	.61	.58	.197	1
2	.75	.71	.67	.72	.68	.65	.67	.63	.61	.62	.59	.57	.57	.55	.53	.51	.194	2
3	.67	.61	.57	.65	.59	.55	.60	.56	.52	.55	.52	.49	.51	.49	.46	.44	.184	3
4	.60	.54	.49	.58	.52	.48	.54	.49	.45	.50	.46	.43	.46	.43	.41	.39	.173	4
5	.54	.47	.43	.52	.46	.42	.49	.43	.40	.45	.41	.38	.42	.39	.36	.34	.162	5
6	.49	.42	.37	.47	.41	.37	.44	.39	.35	.41	.37	.33	.38	.35	.32	.30	.151	6
7	.44	.38	.33	.43	.37	.32	.40	.35	.31	.38	.33	.30	.35	.31	.28	.27	.141	7
8	.40	.34	.29	.39	.33	.29	.37	.31	.28	.34	.30	.27	.32	.28	.26	.24	.132	8
9	.37	.31	.26	.36	.30	.26	.34	.29	.25	.32	.27	.24	.30	.26	.23	.21	.124	9
10	.34	.28	.24	.33	.27	.23	.31	.26	.23	.29	.25	.22	.28	.24	.21	.19	.117	10

27 — Porcelain-enameled reflector with 30°CW × 30°LW shielding

Maint. Cat. II — SC 1.0 — 23½%↓, 57%↓

ρ_{CC}	80			70			50			30			10			0	WDRC	RCR
ρ_W	50	30	10	50	30	10	50	30	10	50	30	10	50	30	10	0		
0	.91	.91	.91	.86	.86	.86	.77	.77	.77	.68	.68	.68	.61	.61	.61	.57		0
1	.80	.77	.75	.76	.74	.71	.69	.67	.65	.62	.60	.59	.55	.54	.53	.50	.182	1
2	.71	.67	.63	.68	.64	.60	.61	.58	.55	.55	.53	.51	.50	.48	.46	.43	.174	2
3	.63	.58	.53	.60	.55	.51	.55	.51	.47	.50	.46	.44	.45	.42	.40	.38	.163	3
4	.57	.51	.46	.54	.49	.44	.49	.45	.41	.45	.41	.38	.41	.38	.35	.33	.151	4
5	.51	.45	.40	.49	.43	.39	.45	.40	.36	.41	.37	.34	.37	.34	.31	.29	.140	5
6	.46	.40	.35	.44	.38	.34	.41	.36	.32	.37	.33	.30	.34	.30	.28	.26	.130	6
7	.42	.36	.31	.40	.35	.30	.37	.32	.29	.34	.29	.26	.31	.27	.24	.23	.121	7
8	.38	.32	.28	.37	.31	.27	.34	.29	.26	.31	.27	.24	.29	.25	.23	.21	.113	8
9	.35	.29	.25	.34	.28	.25	.31	.27	.23	.29	.25	.22	.27	.23	.21	.19	.106	9
10	.33	.27	.23	.31	.26	.22	.29	.24	.21	.27	.23	.20	.25	.21	.19	.17	.099	10

28 — Diffuse aluminum reflector with 35°CW × 35°LW shielding

Maint. Cat. II — SC 1.5/1.1 — 17%↑, 56½%↓

ρ_{CC}	80			70			50			30			10			0	WDRC	RCR
ρ_W	50	30	10	50	30	10	50	30	10	50	30	10	50	30	10	0		
0	.83	.83	.83	.79	.79	.79	.72	.72	.72	.65	.65	.65	.59	.59	.59	.56		0
1	.74	.72	.70	.71	.69	.67	.65	.63	.62	.59	.58	.57	.54	.53	.52	.50	.160	1
2	.66	.62	.59	.64	.60	.57	.58	.56	.53	.54	.51	.49	.49	.47	.46	.44	.158	2
3	.59	.54	.50	.57	.53	.49	.53	.49	.46	.48	.46	.43	.45	.42	.40	.38	.150	3
4	.53	.48	.44	.51	.46	.42	.47	.43	.40	.44	.41	.38	.40	.38	.36	.34	.141	4
5	.48	.42	.38	.46	.41	.37	.43	.39	.35	.40	.36	.33	.37	.34	.32	.30	.132	5
6	.44	.38	.34	.42	.37	.33	.39	.35	.31	.36	.33	.30	.34	.31	.28	.27	.124	6
7	.40	.34	.30	.38	.33	.29	.36	.31	.28	.33	.30	.27	.31	.28	.25	.24	.116	7
8	.36	.31	.27	.35	.30	.26	.33	.28	.25	.31	.27	.24	.29	.25	.23	.21	.109	8
9	.33	.28	.24	.32	.27	.24	.30	.26	.23	.28	.24	.22	.26	.23	.21	.19	.102	9
10	.31	.25	.22	.30	.25	.22	.28	.24	.21	.26	.22	.20	.25	.21	.19	.18	.096	10

29 — Metal or dense diffusing sides with 45°CW × 45°LW shielding

Maint. Cat. II — SC 1.1 — 39%↓, 32%↓

ρ_{CC}	80			70			50			30			10			0	WDRC	RCR
ρ_W	50	30	10	50	30	10	50	30	10	50	30	10	50	30	10	0		
0	.75	.75	.75	.69	.69	.69	.57	.57	.57	.46	.46	.46	.37	.37	.37	.32		0
1	.66	.64	.62	.61	.59	.57	.51	.50	.48	.42	.41	.40	.33	.33	.32	.28	.094	1
2	.59	.55	.52	.54	.51	.48	.46	.43	.41	.38	.36	.34	.30	.28	.25	.25	.091	2
3	.52	.48	.44	.48	.44	.41	.41	.38	.35	.34	.32	.30	.27	.26	.25	.22	.085	3
4	.47	.42	.38	.43	.39	.35	.37	.33	.31	.31	.28	.26	.25	.23	.22	.19	.079	4
5	.42	.37	.33	.39	.34	.31	.33	.30	.27	.28	.25	.23	.23	.21	.20	.17	.073	5
6	.38	.33	.29	.35	.31	.27	.30	.27	.24	.25	.23	.21	.21	.19	.18	.16	.068	6
7	.35	.29	.26	.32	.28	.24	.28	.24	.21	.23	.21	.19	.19	.17	.16	.14	.063	7
8	.32	.26	.23	.29	.25	.22	.25	.22	.19	.22	.19	.17	.18	.16	.15	.13	.059	8
9	.29	.24	.21	.27	.23	.20	.23	.20	.17	.20	.17	.15	.17	.15	.13	.12	.056	9
10	.27	.22	.19	.25	.21	.18	.22	.18	.16	.19	.16	.14	.16	.14	.12	.11	.052	10

30 — Same as unit #29 except with top reflectors

Maint. Cat. IV — SC 1.0 — 6%↑, 46%↓

ρ_{CC}	80			70			50			30			10			0	WDRC	RCR
ρ_W	50	30	10	50	30	10	50	30	10	50	30	10	50	30	10	0		
0	.61	.61	.61	.58	.58	.58	.55	.55	.55	.51	.51	.51	.48	.48	.48	.46		0
1	.54	.52	.50	.52	.50	.49	.49	.47	.46	.46	.45	.43	.43	.42	.41	.40	.159	1
2	.48	.45	.42	.46	.44	.41	.44	.41	.39	.41	.39	.38	.39	.37	.36	.34	.145	2
3	.43	.39	.36	.42	.38	.35	.39	.36	.34	.37	.35	.33	.35	.33	.31	.30	.132	3
4	.39	.35	.32	.38	.34	.31	.36	.32	.30	.34	.31	.29	.32	.30	.28	.27	.121	4
5	.35	.31	.28	.34	.30	.27	.32	.29	.27	.31	.28	.26	.29	.27	.25	.24	.111	5
6	.32	.28	.25	.31	.27	.25	.30	.26	.24	.28	.25	.23	.27	.25	.23	.22	.102	6
7	.29	.25	.22	.29	.25	.22	.27	.24	.21	.26	.23	.21	.25	.23	.21	.20	.095	7
8	.27	.23	.20	.27	.23	.20	.25	.22	.20	.24	.21	.19	.23	.21	.19	.18	.088	8
9	.25	.21	.19	.25	.21	.18	.24	.20	.18	.23	.20	.18	.22	.19	.17	.16	.083	9
10	.23	.20	.17	.23	.19	.17	.22	.19	.17	.21	.18	.16	.20	.18	.16	.15	.077	10

80			70			50			30			10			80			70			50			30			10		
50	30	10	50	30	10	50	30	10	50	30	10	50	30	10	50	30	10	50	30	10	50	30	10	50	30	10	50	30	10
Wall Exitance Coefficients for 20 Per Cent Effective Floor Cavity Reflectance (ρ_{FC} = 20)															Ceiling Cavity Exitance Coefficients for 20 Per Cent Floor Cavity Reflectance (ρ_{FC} = 20)														
															.339	.339	.339	.290	.290	.290	.198	.198	.198	.114	.114	.114	.036	.036	.036
.243	.138	.044	.230	.131	.042	.206	.118	.038	.184	.106	.034	.163	.095	.031	.329	.311	.293	.282	.267	.253	.193	.183	.175	.111	.106	.102	.036	.034	.033
.228	.125	.038	.216	.119	.037	.195	.108	.034	.174	.098	.031	.156	.088	.028	.322	.290	.264	.276	.250	.228	.189	.173	.159	.109	.101	.093	.035	.033	.030
.212	.113	.034	.202	.108	.032	.182	.098	.030	.163	.090	.027	.146	.081	.025	.315	.275	.244	.270	.238	.212	.185	.165	.148	.107	.096	.087	.034	.031	.029
.197	.102	.030	.187	.098	.029	.169	.090	.027	.153	.082	.025	.137	.074	.023	.308	.264	.231	.265	.228	.200	.182	.159	.141	.105	.093	.083	.034	.030	.027
.183	.093	.027	.175	.090	.026	.158	.082	.024	.143	.075	.022	.129	.069	.021	.302	.255	.221	.260	.221	.192	.179	.154	.136	.104	.091	.081	.033	.030	.027
.171	.086	.025	.163	.082	.024	.148	.076	.022	.134	.070	.020	.121	.064	.019	.297	.248	.214	.255	.215	.186	.176	.151	.132	.102	.089	.078	.033	.029	.026
.160	.079	.022	.153	.076	.022	.139	.070	.020	.126	.065	.019	.114	.059	.017	.291	.243	.209	.250	.210	.182	.173	.148	.129	.101	.087	.077	.032	.028	.025
.150	.073	.021	.144	.071	.020	.131	.065	.019	.119	.060	.017	.107	.055	.016	.286	.238	.205	.246	.206	.179	.170	.145	.127	.099	.085	.076	.032	.028	.025
.141	.068	.019	.135	.066	.018	.123	.061	.017	.112	.056	.016	.101	.052	.015	.281	.234	.202	.242	.203	.176	.168	.143	.125	.098	.084	.075	.032	.028	.025
.133	.064	.018	.128	.062	.017	.117	.057	.016	.106	.053	.015	.096	.048	.014	.277	.230	.199	.239	.200	.174	.165	.141	.124	.097	.083	.074	.031	.027	.024
															.286	.286	.286	.244	.244	.244	.167	.167	.167	.096	.096	.096	.031	.031	.031
.209	.119	.038	.198	.113	.036	.178	.102	.033	.159	.092	.029	.142	.082	.027	.275	.259	.244	.235	.222	.210	.161	.153	.145	.093	.088	.084	.030	.028	.027
.200	.110	.034	.191	.105	.032	.173	.096	.030	.156	.087	.027	.140	.079	.025	.267	.239	.216	.229	.206	.187	.157	.143	.130	.090	.083	.076	.029	.027	.025
.190	.101	.030	.181	.097	.029	.164	.089	.027	.149	.082	.025	.135	.075	.023	.260	.225	.197	.223	.194	.171	.153	.135	.120	.088	.079	.071	.028	.026	.023
.178	.093	.027	.170	.089	.026	.156	.083	.025	.142	.076	.023	.129	.070	.021	.254	.214	.184	.218	.185	.159	.150	.129	.112	.087	.076	.066	.028	.025	.022
.168	.085	.025	.161	.082	.024	.147	.077	.023	.134	.071	.021	.123	.065	.020	.248	.206	.174	.213	.178	.151	.147	.124	.107	.085	.073	.063	.027	.024	.021
.158	.079	.023	.151	.076	.022	.139	.071	.021	.127	.066	.019	.116	.061	.018	.243	.199	.167	.209	.172	.145	.144	.121	.103	.084	.071	.061	.027	.023	.020
.148	.073	.021	.142	.071	.020	.131	.066	.019	.120	.062	.018	.110	.057	.017	.238	.193	.162	.205	.168	.141	.142	.118	.100	.082	.069	.060	.027	.023	.020
.140	.068	.019	.135	.066	.019	.124	.062	.018	.114	.058	.017	.105	.054	.016	.234	.189	.158	.201	.164	.138	.139	.115	.098	.081	.068	.058	.026	.022	.019
.132	.064	.018	.127	.062	.017	.118	.058	.016	.108	.054	.016	.100	.051	.015	.229	.185	.155	.197	.160	.135	.137	.113	.096	.080	.067	.057	.026	.022	.019
.125	.060	.017	.121	.058	.016	.112	.055	.015	.103	.051	.015	.095	.048	.014	.225	.182	.153	.194	.158	.133	.135	.111	.095	.079	.066	.056	.025	.022	.019
															.334	.334	.334	.286	.286	.286	.195	.195	.195	.112	.112	.112	.036	.036	.036
.210	.119	.038	.197	.113	.036	.173	.099	.032	.151	.087	.028	.131	.076	.025	.325	.308	.294	.278	.265	.253	.190	.182	.175	.109	.105	.102	.035	.034	.033
.199	.109	.033	.187	.103	.032	.166	.092	.029	.146	.082	.026	.127	.072	.023	.317	.290	.267	.272	.249	.231	.186	.173	.161	.107	.100	.094	.034	.032	.031
.186	.099	.030	.176	.094	.028	.156	.085	.026	.138	.076	.023	.121	.067	.021	.311	.276	.249	.266	.238	.215	.183	.165	.151	.106	.097	.089	.034	.031	.029
.174	.090	.027	.164	.086	.025	.146	.078	.023	.130	.070	.021	.115	.062	.019	.305	.266	.236	.261	.230	.205	.180	.160	.144	.104	.094	.085	.033	.031	.028
.162	.083	.024	.154	.079	.023	.137	.072	.021	.122	.065	.019	.108	.058	.017	.299	.258	.227	.257	.223	.198	.177	.156	.139	.103	.091	.083	.033	.030	.027
.152	.076	.022	.144	.073	.021	.129	.066	.019	.115	.060	.018	.102	.054	.016	.294	.251	.221	.253	.218	.192	.174	.152	.136	.101	.090	.081	.033	.029	.027
.143	.071	.020	.135	.067	.019	.122	.062	.018	.109	.056	.016	.097	.050	.015	.289	.246	.216	.249	.213	.188	.172	.149	.133	.100	.088	.079	.032	.029	.026
.134	.066	.018	.128	.063	.018	.115	.057	.016	.103	.052	.015	.091	.047	.014	.284	.241	.212	.245	.209	.185	.169	.147	.131	.099	.087	.078	.032	.028	.026
.127	.061	.017	.120	.059	.016	.109	.054	.015	.097	.049	.014	.087	.044	.013	.280	.238	.209	.241	.206	.182	.167	.145	.129	.097	.086	.077	.032	.028	.026
.120	.057	.016	.114	.055	.015	.103	.050	.014	.092	.046	.013	.082	.041	.012	.276	.234	.207	.238	.204	.180	.165	.143	.128	.096	.085	.077	.031	.028	.025
															.268	.268	.268	.229	.229	.229	.156	.156	.156	.090	.090	.090	.029	.029	.029
.180	.103	.032	.170	.097	.031	.151	.087	.028	.134	.077	.025	.118	.068	.022	.259	.245	.232	.221	.210	.200	.151	.144	.138	.087	.084	.080	.028	.027	.026
.173	.095	.029	.164	.090	.028	.146	.081	.025	.131	.073	.023	.116	.066	.021	.251	.227	.207	.215	.196	.179	.148	.135	.125	.085	.079	.073	.027	.025	.024
.163	.087	.026	.155	.083	.025	.139	.076	.023	.125	.069	.021	.112	.062	.019	.245	.215	.191	.210	.185	.165	.144	.129	.116	.083	.075	.068	.027	.024	.022
.153	.080	.023	.146	.076	.023	.132	.070	.021	.119	.064	.019	.107	.058	.018	.240	.205	.179	.206	.177	.156	.141	.124	.110	.082	.072	.065	.026	.024	.021
.144	.074	.021	.138	.071	.021	.125	.065	.019	.113	.060	.018	.102	.054	.016	.235	.198	.171	.202	.171	.148	.139	.120	.105	.080	.070	.062	.026	.023	.020
.136	.068	.019	.130	.065	.019	.118	.060	.018	.107	.056	.016	.097	.051	.015	.230	.192	.165	.198	.166	.143	.136	.116	.101	.079	.068	.060	.026	.022	.020
.128	.063	.018	.122	.061	.017	.111	.056	.016	.101	.052	.015	.092	.048	.014	.226	.187	.160	.194	.162	.139	.134	.114	.099	.078	.067	.059	.025	.022	.019
.121	.059	.017	.116	.057	.016	.106	.053	.015	.096	.049	.014	.087	.045	.013	.222	.183	.156	.191	.159	.136	.132	.111	.097	.077	.066	.058	.025	.022	.019
.114	.055	.015	.109	.053	.015	.100	.050	.014	.091	.046	.013	.083	.042	.012	.218	.180	.154	.188	.156	.134	.130	.110	.095	.076	.065	.057	.025	.021	.019
.108	.052	.014	.104	.050	.014	.095	.047	.013	.087	.043	.012	.079	.040	.011	.214	.177	.152	.185	.153	.132	.128	.108	.094	.075	.064	.056	.024	.021	.019
															.433	.433	.433	.370	.370	.370	.253	.253	.253	.145	.145	.145	.046	.046	.046
.180	.102	.032	.163	.093	.030	.132	.076	.024	.103	.060	.019	.077	.044	.014	.426	.411	.399	.364	.353	.343	.249	.243	.237	.143	.141	.138	.046	.045	.045
.168	.092	.028	.153	.084	.026	.125	.069	.022	.098	.055	.017	.074	.042	.013	.419	.396	.377	.359	.341	.326	.246	.236	.227	.142	.137	.133	.046	.044	.043
.157	.083	.025	.143	.077	.023	.117	.063	.019	.093	.051	.016	.070	.039	.012	.414	.385	.362	.355	.332	.314	.244	.231	.220	.141	.135	.130	.045	.044	.042
.146	.076	.022	.133	.070	.021	.109	.058	.017	.087	.047	.014	.066	.036	.011	.409	.376	.351	.351	.325	.305	.241	.227	.215	.140	.133	.127	.045	.043	.042
.136	.069	.020	.125	.064	.019	.102	.053	.016	.082	.043	.013	.063	.034	.010	.404	.370	.344	.347	.320	.299	.239	.223	.211	.139	.131	.125	.045	.043	.041
.127	.064	.018	.117	.059	.017	.096	.049	.014	.077	.040	.012	.059	.031	.009	.400	.364	.339	.344	.315	.295	.237	.221	.209	.138	.130	.124	.044	.042	.041
.119	.059	.017	.109	.055	.016	.091	.046	.013	.073	.037	.011	.056	.029	.009	.396	.360	.335	.341	.312	.292	.235	.219	.207	.137	.129	.123	.044	.042	.041
.112	.055	.015	.103	.051	.014	.085	.043	.012	.069	.035	.010	.053	.027	.008	.392	.356	.331	.338	.309	.289	.234	.217	.205	.136	.128	.122	.044	.042	.040
.106	.051	.014	.097	.047	.013	.081	.040	.011	.065	.033	.009	.051	.026	.007	.388	.353	.329	.335	.306	.287	.232	.215	.204	.135	.127	.122	.044	.042	.040
.100	.048	.013	.092	.044	.012	.077	.037	.011	.062	.031	.009	.048	.024	.007	.385	.350	.327	.332	.304	.286	.230	.214	.203	.134	.127	.121	.044	.042	.040
															.145	.145	.145	.124	.124	.124	.085	.085	.085	.049	.049	.049	.016	.016	.016
.142	.081	.026	.137	.078	.025	.127	.073	.023	.117	.068	.022	.108	.063	.020	.139	.128	.118	.119	.110	.102	.081	.076	.070	.047	.044	.041	.015	.014	.013
.132	.072	.022	.127	.070	.022	.118	.066	.020	.109	.061	.019	.102	.058	.018	.134	.116	.100	.115	.099	.087	.079	.069	.060	.045	.040	.035	.015	.013	.012
.122	.065	.019	.118	.063	.019	.109	.059	.018	.102	.056	.017	.095	.053	.016	.129	.106	.088	.111	.092	.077	.076	.064	.054	.044	.037	.032	.014	.012	.010
.113	.059	.017	.109	.057	.017	.102	.054	.016	.095	.051	.015	.089	.048	.015	.125	.099	.080	.107	.086	.070	.074	.060	.049	.043	.035	.029	.014	.011	.010
.105	.054	.016	.102	.052	.015	.095	.050	.015	.089	.047	.014	.083	.044	.013	.121	.094	.074	.104	.081	.065	.071	.057	.046	.041	.033	.027	.013	.011	.009
.098	.049	.014	.095	.048	.014	.089	.046	.013	.083	.043	.013	.078	.041	.012	.117	.090	.070	.101	.078	.061	.069	.054	.043	.040	.032	.026	.013	.010	.008
.092	.045	.013	.089	.044	.013	.084	.042	.012	.079	.040	.012	.074	.038	.011	.114	.086	.067	.098	.075	.058	.068	.052	.041	.039	.031	.024	.013	.010	.008
.086	.042	.012	.084	.041	.012	.079	.039	.011	.074	.038	.011	.070	.036	.010	.111	.083	.064	.095	.072	.056	.066	.051	.040	.038	.030	.024	.012	.010	.008
.081	.039	.011	.079	.039	.011	.074	.037	.010	.070	.035	.010	.066	.034	.010	.108	.080	.062	.093	.070	.054	.064	.049	.038	.037	.029	.023	.012	.010	.008
.077	.037	.010	.075	.036	.010	.071	.035	.010	.067	.033	.009	.063	.032	.009	.105	.078	.060	.091	.068	.053	.063	.048	.037	.037	.028	.022	.012	.009	.007

Fig. 61 *Continued (See page 119 for instructions and notes)*

Typical Luminaire	Typical Intensity Distribution and Per Cent Lamp Lumens			ρcc →	80			70			50			30			10			0	WDRC	ρcc →
				ρw →	50	30	10	50	30	10	50	30	10	50	30	10	50	30	10	0		ρw →
	Maint. Cat.	SC		RCR ↓	Coefficients of Utilization for 20 Per Cent Effective Floor Cavity Reflectance (ρ_{FC} = 20)																	RCR ↓
31 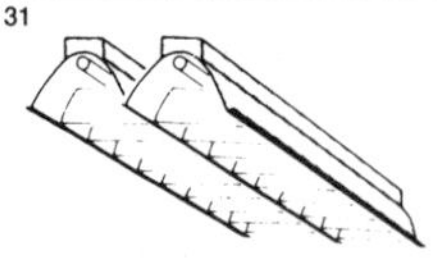	IV	1.5/1.2	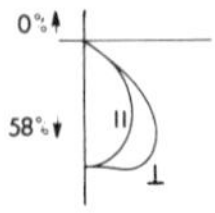	0	.69	.69	.69	.67	.67	.67	.64	.64	.64	.62	.62	.62	.59	.59	.59	.58		
				1	.62	.61	.59	.61	.59	.58	.59	.57	.56	.57	.55	.54	.55	.54	.53	.52	.159	1
				2	.56	.53	.50	.55	.52	.50	.53	.50	.48	.51	.49	.47	.49	.48	.46	.45	.160	2
				3	.50	.46	.43	.49	.46	.43	.48	.44	.42	.46	.43	.41	.45	.42	.41	.39	.155	3
				4	.45	.41	.37	.44	.40	.37	.43	.39	.36	.42	.38	.36	.40	.38	.36	.34	.147	4
				5	.40	.36	.32	.40	.36	.32	.39	.35	.32	.38	.34	.32	.37	.34	.31	.30	.139	5
				6	.37	.32	.29	.36	.32	.28	.35	.31	.28	.34	.31	.28	.33	.30	.28	.27	.131	6
				7	.33	.29	.25	.33	.28	.25	.32	.28	.25	.31	.28	.25	.30	.27	.25	.24	.123	7
150 mm × 150 mm (6 × 6") cell parabolic				8	.30	.26	.23	.30	.26	.22	.29	.25	.22	.28	.25	.22	.28	.25	.22	.21	.115	8
wedge louver—multiply by 1.1 for 250				9	.28	.23	.20	.27	.23	.20	.27	.23	.20	.26	.23	.20	.26	.22	.20	.19	.109	9
× 250 mm (10 × 10") cells				10	.26	.21	.18	.25	.21	.18	.25	.21	.18	.24	.21	.18	.24	.20	.18	.17	.102	10
32 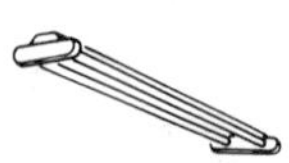	I	1.3	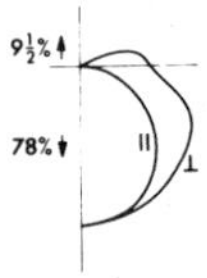	0	1.02	1.02	1.02	.99	.99	.99	.92	.92	.92	.86	.86	.86	.81	.81	.81	.78		
				1	.85	.80	.76	.82	.78	.74	.76	.73	.70	.71	.68	.66	.67	.64	.62	.60	.467	1
				2	.72	.65	.59	.70	.63	.58	.65	.60	.55	.61	.56	.52	.57	.53	.50	.47	.387	2
				3	.63	.55	.48	.60	.53	.47	.56	.50	.45	.53	.47	.43	.49	.45	.41	.38	.331	3
				4	.55	.46	.40	.53	.45	.39	.50	.43	.37	.46	.41	.36	.43	.38	.34	.32	.289	4
				5	.49	.40	.34	.47	.39	.33	.44	.37	.32	.41	.35	.31	.39	.34	.29	.27	.255	5
2-lamp, surface mounted, bare lamp				6	.43	.35	.29	.42	.34	.29	.40	.33	.28	.37	.31	.27	.35	.30	.26	.23	.228	6
unit—photometry with 460 mm (18")				7	.39	.31	.25	.38	.30	.25	.36	.29	.24	.34	.28	.23	.32	.26	.22	.20	.206	7
wide panel above luminaire—lamps on				8	.36	.28	.22	.35	.27	.22	.33	.26	.21	.31	.25	.21	.29	.24	.20	.18	.188	8
150 mm (6") centers				9	.33	.25	.20	.32	.25	.20	.30	.24	.19	.28	.23	.18	.27	.22	.18	.16	.173	9
				10	.30	.23	.18	.29	.22	.18	.28	.21	.17	.26	.21	.17	.25	.20	.16	.14	.159	10
33 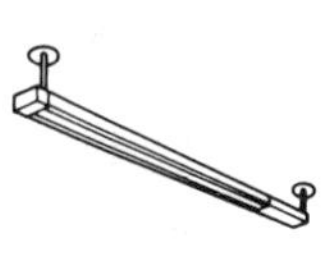	VI	N.A.	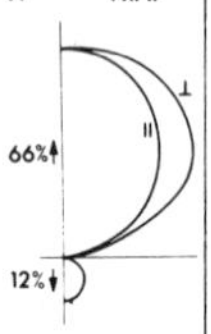	0	.77	.77	.77	.68	.68	.68	.50	.50	.50	.34	.34	.34	.19	.19	.19	.12		
				1	.67	.64	.61	.59	.56	.54	.43	.42	.41	.29	.29	.28	.17	.16	.16	.10	.048	1
				2	.58	.54	.50	.51	.48	.44	.38	.36	.34	.26	.24	.23	.14	.14	.13	.08	.045	2
				3	.51	.46	.42	.45	.41	.37	.33	.30	.28	.23	.21	.19	.13	.12	.11	.07	.041	3
				4	.45	.39	.35	.40	.35	.31	.30	.26	.24	.20	.18	.17	.11	.10	.10	.06	.037	4
				5	.40	.34	.30	.35	.30	.26	.26	.23	.20	.18	.16	.14	.10	.09	.08	.05	.034	5
				6	.36	.30	.25	.31	.26	.23	.24	.20	.17	.16	.14	.12	.09	.08	.07	.04	.031	6
				7	.32	.26	.22	.28	.23	.20	.21	.18	.15	.15	.12	.11	.08	.07	.06	.04	.028	7
				8	.29	.23	.19	.26	.21	.17	.19	.16	.13	.13	.11	.09	.08	.06	.06	.03	.026	8
Luminous bottom suspended unit with				9	.26	.21	.17	.23	.18	.15	.17	.14	.12	.12	.10	.08	.07	.06	.05	.03	.024	9
extra-high output lamp				10	.24	.19	.15	.21	.17	.13	.16	.13	.10	.11	.09	.07	.06	.05	.04	.03	.022	10
34 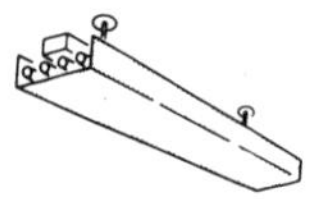	VI	1.4/1.2	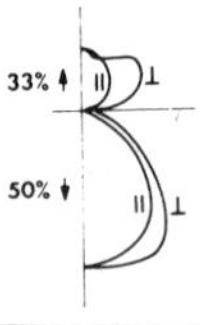	0	.91	.91	.91	.85	.85	.85	.74	.74	.74	.64	.64	.64	.54	.54	.54	.50		
				1	.80	.77	.74	.75	.72	.70	.65	.63	.61	.57	.55	.54	.49	.47	.47	.43	.179	1
				2	.70	.65	.61	.66	.62	.58	.58	.54	.52	.50	.48	.46	.43	.42	.40	.37	.166	2
				3	.62	.56	.51	.58	.53	.49	.51	.47	.44	.45	.42	.39	.39	.37	.35	.32	.153	3
				4	.55	.49	.44	.52	.46	.42	.46	.41	.38	.40	.37	.34	.35	.32	.30	.27	.140	4
				5	.50	.43	.38	.47	.41	.36	.41	.37	.33	.36	.33	.30	.32	.29	.26	.24	.129	5
				6	.45	.38	.33	.42	.36	.32	.37	.33	.29	.33	.29	.26	.29	.26	.23	.21	.119	6
				7	.40	.34	.29	.38	.32	.28	.34	.29	.26	.30	.26	.23	.26	.23	.21	.19	.111	7
				8	.37	.30	.26	.35	.29	.25	.31	.26	.23	.28	.24	.21	.24	.21	.19	.17	.103	8
Prismatic bottom and sides, open top, 4-				9	.34	.27	.23	.32	.26	.22	.29	.24	.21	.25	.22	.19	.22	.19	.17	.15	.096	9
lamp suspended unit—see note 7				10	.31	.25	.21	.29	.24	.20	.26	.22	.19	.23	.20	.17	.21	.18	.15	.14	.090	10
35 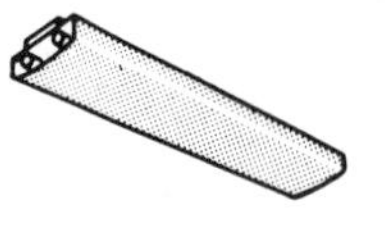	V	1.5/1.2	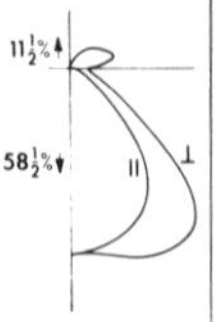	0	.81	.81	.81	.78	.78	.78	.72	.72	.72	.66	.66	.66	.61	.61	.61	.59		
				1	.71	.68	.66	.68	.66	.63	.63	.61	.59	.58	.57	.56	.54	.53	.52	.50	.223	1
				2	.63	.58	.55	.60	.56	.53	.56	.53	.50	.52	.50	.47	.48	.46	.45	.43	.201	2
				3	.56	.50	.46	.54	.49	.45	.50	.46	.43	.47	.43	.41	.43	.41	.39	.37	.183	3
				4	.50	.44	.40	.48	.43	.39	.45	.40	.37	.42	.38	.35	.39	.36	.34	.32	.167	4
				5	.45	.39	.34	.43	.38	.34	.40	.36	.32	.38	.34	.31	.35	.32	.30	.28	.153	5
				6	.40	.34	.30	.39	.34	.30	.37	.32	.28	.34	.30	.27	.32	.29	.26	.25	.142	6
				7	.37	.31	.27	.35	.30	.26	.33	.29	.25	.31	.27	.24	.30	.26	.23	.22	.131	7
				8	.33	.28	.24	.32	.27	.23	.30	.26	.23	.29	.25	.22	.27	.24	.21	.20	.122	8
2-lamp prismatic wraparound—see				9	.31	.25	.21	.30	.25	.21	.28	.24	.20	.26	.23	.20	.25	.22	.19	.18	.114	9
note 7				10	.28	.23	.19	.27	.22	.19	.26	.21	.18	.24	.21	.18	.23	.20	.17	.16	.107	10
36 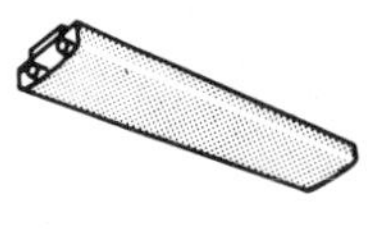	V	1.2	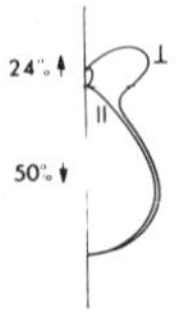	0	.82	.82	.82	.77	.77	.77	.69	.69	.69	.61	.61	.61	.53	.53	.53	.50		
				1	.71	.67	.65	.67	.64	.61	.59	.57	.55	.52	.51	.49	.46	.45	.44	.40	.234	1
				2	.62	.57	.53	.59	.54	.51	.52	.49	.46	.46	.44	.41	.41	.39	.37	.34	.194	2
				3	.55	.49	.45	.52	.47	.43	.46	.42	.39	.41	.38	.36	.37	.34	.32	.30	.168	3
				4	.49	.43	.39	.47	.41	.37	.42	.37	.34	.37	.34	.31	.33	.30	.28	.26	.150	4
				5	.44	.38	.34	.42	.36	.32	.38	.33	.30	.34	.30	.27	.30	.27	.25	.23	.135	5
				6	.40	.34	.29	.38	.32	.28	.34	.30	.26	.31	.27	.24	.28	.25	.22	.20	.123	6
				7	.36	.30	.26	.35	.29	.25	.31	.27	.23	.28	.25	.22	.25	.22	.20	.18	.112	7
				8	.33	.27	.23	.32	.26	.23	.29	.24	.21	.26	.22	.20	.23	.20	.18	.16	.104	8
2-lamp prismatic wraparound—see				9	.30	.25	.21	.29	.24	.20	.26	.22	.19	.24	.20	.18	.22	.19	.16	.15	.097	9
note 7				10	.28	.23	.19	.27	.22	.18	.25	.20	.17	.22	.19	.16	.20	.17	.15	.14	.090	10

Wall Exitance Coefficients for 20 Per Cent Effective Floor Cavity Reflectance (ρ_{FC} = 20)

80			70			50			30			10		
50	30	10	50	30	10	50	30	10	50	30	10	50	30	10
.138	.078	.025	.134	.076	.024	.126	.072	.023	.119	.069	.022	.113	.065	.021
.136	.074	.023	.132	.073	.022	.126	.070	.022	.120	.067	.021	.114	.065	.020
.130	.069	.021	.127	.068	.021	.122	.066	.020	.117	.064	.020	.112	.062	.019
.124	.065	.019	.122	.064	.019	.117	.062	.018	.112	.060	.018	.108	.058	.018
.118	.060	.017	.115	.059	.017	.111	.058	.017	.107	.056	.017	.103	.055	.017
.112	.056	.016	.109	.055	.016	.105	.054	.016	.102	.053	.016	.098	.052	.015
.106	.052	.015	.104	.052	.015	.100	.051	.015	.097	.050	.014	.094	.049	.014
.100	.049	.014	.098	.048	.014	.095	.047	.014	.092	.047	.013	.089	.046	.013
.095	.046	.013	.093	.045	.013	.090	.045	.013	.087	.044	.013	.085	.043	.012
.090	.043	.012	.088	.043	.012	.086	.042	.012	.083	.041	.012	.081	.041	.012
.345	.196	.062	.335	.191	.061	.318	.182	.058	.302	.174	.056	.287	.166	.054
.300	.164	.050	.292	.161	.049	.276	.153	.048	.262	.147	.046	.248	.140	.044
.267	.142	.043	.259	.139	.042	.245	.133	.040	.232	.127	.039	.220	.122	.038
.240	.125	.037	.233	.122	.036	.220	.117	.035	.209	.112	.034	.198	.107	.033
.218	.111	.032	.212	.109	.032	.200	.104	.031	.190	.100	.030	.180	.096	.029
.199	.100	.029	.194	.098	.028	.183	.094	.027	.174	.090	.027	.165	.087	.026
.184	.091	.026	.179	.089	.025	.169	.086	.025	.160	.082	.024	.152	.079	.023
.170	.083	.023	.166	.082	.023	.157	.078	.022	.149	.075	.022	.141	.073	.021
.158	.077	.021	.154	.075	.021	.146	.072	.021	.139	.070	.020	.132	.067	.019
.148	.071	.020	.144	.070	.019	.137	.067	.019	.130	.065	.018	.124	.062	.018
.206	.117	.037	.181	.103	.033	.133	.077	.024	.090	.052	.017	.049	.029	.009
.191	.104	.032	.167	.092	.028	.124	.069	.021	.084	.047	.015	.047	.026	.008
.176	.094	.028	.155	.083	.025	.115	.062	.019	.078	.043	.013	.043	.024	.007
.163	.085	.025	.144	.075	.022	.107	.057	.017	.072	.039	.012	.040	.022	.007
.152	.077	.022	.133	.069	.020	.099	.052	.015	.067	.036	.011	.038	.020	.006
.141	.071	.020	.124	.063	.018	.093	.047	.014	.063	.033	.010	.035	.019	.006
.132	.065	.018	.116	.058	.017	.087	.044	.013	.059	.030	.009	.033	.017	.005
.124	.060	.017	.109	.054	.015	.081	.041	.012	.055	.028	.008	.031	.016	.005
.116	.056	.016	.103	.050	.014	.077	.038	.011	.052	.026	.007	.029	.015	.004
.109	.052	.015	.097	.047	.013	.072	.035	.010	.049	.025	.007	.028	.014	.004
.226	.129	.041	.210	.120	.038	.181	.104	.033	.154	.089	.028	.129	.075	.024
.210	.115	.035	.196	.108	.033	.169	.094	.029	.145	.081	.025	.122	.069	.022
.195	.104	.031	.182	.098	.029	.158	.086	.026	.135	.074	.023	.115	.063	.020
.182	.094	.028	.170	.089	.026	.147	.078	.023	.127	.068	.021	.107	.058	.018
.169	.086	.025	.158	.081	.024	.138	.072	.021	.119	.063	.019	.101	.054	.016
.158	.079	.023	.148	.075	.022	.129	.066	.019	.111	.058	.017	.095	.050	.015
.148	.073	.021	.139	.069	.020	.121	.061	.018	.105	.054	.016	.089	.046	.014
.139	.068	.019	.130	.064	.018	.114	.057	.016	.099	.050	.014	.084	.043	.013
.131	.063	.018	.123	.060	.017	.108	.053	.015	.093	.047	.013	.080	.041	.012
.123	.059	.016	.116	.056	.016	.102	.050	.014	.089	.044	.012	.076	.038	.011
.202	.115	.036	.193	.110	.035	.178	.102	.033	.163	.094	.030	.150	.087	.028
.186	.102	.031	.178	.098	.030	.164	.091	.028	.151	.085	.027	.139	.079	.025
.172	.091	.027	.165	.088	.027	.153	.083	.025	.141	.077	.024	.130	.072	.022
.159	.083	.024	.153	.080	.024	.142	.075	.022	.131	.071	.021	.121	.066	.020
.148	.076	.022	.143	.073	.021	.133	.069	.020	.123	.065	.019	.114	.061	.018
.139	.069	.020	.134	.068	.019	.124	.064	.019	.115	.060	.018	.107	.056	.017
.130	.064	.018	.125	.062	.018	.117	.059	.017	.108	.056	.016	.101	.052	.015
.122	.060	.017	.118	.058	.016	.110	.055	.016	.102	.052	.015	.095	.049	.014
.115	.056	.016	.111	.054	.015	.104	.051	.015	.097	.048	.014	.090	.046	.013
.108	.052	.014	.105	.051	.014	.098	.048	.014	.092	.046	.013	.086	.043	.012
.232	.132	.042	.219	.125	.040	.196	.112	.036	.175	.101	.032	.155	.090	.029
.204	.112	.034	.193	.106	.033	.172	.096	.030	.152	.086	.027	.134	.076	.024
.185	.098	.029	.174	.093	.028	.155	.084	.026	.137	.075	.023	.121	.067	.021
.169	.088	.026	.160	.083	.025	.142	.075	.022	.126	.067	.020	.111	.060	.018
.156	.079	.023	.147	.076	.022	.131	.068	.020	.116	.061	.018	.102	.055	.016
.144	.072	.021	.136	.069	.020	.122	.062	.018	.108	.056	.016	.095	.050	.015
.134	.066	.019	.127	.063	.018	.114	.057	.017	.101	.052	.015	.089	.046	.014
.126	.061	.017	.119	.059	.017	.107	.053	.015	.095	.048	.014	.084	.043	.012
.118	.057	.016	.112	.055	.015	.100	.050	.014	.089	.045	.013	.079	.040	.012
.111	.053	.015	.106	.051	.014	.095	.046	.013	.084	.042	.012	.075	.038	.011

Ceiling Cavity Exitance Coefficients for 20 Per Cent Floor Cavity Reflectance (ρ_{FC} = 20)

80			70			50			30			10		
50	30	10	50	30	10	50	30	10	50	30	10	50	30	10
.111	.111	.111	.094	.094	.094	.064	.064	.064	.037	.037	.037	.012	.012	.012
.102	.091	.081	.087	.078	.070	.060	.054	.048	.034	.031	.028	.011	.010	.009
.095	.077	.061	.082	.066	.053	.056	.046	.037	.032	.027	.022	.010	.009	.007
.090	.066	.047	.077	.057	.041	.053	.040	.028	.031	.023	.017	.010	.008	.005
.086	.058	.037	.074	.050	.032	.051	.035	.023	.029	.021	.013	.009	.007	.004
.082	.052	.030	.070	.045	.026	.049	.031	.018	.028	.018	.011	.009	.006	.004
.078	.047	.024	.067	.041	.021	.046	.028	.015	.027	.017	.009	.009	.005	.003
.075	.043	.021	.064	.037	.018	.044	.026	.013	.026	.015	.008	.008	.005	.003
.072	.040	.018	.062	.034	.015	.043	.024	.011	.025	.014	.007	.008	.005	.002
.068	.037	.015	.059	.032	.013	.041	.022	.010	.024	.013	.006	.008	.004	.002
.066	.034	.014	.057	.030	.012	.039	.021	.008	.023	.012	.005	.007	.004	.002
.239	.239	.239	.205	.205	.205	.140	.140	.140	.080	.080	.080	.026	.026	.026
.236	.209	.185	.202	.180	.159	.138	.123	.110	.080	.071	.064	.025	.023	.021
.230	.189	.154	.197	.163	.133	.135	.112	.093	.078	.065	.054	.025	.021	.018
.224	.174	.135	.192	.150	.117	.132	.104	.082	.076	.061	.048	.024	.020	.016
.217	.163	.122	.186	.141	.106	.128	.098	.075	.074	.058	.044	.024	.019	.015
.210	.154	.113	.180	.134	.099	.124	.093	.070	.072	.055	.041	.023	.018	.014
.203	.147	.107	.175	.128	.093	.121	.089	.066	.070	.053	.039	.023	.017	.013
.197	.141	.103	.169	.123	.089	.117	.086	.063	.068	.051	.038	.022	.017	.012
.191	.137	.099	.164	.118	.086	.114	.083	.061	.066	.049	.037	.021	.016	.012
.185	.132	.096	.160	.115	.084	.111	.081	.060	.064	.048	.036	.021	.016	.012
.180	.129	.094	.155	.112	.082	.108	.079	.058	.063	.046	.035	.020	.015	.012
.653	.653	.653	.558	.558	.558	.381	.381	.381	.219	.219	.219	.070	.070	.070
.647	.631	.616	.553	.541	.530	.378	.372	.367	.218	.215	.213	.070	.069	.069
.641	.615	.593	.549	.529	.512	.376	.366	.357	.217	.213	.209	.070	.069	.068
.636	.603	.577	.545	.521	.501	.374	.362	.351	.216	.211	.207	.069	.069	.068
.631	.595	.567	.542	.514	.493	.373	.358	.347	.216	.210	.205	.069	.068	.067
.627	.588	.560	.538	.509	.487	.371	.356	.344	.215	.209	.204	.069	.068	.067
.623	.583	.554	.535	.505	.483	.369	.353	.342	.214	.208	.203	.069	.068	.067
.618	.578	.551	.532	.501	.480	.367	.352	.340	.214	.207	.202	.069	.068	.067
.614	.575	.548	.529	.499	.478	.366	.350	.339	.213	.206	.202	.069	.068	.067
.611	.572	.545	.526	.496	.476	.364	.349	.338	.212	.206	.201	.069	.068	.067
.607	.569	.544	.523	.494	.474	.363	.348	.337	.212	.206	.201	.069	.068	.067
.409	.409	.409	.350	.350	.350	.239	.239	.239	.137	.137	.137	.044	.044	.044
.401	.383	.367	.343	.329	.316	.235	.226	.219	.135	.131	.127	.043	.042	.041
.394	.365	.340	.337	.314	.294	.231	.217	.205	.133	.126	.120	.043	.041	.039
.388	.351	.322	.332	.303	.279	.228	.211	.196	.132	.123	.116	.042	.040	.038
.382	.341	.310	.328	.295	.269	.225	.205	.190	.130	.120	.112	.042	.039	.037
.376	.333	.301	.323	.288	.262	.223	.201	.185	.129	.118	.110	.042	.039	.036
.371	.327	.295	.319	.283	.257	.220	.198	.182	.128	.116	.108	.041	.038	.036
.366	.321	.290	.315	.279	.253	.218	.195	.179	.126	.115	.107	.041	.038	.035
.361	.317	.286	.311	.275	.250	.215	.193	.177	.125	.114	.106	.041	.037	.035
.357	.313	.284	.308	.272	.247	.213	.191	.176	.124	.113	.105	.040	.037	.035
.353	.310	.281	.304	.269	.246	.211	.189	.174	.123	.112	.104	.040	.037	.035
.221	.221	.221	.189	.189	.189	.129	.129	.129	.074	.074	.074	.024	.024	.024
.213	.198	.183	.183	.170	.158	.125	.117	.109	.072	.068	.064	.023	.022	.021
.207	.181	.160	.177	.156	.138	.121	.108	.096	.070	.063	.056	.022	.020	.018
.201	.169	.144	.172	.146	.125	.118	.101	.087	.068	.059	.051	.022	.019	.017
.196	.160	.133	.168	.138	.115	.115	.096	.081	.067	.056	.048	.021	.018	.016
.191	.153	.125	.164	.132	.109	.113	.092	.077	.065	.054	.045	.021	.018	.015
.186	.147	.119	.160	.127	.104	.110	.089	.073	.064	.052	.044	.021	.017	.014
.182	.142	.115	.156	.123	.100	.108	.087	.071	.063	.051	.042	.020	.017	.014
.178	.138	.112	.153	.120	.097	.106	.084	.069	.062	.050	.041	.020	.016	.014
.174	.135	.109	.150	.117	.095	.104	.082	.068	.060	.049	.040	.020	.016	.013
.170	.132	.107	.147	.115	.094	.102	.081	.066	.059	.048	.040	.019	.016	.013
.324	.324	.324	.277	.277	.277	.189	.189	.189	.108	.108	.108	.035	.035	.035
.318	.300	.284	.272	.257	.244	.186	.177	.169	.107	.102	.098	.034	.033	.032
.312	.283	.260	.267	.244	.224	.183	.169	.157	.105	.098	.092	.034	.032	.030
.305	.271	.244	.262	.234	.211	.180	.163	.148	.104	.095	.087	.033	.031	.029
.300	.262	.233	.257	.226	.202	.177	.158	.143	.102	.092	.084	.033	.030	.028
.294	.255	.225	.253	.220	.196	.174	.154	.138	.101	.090	.082	.033	.029	.027
.289	.249	.220	.249	.216	.191	.172	.151	.135	.100	.089	.080	.032	.029	.027
.285	.244	.216	.245	.212	.188	.169	.148	.133	.098	.087	.079	.032	.029	.026
.280	.240	.212	.241	.208	.185	.167	.146	.131	.097	.086	.078	.031	.028	.026
.276	.237	.210	.238	.205	.183	.165	.144	.130	.096	.085	.078	.031	.028	.026
.272	.234	.208	.235	.203	.181	.163	.143	.129	.095	.084	.077	.031	.028	.026

Fig. 61 Continued (See page 119 for instructions and notes)

Table header:

Typical Luminaire	Typical Intensity Distribution and Per Cent Lamp Lumens	Maint. Cat.	SC	RCR ↓	ρcc → 80			70			50			30			10			0	WDRC	RCR ↓
					ρw → 50	30	10	50	30	10	50	30	10	50	30	10	50	30	10	0		

Coefficients of Utilization for 20 Per Cent Effective Floor Cavity Reflectance (ρFC = 20)

37 — 2-lamp diffuse wraparound—see note 7. V, SC 1.3, 8%↑, 37½%↓

RCR	80/50	80/30	80/10	70/50	70/30	70/10	50/50	50/30	50/10	30/50	30/30	30/10	10/50	10/30	10/10	0	WDRC
0	.52	.52	.52	.50	.50	.50	.46	.46	.46	.43	.43	.43	.39	.39	.39	.38	
1	.44	.42	.40	.42	.40	.39	.39	.37	.36	.36	.35	.33	.33	.32	.31	.30	.201
2	.38	.35	.32	.37	.33	.31	.34	.31	.29	.31	.29	.27	.28	.27	.25	.24	.171
3	.33	.29	.26	.32	.28	.25	.29	.26	.24	.27	.25	.22	.25	.23	.21	.20	.149
4	.29	.25	.22	.28	.24	.21	.26	.23	.20	.24	.21	.19	.22	.20	.18	.17	.132
5	.26	.22	.19	.25	.21	.18	.23	.20	.17	.21	.18	.16	.20	.17	.15	.14	.117
6	.23	.19	.16	.22	.18	.16	.21	.17	.15	.19	.16	.14	.18	.15	.13	.12	.106
7	.21	.17	.14	.20	.16	.14	.19	.15	.13	.17	.15	.12	.16	.14	.12	.11	.096
8	.19	.15	.12	.18	.15	.12	.17	.14	.12	.16	.13	.11	.15	.12	.11	.10	.088
9	.17	.14	.11	.17	.13	.11	.16	.13	.10	.15	.12	.10	.14	.11	.09	.09	.081
10	.16	.12	.10	.15	.12	.10	.14	.11	.09	.14	.11	.09	.13	.10	.09	.08	.075

38 — 4-lamp, 610 mm (2') wide troffer with 45° plastic louver—see note 7. IV, SC 1.0, 0%↑, 50%↓

RCR	80/50	80/30	80/10	70/50	70/30	70/10	50/50	50/30	50/10	30/50	30/30	30/10	10/50	10/30	10/10	0	WDRC
0	.60	.60	.60	.58	.58	.58	.56	.56	.56	.53	.53	.53	.51	.51	.51	.50	
1	.53	.51	.49	.52	.50	.49	.50	.48	.47	.48	.47	.46	.46	.45	.44	.43	.168
2	.47	.44	.42	.46	.43	.41	.44	.42	.40	.43	.41	.39	.41	.40	.38	.37	.159
3	.42	.38	.36	.41	.38	.35	.40	.37	.35	.39	.36	.34	.37	.35	.34	.32	.146
4	.38	.34	.31	.37	.34	.31	.36	.33	.30	.35	.32	.30	.34	.32	.30	.29	.135
5	.34	.30	.27	.34	.30	.27	.33	.29	.27	.32	.29	.27	.31	.28	.26	.25	.124
6	.31	.27	.24	.31	.27	.24	.30	.27	.24	.29	.26	.24	.28	.26	.24	.23	.114
7	.29	.25	.22	.28	.24	.22	.28	.24	.22	.27	.24	.21	.26	.23	.21	.20	.106
8	.26	.22	.20	.26	.22	.20	.25	.22	.20	.25	.22	.20	.24	.21	.19	.19	.099
9	.24	.21	.18	.24	.21	.18	.24	.20	.18	.23	.20	.18	.23	.20	.18	.17	.092
10	.23	.19	.17	.22	.19	.17	.22	.19	.16	.22	.19	.16	.21	.18	.16	.16	.086

39 — 4-lamp, 610 mm (2') wide troffer with 45° white metal louver—see note 7. IV, SC 0.9, 0%↑, 46%↓

RCR	80/50	80/30	80/10	70/50	70/30	70/10	50/50	50/30	50/10	30/50	30/30	30/10	10/50	10/30	10/10	0	WDRC
0	.55	.55	.55	.54	.54	.54	.51	.51	.51	.49	.49	.49	.47	.47	.47	.46	
1	.49	.48	.46	.48	.47	.46	.46	.45	.44	.45	.44	.43	.43	.42	.42	.41	.137
2	.44	.42	.40	.43	.41	.39	.42	.40	.38	.40	.39	.37	.39	.38	.37	.36	.131
3	.40	.37	.34	.39	.36	.34	.38	.36	.33	.37	.35	.33	.36	.34	.32	.32	.122
4	.36	.33	.30	.36	.33	.30	.35	.32	.30	.34	.31	.29	.33	.31	.29	.28	.113
5	.33	.30	.27	.33	.29	.27	.32	.29	.27	.31	.28	.26	.30	.28	.26	.25	.104
6	.30	.27	.24	.30	.27	.24	.29	.26	.24	.29	.26	.24	.28	.25	.24	.23	.097
7	.28	.25	.22	.28	.24	.22	.27	.24	.22	.26	.24	.22	.26	.23	.22	.21	.090
8	.26	.23	.20	.26	.22	.20	.25	.22	.20	.25	.22	.20	.24	.22	.20	.19	.085
9	.24	.21	.19	.24	.21	.19	.23	.20	.18	.23	.20	.18	.23	.20	.18	.18	.079
10	.22	.19	.17	.22	.19	.17	.22	.19	.17	.22	.19	.17	.21	.19	.17	.16	.075

40 — Fluorescent unit dropped diffuser, 4-lamp 610 mm (2') wide—see note 7. V, SC 1.2, 1%↑, 60½%↓

RCR	80/50	80/30	80/10	70/50	70/30	70/10	50/50	50/30	50/10	30/50	30/30	30/10	10/50	10/30	10/10	0	WDRC
0	.73	.73	.73	.71	.71	.71	.68	.68	.68	.65	.65	.65	.62	.62	.62	.60	
1	.63	.60	.58	.62	.59	.57	.59	.57	.55	.56	.55	.53	.54	.53	.51	.50	.259
2	.55	.51	.47	.54	.50	.46	.51	.48	.45	.49	.46	.44	.47	.45	.43	.42	.236
3	.48	.43	.39	.47	.42	.39	.45	.41	.38	.43	.40	.37	.42	.39	.36	.35	.212
4	.43	.37	.33	.42	.37	.33	.40	.36	.32	.39	.35	.32	.37	.34	.31	.30	.191
5	.38	.33	.29	.37	.32	.28	.36	.31	.28	.35	.31	.28	.33	.30	.27	.26	.173
6	.34	.29	.25	.34	.29	.25	.33	.28	.24	.31	.27	.24	.30	.27	.24	.23	.158
7	.31	.26	.22	.31	.26	.22	.30	.25	.22	.28	.24	.21	.28	.24	.21	.20	.144
8	.28	.23	.20	.28	.23	.20	.27	.23	.19	.26	.22	.19	.25	.22	.19	.18	.133
9	.26	.21	.18	.26	.21	.18	.25	.21	.17	.24	.20	.17	.24	.20	.17	.16	.123
10	.24	.19	.16	.24	.19	.16	.23	.19	.16	.22	.18	.16	.22	.18	.16	.15	.115

41 — Fluorescent unit with flat bottom diffuser, 4-lamp 610 mm (2') wide—see note 7. V, SC 1.2, 0%↑, 57½%↓

RCR	80/50	80/30	80/10	70/50	70/30	70/10	50/50	50/30	50/10	30/50	30/30	30/10	10/50	10/30	10/10	0	WDRC
0	.69	.69	.69	.67	.67	.67	.64	.64	.64	.61	.61	.61	.59	.59	.59	.58	
1	.60	.58	.55	.59	.57	.55	.56	.55	.53	.54	.53	.51	.52	.51	.50	.49	.227
2	.52	.49	.45	.51	.48	.45	.49	.46	.44	.47	.45	.43	.45	.43	.42	.41	.214
3	.46	.41	.38	.45	.41	.37	.43	.40	.37	.42	.39	.36	.40	.38	.35	.34	.196
4	.41	.36	.32	.40	.35	.32	.39	.34	.31	.37	.34	.31	.36	.33	.30	.29	.178
5	.36	.31	.28	.36	.31	.27	.35	.30	.27	.33	.30	.27	.32	.29	.26	.25	.162
6	.33	.28	.24	.32	.27	.24	.31	.27	.24	.30	.26	.23	.29	.26	.23	.22	.148
7	.29	.25	.21	.29	.25	.21	.28	.24	.21	.28	.24	.21	.27	.23	.21	.20	.136
8	.27	.22	.19	.27	.22	.19	.26	.22	.19	.25	.21	.19	.25	.21	.19	.17	.126
9	.25	.20	.17	.25	.20	.17	.24	.20	.17	.23	.20	.17	.23	.19	.17	.16	.116
10	.23	.18	.15	.23	.18	.15	.22	.18	.15	.22	.18	.15	.21	.18	.15	.14	.108

42 — Fluorescent unit with flat prismatic lens, 4-lamp 610 mm (2') wide—see note 7. V, SC 1.4/1.2, 0%↑, 63%↓, 60°

RCR	80/50	80/30	80/10	70/50	70/30	70/10	50/50	50/30	50/10	30/50	30/30	30/10	10/50	10/30	10/10	0	WDRC
0	.75	.75	.75	.73	.73	.73	.70	.70	.70	.67	.67	.67	.64	.64	.64	.63	
1	.67	.64	.62	.65	.63	.61	.63	.61	.59	.60	.59	.58	.58	.57	.56	.55	.208
2	.59	.56	.52	.58	.55	.52	.56	.53	.51	.54	.52	.49	.52	.50	.48	.47	.199
3	.53	.48	.45	.52	.48	.44	.50	.46	.43	.48	.45	.43	.47	.44	.42	.41	.186
4	.47	.42	.38	.46	.42	.38	.45	.41	.38	.44	.40	.37	.42	.39	.37	.35	.172
5	.43	.37	.34	.42	.37	.33	.41	.36	.33	.39	.36	.33	.38	.35	.32	.31	.160
6	.39	.33	.30	.38	.33	.29	.37	.32	.29	.36	.32	.29	.35	.31	.29	.27	.148
7	.35	.30	.26	.35	.30	.26	.34	.29	.26	.33	.28	.26	.32	.28	.26	.24	.138
8	.32	.27	.24	.32	.27	.23	.31	.26	.23	.30	.26	.23	.29	.26	.23	.22	.128
9	.30	.25	.21	.29	.24	.21	.28	.24	.21	.28	.24	.21	.27	.24	.21	.20	.120
10	.27	.22	.19	.27	.22	.19	.26	.22	.19	.26	.22	.19	.25	.22	.19	.18	.113

80			70			50			30			10			80			70			50			30			10		
50	30	10	50	30	10	50	30	10	50	30	10	50	30	10	50	30	10	50	30	10	50	30	10	50	30	10	50	30	10
Wall Exitance Coefficients for 20 Per Cent Effective Floor Cavity Reflectance (ρ_{FC} = 20)															Ceiling Cavity Exitance Coefficients for 20 Per Cent Floor Cavity Reflectance (ρ_{FC} = 20)														
															.147	.147	.147	.125	.125	.125	.085	.085	.085	.049	.049	.049	.016	.016	.016
.162	.092	.029	.156	.089	.028	.145	.083	.027	.136	.078	.025	.127	.073	.024	.144	.131	.120	.123	.113	.103	.084	.077	.071	.048	.045	.041	.016	.014	.013
.144	.079	.024	.139	.076	.024	.129	.072	.022	.120	.068	.021	.112	.064	.020	.141	.121	.104	.120	.104	.090	.083	.072	.063	.048	.042	.037	.015	.014	.012
.130	.069	.021	.125	.067	.020	.116	.063	.019	.108	.059	.018	.101	.056	.017	.137	.113	.094	.118	.098	.081	.081	.068	.057	.047	.040	.034	.015	.013	.011
.118	.061	.018	.114	.059	.018	.106	.056	.017	.098	.053	.016	.092	.050	.015	.134	.107	.087	.115	.093	.076	.079	.065	.053	.046	.038	.032	.015	.012	.010
.108	.055	.016	.104	.053	.016	.097	.050	.015	.090	.048	.014	.084	.045	.013	.130	.103	.083	.112	.089	.072	.077	.062	.051	.045	.037	.030	.014	.012	.010
.099	.050	.014	.096	.048	.014	.089	.046	.013	.083	.043	.013	.077	.041	.012	.127	.099	.079	.109	.086	.069	.075	.060	.049	.044	.035	.029	.014	.012	.010
.092	.045	.013	.088	.044	.013	.083	.042	.012	.077	.039	.011	.072	.037	.011	.124	.096	.077	.107	.083	.067	.074	.059	.047	.043	.034	.028	.014	.011	.009
.085	.042	.012	.082	.041	.011	.077	.038	.011	.072	.036	.010	.067	.034	.010	.121	.094	.075	.104	.081	.065	.072	.057	.046	.042	.034	.028	.014	.011	.009
.080	.038	.011	.077	.037	.011	.072	.036	.010	.067	.034	.010	.063	.032	.009	.118	.092	.074	.102	.079	.064	.071	.056	.046	.041	.033	.027	.013	.011	.009
.075	.036	.010	.072	.035	.010	.067	.033	.009	.063	.031	.009	.059	.030	.009	.116	.090	.072	.100	.078	.063	.069	.055	.045	.040	.032	.027	.013	.011	.009
															.095	.095	.095	.082	.082	.082	.056	.056	.056	.032	.032	.032	.010	.010	.010
.135	.077	.024	.132	.075	.024	.125	.072	.023	.119	.069	.022	.114	.066	.021	.089	.078	.069	.076	.067	.059	.052	.046	.041	.030	.027	.024	.010	.009	.008
.128	.070	.022	.125	.069	.021	.120	.066	.021	.114	.064	.020	.109	.062	.020	.084	.066	.051	.072	.057	.044	.049	.039	.031	.028	.023	.018	.009	.007	.006
.120	.064	.019	.117	.063	.019	.112	.061	.018	.108	.059	.018	.103	.057	.018	.079	.057	.039	.068	.049	.034	.047	.034	.024	.027	.020	.014	.009	.006	.005
.112	.058	.017	.109	.057	.017	.105	.056	.017	.101	.054	.016	.097	.053	.016	.075	.050	.031	.065	.043	.027	.044	.030	.019	.026	.018	.011	.008	.006	.004
.104	.053	.015	.102	.052	.015	.098	.051	.015	.094	.050	.015	.091	.049	.015	.071	.045	.025	.061	.039	.022	.042	.027	.016	.024	.016	.009	.008	.005	.003
.097	.049	.014	.095	.048	.014	.092	.047	.014	.089	.046	.014	.086	.045	.013	.068	.041	.021	.058	.035	.018	.040	.025	.013	.023	.014	.008	.008	.005	.003
.091	.045	.013	.089	.045	.013	.086	.044	.013	.083	.043	.012	.081	.042	.012	.065	.037	.018	.056	.032	.015	.038	.023	.011	.022	.013	.007	.007	.004	.002
.086	.042	.012	.084	.041	.012	.081	.041	.012	.079	.040	.012	.076	.039	.011	.062	.034	.015	.053	.030	.013	.037	.021	.009	.021	.012	.006	.007	.004	.002
.081	.039	.011	.079	.039	.011	.077	.038	.011	.075	.037	.011	.072	.037	.011	.059	.032	.013	.051	.027	.012	.035	.019	.008	.020	.011	.005	.007	.004	.002
.076	.037	.010	.075	.036	.010	.073	.036	.010	.071	.035	.010	.069	.035	.010	.056	.029	.012	.048	.026	.010	.033	.018	.007	.020	.011	.004	.006	.003	.001
															.088	.088	.088	.075	.075	.075	.051	.051	.051	.029	.029	.029	.009	.009	.009
.115	.065	.021	.112	.064	.020	.106	.061	.019	.100	.058	.019	.095	.055	.018	.081	.072	.064	.069	.062	.055	.048	.043	.038	.027	.025	.022	.009	.008	.007
.109	.060	.018	.107	.059	.018	.102	.056	.018	.097	.054	.017	.092	.052	.016	.076	.061	.048	.065	.052	.042	.045	.036	.029	.026	.021	.017	.008	.007	.006
.103	.055	.016	.100	.054	.016	.096	.052	.016	.092	.050	.015	.088	.049	.015	.072	.053	.037	.061	.045	.032	.042	.031	.023	.024	.018	.013	.008	.006	.004
.096	.050	.015	.094	.049	.015	.090	.048	.014	.086	.046	.014	.083	.045	.014	.068	.046	.030	.058	.040	.026	.040	.028	.018	.023	.016	.011	.007	.005	.004
.090	.046	.013	.088	.045	.013	.085	.044	.013	.081	.043	.013	.078	.042	.013	.064	.041	.024	.055	.036	.021	.038	.025	.015	.022	.015	.009	.007	.005	.003
.084	.042	.012	.083	.042	.012	.080	.041	.012	.077	.040	.012	.074	.039	.012	.061	.037	.020	.052	.032	.017	.036	.022	.012	.021	.013	.007	.007	.004	.002
.079	.039	.011	.078	.039	.011	.075	.038	.011	.073	.037	.011	.070	.036	.011	.058	.034	.017	.050	.029	.015	.034	.021	.010	.020	.012	.006	.006	.004	.002
.075	.037	.010	.074	.036	.010	.071	.035	.010	.069	.035	.010	.067	.034	.010	.055	.031	.015	.047	.027	.013	.033	.019	.009	.019	.011	.005	.006	.004	.002
.071	.034	.010	.070	.034	.010	.067	.033	.009	.065	.033	.009	.063	.032	.009	.052	.029	.013	.045	.025	.011	.031	.018	.008	.018	.010	.005	.006	.003	.002
.067	.032	.009	.066	.032	.009	.064	.031	.009	.062	.031	.009	.060	.030	.009	.050	.027	.011	.043	.023	.010	.030	.016	.007	.017	.010	.004	.006	.003	.001
															.123	.123	.123	.105	.105	.105	.072	.072	.072	.041	.041	.041	.013	.013	.013
.196	.111	.035	.191	.109	.035	.182	.105	.033	.174	.101	.032	.167	.097	.031	.118	.102	.089	.101	.088	.076	.069	.060	.053	.040	.035	.031	.013	.011	.010
.181	.099	.030	.177	.098	.030	.170	.094	.029	.163	.091	.029	.156	.088	.028	.113	.087	.066	.096	.075	.057	.066	.052	.040	.038	.030	.023	.012	.010	.008
.167	.089	.027	.163	.087	.026	.156	.085	.026	.150	.082	.025	.144	.080	.025	.108	.077	.052	.092	.066	.045	.063	.046	.032	.037	.027	.019	.012	.009	.006
.153	.080	.023	.150	.079	.023	.144	.076	.023	.139	.074	.022	.133	.072	.022	.103	.069	.042	.088	.059	.037	.061	.041	.026	.035	.024	.015	.011	.008	.005
.141	.072	.021	.139	.071	.021	.133	.069	.020	.128	.068	.020	.124	.066	.020	.098	.062	.036	.085	.054	.031	.058	.038	.022	.034	.022	.013	.011	.007	.004
.131	.066	.019	.128	.065	.019	.124	.063	.018	.119	.062	.018	.115	.061	.018	.094	.057	.031	.081	.050	.027	.056	.035	.019	.032	.020	.011	.010	.007	.004
.122	.060	.017	.119	.060	.017	.115	.058	.017	.111	.057	.017	.107	.056	.016	.090	.053	.027	.077	.046	.024	.053	.032	.017	.031	.019	.010	.010	.006	.003
.114	.056	.016	.112	.055	.016	.108	.054	.015	.104	.053	.015	.101	.052	.015	.086	.049	.024	.074	.043	.021	.051	.030	.015	.030	.018	.009	.010	.006	.003
.106	.051	.014	.105	.051	.014	.101	.050	.014	.098	.049	.014	.095	.048	.014	.082	.046	.022	.071	.040	.019	.049	.028	.014	.029	.017	.008	.009	.005	.003
.100	.048	.013	.098	.048	.013	.095	.047	.013	.092	.046	.013	.089	.045	.013	.079	.044	.021	.068	.038	.018	.047	.027	.013	.027	.016	.008	.009	.005	.003
															.110	.110	.110	.094	.094	.094	.064	.064	.064	.037	.037	.037	.012	.012	.012
.174	.099	.031	.170	.097	.031	.162	.093	.030	.155	.089	.029	.149	.086	.028	.104	.090	.078	.089	.077	.067	.061	.053	.046	.035	.031	.027	.011	.010	.009
.165	.090	.028	.161	.089	.027	.155	.086	.027	.149	.083	.026	.143	.081	.025	.099	.076	.057	.085	.065	.049	.058	.045	.034	.033	.026	.020	.011	.009	.007
.153	.082	.024	.150	.080	.024	.144	.078	.024	.139	.076	.023	.134	.074	.023	.094	.066	.043	.081	.057	.037	.056	.039	.026	.032	.023	.015	.010	.007	.005
.142	.074	.022	.139	.073	.022	.134	.071	.021	.129	.069	.021	.124	.068	.021	.090	.058	.034	.077	.050	.029	.053	.035	.021	.031	.021	.012	.010	.007	.004
.131	.067	.019	.129	.066	.019	.124	.065	.019	.120	.063	.019	.116	.062	.019	.086	.052	.027	.074	.045	.024	.051	.032	.017	.029	.018	.010	.009	.006	.003
.122	.061	.018	.120	.061	.017	.116	.059	.017	.112	.058	.017	.108	.057	.017	.082	.047	.023	.070	.041	.020	.048	.029	.014	.028	.017	.008	.009	.006	.003
.114	.056	.016	.112	.056	.016	.108	.055	.016	.104	.054	.016	.101	.053	.015	.078	.043	.019	.067	.038	.017	.046	.026	.012	.027	.016	.007	.009	.005	.002
.106	.052	.015	.104	.051	.015	.101	.051	.014	.098	.050	.014	.095	.049	.014	.074	.040	.017	.064	.035	.014	.044	.024	.010	.026	.014	.006	.008	.005	.002
.100	.048	.013	.098	.048	.013	.095	.047	.013	.092	.046	.013	.090	.046	.013	.070	.037	.015	.061	.032	.013	.042	.023	.009	.025	.013	.005	.008	.004	.002
.094	.045	.012	.092	.045	.012	.090	.044	.012	.087	.043	.012	.085	.043	.012	.067	.035	.013	.058	.030	.011	.040	.021	.008	.023	.013	.005	.008	.004	.002
															.120	.120	.120	.103	.103	.103	.070	.070	.070	.040	.040	.040	.013	.013	.013
.168	.096	.030	.164	.093	.030	.156	.089	.029	.148	.085	.027	.141	.082	.026	.112	.099	.087	.096	.085	.075	.065	.058	.052	.038	.034	.030	.012	.011	.010
.161	.088	.027	.157	.087	.027	.150	.083	.026	.143	.080	.025	.137	.078	.024	.105	.083	.064	.090	.072	.056	.062	.050	.039	.036	.029	.023	.011	.009	.007
.152	.081	.024	.148	.079	.024	.142	.077	.023	.136	.075	.023	.131	.072	.022	.100	.072	.049	.086	.062	.043	.059	.043	.030	.034	.025	.018	.011	.008	.006
.142	.074	.022	.139	.073	.022	.134	.071	.021	.128	.069	.021	.124	.067	.020	.095	.063	.039	.082	.055	.034	.056	.038	.024	.032	.022	.014	.010	.007	.005
.133	.068	.020	.131	.067	.020	.126	.065	.019	.121	.064	.019	.117	.062	.019	.091	.057	.031	.078	.049	.027	.054	.034	.019	.031	.020	.011	.010	.007	.004
.125	.063	.018	.123	.062	.018	.118	.061	.018	.114	.059	.017	.110	.058	.017	.086	.051	.026	.074	.044	.023	.051	.031	.016	.030	.018	.010	.010	.006	.003
.117	.058	.016	.115	.057	.016	.111	.056	.016	.108	.055	.016	.104	.054	.016	.082	.047	.022	.071	.041	.019	.049	.028	.014	.028	.017	.008	.009	.005	.003
.110	.054	.015	.109	.053	.015	.105	.052	.015	.102	.052	.015	.099	.051	.015	.079	.043	.019	.068	.037	.016	.047	.026	.012	.027	.016	.007	.009	.005	.002
.104	.050	.014	.102	.050	.014	.099	.049	.014	.096	.048	.014	.093	.047	.014	.075	.040	.017	.065	.035	.014	.045	.024	.010	.026	.014	.006	.008	.005	.002
.098	.047	.013	.097	.047	.013	.094	.046	.013	.091	.045	.013	.089	.045	.013	.072	.037	.015	.062	.032	.013	.043	.023	.009	.025	.014	.005	.008	.004	.002

Fig. 61 *Continued (See page 119 for instructions and notes)*

	Typical Intensity Distribution and Per Cent Lamp Lumens	ρCC →	80			70			50			30			10			0		ρCC →
Typical Luminaire		ρW →	50	30	10	50	30	10	50	30	10	50	30	10	50	30	10	0	WDRC	ρW →
	Maint. Cat. / SC	RCR ↓	Coefficients of Utilization for 20 Per Cent Effective Floor Cavity Reflectance (ρFC = 20)																	RCR ↓

43 — 4-lamp, 610 mm (2') wide unit with sharp cutoff (high angle—low luminance) flat prismatic lens—see note 7 — V, SC 1.4/1.3

RCR	80/50	80/30	80/10	70/50	70/30	70/10	50/50	50/30	50/10	30/50	30/30	30/10	10/50	10/30	10/10	0	WDRC
0	.78	.78	.78	.76	.76	.76	.73	.73	.73	.70	.70	.70	.67	.67	.67	.66	
1	.71	.68	.66	.69	.67	.65	.66	.65	.63	.64	.63	.61	.62	.61	.60	.58	.181
2	.63	.60	.57	.62	.59	.56	.60	.57	.55	.58	.56	.54	.56	.54	.52	.51	.180
3	.57	.52	.49	.56	.52	.48	.54	.51	.48	.52	.49	.47	.51	.48	.46	.45	.173
4	.51	.46	.43	.50	.46	.42	.49	.45	.42	.47	.44	.41	.46	.43	.41	.39	.164
5	.46	.41	.37	.46	.41	.37	.44	.40	.37	.43	.39	.36	.42	.39	.36	.35	.154
6	.42	.37	.33	.41	.37	.33	.40	.36	.33	.39	.35	.32	.38	.35	.32	.31	.145
7	.38	.33	.29	.38	.33	.29	.37	.32	.29	.36	.32	.29	.35	.32	.29	.28	.136
8	.35	.30	.26	.35	.30	.26	.34	.29	.26	.33	.29	.26	.32	.29	.26	.25	.127
9	.32	.27	.24	.32	.27	.24	.31	.27	.24	.31	.27	.24	.30	.26	.24	.22	.120
10	.30	.25	.22	.30	.25	.22	.29	.25	.22	.28	.24	.22	.28	.24	.21	.20	.113

44 — Bilateral batwing distribution—louvered fluorescent unit — IV, SC N.A.

RCR	80/50	80/30	80/10	70/50	70/30	70/10	50/50	50/30	50/10	30/50	30/30	30/10	10/50	10/30	10/10	0	WDRC
0	.71	.71	.71	.70	.70	.70	.66	.66	.66	.64	.64	.64	.61	.61	.61	.60	
1	.64	.62	.60	.63	.61	.60	.60	.59	.58	.58	.57	.56	.56	.55	.54	.53	.167
2	.57	.54	.51	.56	.53	.51	.54	.52	.50	.52	.50	.48	.51	.49	.47	.46	.170
3	.51	.47	.44	.50	.46	.43	.49	.45	.43	.47	.44	.42	.46	.43	.41	.40	.165
4	.46	.41	.38	.45	.41	.37	.44	.40	.37	.42	.39	.36	.41	.38	.36	.35	.157
5	.41	.36	.33	.40	.36	.32	.39	.35	.32	.38	.35	.32	.37	.34	.31	.30	.148
6	.37	.32	.28	.36	.32	.28	.35	.31	.28	.34	.31	.28	.34	.30	.28	.27	.139
7	.33	.29	.25	.33	.28	.25	.32	.28	.25	.31	.27	.25	.30	.27	.25	.23	.130
8	.30	.26	.22	.30	.25	.22	.29	.25	.22	.28	.25	.22	.28	.24	.22	.21	.122
9	.28	.23	.20	.27	.23	.20	.27	.23	.20	.26	.22	.20	.25	.22	.19	.18	.115
10	.25	.21	.18	.25	.21	.18	.25	.20	.18	.24	.20	.18	.24	.20	.18	.17	.108

45 — Bilateral batwing distribution—4-lamp, 610 mm (2') wide fluorescent unit with flat prismatic lens and overlay—see note 7 — V, SC N.A.

RCR	80/50	80/30	80/10	70/50	70/30	70/10	50/50	50/30	50/10	30/50	30/30	30/10	10/50	10/30	10/10	0	WDRC
0	.57	.57	.57	.56	.56	.56	.53	.53	.53	.51	.51	.51	.49	.49	.49	.48	
1	.50	.48	.46	.49	.47	.45	.47	.45	.44	.45	.43	.42	.43	.42	.41	.40	.204
2	.43	.40	.37	.42	.39	.36	.40	.38	.35	.39	.37	.35	.37	.36	.34	.33	.192
3	.37	.33	.30	.37	.33	.30	.35	.32	.29	.34	.31	.29	.33	.30	.28	.27	.175
4	.33	.28	.25	.32	.28	.25	.31	.27	.24	.30	.27	.24	.29	.26	.24	.23	.159
5	.29	.24	.21	.28	.24	.21	.27	.24	.21	.26	.23	.20	.25	.23	.20	.19	.145
6	.26	.21	.18	.25	.21	.18	.24	.21	.18	.24	.20	.18	.23	.20	.17	.16	.132
7	.23	.19	.16	.23	.18	.15	.22	.18	.15	.21	.18	.15	.21	.17	.15	.14	.122
8	.21	.17	.14	.21	.16	.14	.20	.16	.13	.19	.16	.13	.19	.16	.13	.12	.112
9	.19	.15	.12	.19	.15	.12	.18	.14	.12	.18	.14	.12	.17	.14	.12	.11	.104
10	.17	.13	.11	.17	.13	.11	.17	.13	.11	.16	.13	.11	.16	.13	.10	.10	.096

46 — Bilateral batwing distribution—one-lamp, surface mounted fluorescent with prismatic wraparound lens — V, SC N.A.

RCR	80/50	80/30	80/10	70/50	70/30	70/10	50/50	50/30	50/10	30/50	30/30	30/10	10/50	10/30	10/10	0	WDRC
0	.87	.87	.87	.84	.84	.84	.77	.77	.77	.72	.72	.72	.66	.66	.66	.64	
1	.75	.72	.69	.72	.69	.66	.67	.64	.62	.62	.60	.58	.57	.56	.54	.52	.296
2	.65	.60	.56	.63	.58	.54	.58	.54	.51	.54	.51	.48	.50	.47	.45	.43	.261
3	.57	.51	.46	.55	.49	.45	.51	.46	.42	.47	.43	.40	.44	.41	.38	.36	.232
4	.50	.44	.39	.48	.42	.38	.45	.40	.36	.42	.38	.34	.39	.35	.32	.30	.209
5	.45	.38	.33	.43	.37	.32	.40	.35	.31	.37	.33	.29	.35	.31	.28	.26	.189
6	.40	.33	.28	.39	.32	.28	.36	.31	.26	.34	.29	.25	.31	.27	.24	.22	.172
7	.36	.29	.25	.35	.29	.24	.32	.27	.23	.30	.26	.22	.28	.24	.21	.19	.158
8	.33	.26	.22	.31	.25	.21	.29	.24	.20	.28	.23	.20	.26	.22	.19	.17	.146
9	.30	.23	.19	.29	.23	.19	.27	.22	.18	.25	.21	.17	.24	.20	.17	.15	.135
10	.27	.21	.17	.26	.21	.17	.25	.20	.16	.23	.19	.16	.22	.18	.15	.13	.126

47 — Radial batwing distribution—4-lamp, 610 mm (2') wide fluorescent unit with flat prismatic lens—see note 7 — V, SC 1.7

RCR	80/50	80/30	80/10	70/50	70/30	70/10	50/50	50/30	50/10	30/50	30/30	30/10	10/50	10/30	10/10	0	WDRC
0	.71	.71	.71	.69	.69	.69	.66	.66	.66	.63	.63	.63	.61	.61	.61	.60	
1	.62	.59	.57	.60	.58	.56	.58	.56	.54	.55	.54	.52	.53	.52	.51	.50	.251
2	.53	.49	.46	.52	.48	.45	.50	.47	.44	.48	.45	.43	.46	.44	.42	.41	.237
3	.46	.41	.37	.45	.41	.37	.44	.40	.36	.42	.39	.36	.40	.38	.35	.34	.216
4	.41	.35	.31	.40	.35	.31	.38	.34	.30	.37	.33	.30	.36	.32	.30	.28	.196
5	.36	.30	.26	.35	.30	.26	.34	.29	.26	.33	.29	.26	.32	.28	.25	.24	.178
6	.32	.27	.23	.32	.26	.23	.31	.26	.22	.29	.25	.22	.29	.25	.22	.21	.162
7	.29	.24	.20	.29	.23	.20	.28	.23	.19	.27	.22	.19	.26	.22	.19	.18	.149
8	.26	.21	.17	.26	.21	.17	.25	.20	.17	.24	.20	.17	.24	.20	.17	.16	.137
9	.24	.19	.15	.24	.19	.15	.23	.18	.15	.22	.18	.15	.22	.18	.15	.14	.127
10	.22	.17	.14	.22	.17	.14	.21	.17	.14	.20	.16	.14	.20	.16	.14	.12	.118

48 — 2-lamp fluorescent strip unit — I, SC 1.6/1.2

RCR	80/50	80/30	80/10	70/50	70/30	70/10	50/50	50/30	50/10	30/50	30/30	30/10	10/50	10/30	10/10	0	WDRC
0	1.01	1.01	1.01	.96	.96	.96	.87	.87	.87	.79	.79	.79	.72	.72	.72	.68	
1	.84	.79	.75	.80	.76	.72	.72	.69	.66	.65	.63	.60	.59	.57	.55	.52	.414
2	.72	.65	.59	.68	.62	.57	.62	.57	.52	.56	.52	.48	.50	.47	.44	.41	.343
3	.62	.54	.48	.59	.52	.46	.53	.47	.42	.48	.43	.39	.43	.39	.36	.33	.293
4	.54	.46	.39	.52	.44	.38	.47	.40	.35	.42	.37	.33	.38	.34	.30	.27	.255
5	.47	.39	.33	.46	.38	.32	.41	.35	.30	.38	.32	.28	.34	.29	.26	.23	.225
6	.43	.35	.29	.41	.33	.28	.37	.31	.26	.34	.28	.24	.30	.26	.22	.20	.202
7	.38	.30	.24	.36	.29	.24	.33	.27	.22	.31	.25	.21	.28	.23	.19	.17	.182
8	.35	.27	.22	.33	.26	.21	.31	.24	.20	.28	.22	.18	.25	.21	.17	.15	.166
9	.32	.24	.19	.30	.24	.19	.28	.22	.18	.26	.20	.16	.23	.19	.15	.13	.152
10	.29	.22	.17	.28	.21	.17	.26	.20	.16	.24	.18	.15	.22	.17	.14	.12	.140

Column reflectance headings (ρCC): 80 / 70 / 50 / 30 / 10, each subdivided by wall reflectance (ρW): 50, 30, 10.

Wall Exitance Coefficients for 20 Per Cent Effective Floor Cavity Reflectance (ρFC = 20)

ρCC = 80 (50 / 30 / 10)	70 (50 / 30 / 10)	50 (50 / 30 / 10)	30 (50 / 30 / 10)	10 (50 / 30 / 10)
.156 .089 .028	.152 .087 .028	.143 .082 .026	.136 .078 .025	.128 .074 .024
.153 .084 .026	.149 .082 .025	.142 .079 .024	.135 .076 .024	.129 .073 .023
.146 .078 .023	.143 .076 .023	.136 .074 .022	.130 .071 .022	.125 .069 .021
.139 .072 .021	.136 .071 .021	.130 .069 .021	.125 .067 .020	.120 .065 .020
.131 .067 .019	.129 .066 .019	.124 .064 .019	.119 .063 .019	.115 .061 .018
.124 .062 .018	.122 .061 .018	.117 .060 .017	.113 .059 .017	.109 .057 .017
.117 .058 .016	.115 .057 .016	.111 .056 .016	.107 .055 .016	.104 .054 .016
.111 .054 .015	.109 .054 .015	.105 .053 .015	.102 .052 .015	.099 .051 .015
.105 .051 .014	.103 .050 .014	.100 .050 .014	.097 .049 .014	.094 .048 .014
.100 .048 .013	.098 .047 .013	.095 .047 .013	.092 .046 .013	.090 .045 .013
.144 .082 .026	.140 .080 .025	.132 .076 .024	.125 .072 .023	.118 .069 .022
.142 .078 .024	.139 .076 .024	.132 .073 .023	.126 .071 .022	.120 .068 .021
.137 .073 .022	.134 .072 .022	.128 .069 .021	.123 .067 .021	.118 .065 .020
.131 .068 .020	.128 .067 .020	.123 .065 .019	.118 .063 .019	.113 .062 .019
.124 .063 .018	.122 .062 .018	.117 .061 .018	.113 .059 .018	.109 .058 .017
.117 .059 .017	.115 .058 .017	.111 .057 .017	.107 .056 .016	.104 .055 .016
.111 .055 .016	.109 .054 .015	.105 .053 .015	.102 .052 .015	.099 .051 .015
.105 .051 .014	.103 .051 .014	.100 .050 .014	.097 .049 .014	.094 .048 .014
.100 .048 .013	.098 .048 .013	.095 .047 .013	.092 .046 .013	.089 .045 .013
.094 .045 .013	.093 .045 .013	.090 .044 .012	.088 .044 .012	.085 .043 .012
.153 .087 .027	.149 .085 .027	.143 .082 .026	.137 .079 .025	.131 .076 .025
.145 .079 .024	.142 .078 .024	.136 .076 .024	.131 .074 .023	.126 .071 .023
.135 .072 .022	.132 .071 .021	.127 .069 .021	.123 .067 .021	.118 .065 .020
.125 .065 .019	.123 .064 .019	.118 .063 .019	.114 .061 .018	.110 .060 .018
.116 .059 .017	.114 .058 .017	.110 .057 .017	.106 .056 .017	.102 .055 .016
.107 .054 .015	.106 .053 .015	.102 .052 .015	.099 .051 .015	.096 .050 .015
.100 .049 .014	.098 .049 .014	.095 .048 .014	.092 .047 .014	.089 .046 .014
.093 .046 .013	.092 .045 .013	.089 .044 .013	.086 .044 .013	.084 .043 .013
.087 .042 .012	.086 .042 .012	.083 .041 .012	.081 .041 .012	.079 .040 .012
.082 .039 .011	.081 .039 .011	.079 .038 .011	.076 .038 .011	.074 .037 .011
.247 .140 .044	.238 .136 .043	.221 .127 .040	.205 .118 .038	.191 .111 .036
.224 .123 .038	.216 .119 .037	.201 .112 .035	.187 .105 .033	.174 .098 .031
.205 .109 .033	.198 .106 .032	.184 .100 .030	.171 .094 .029	.160 .088 .027
.188 .098 .029	.182 .095 .028	.169 .090 .027	.158 .085 .026	.147 .080 .025
.174 .089 .026	.168 .086 .025	.157 .082 .024	.146 .077 .023	.136 .073 .022
.161 .081 .023	.156 .079 .023	.146 .075 .022	.136 .071 .021	.127 .067 .020
.150 .074 .021	.145 .072 .021	.136 .069 .020	.127 .065 .019	.119 .062 .019
.140 .069 .019	.136 .067 .019	.127 .063 .018	.119 .060 .017	.111 .057 .017
.132 .064 .018	.127 .062 .017	.119 .059 .017	.112 .056 .016	.105 .053 .015
.124 .059 .016	.120 .058 .016	.112 .055 .016	.106 .052 .015	.099 .050 .014
.188 .107 .034	.184 .105 .033	.176 .101 .032	.169 .097 .031	.162 .094 .030
.179 .098 .030	.176 .097 .030	.169 .094 .029	.162 .091 .028	.156 .088 .028
.167 .089 .027	.163 .087 .026	.157 .085 .026	.151 .083 .025	.146 .081 .025
.154 .080 .024	.151 .079 .023	.145 .077 .023	.140 .075 .023	.136 .074 .022
.142 .073 .021	.140 .072 .021	.135 .070 .021	.130 .069 .020	.126 .067 .020
.132 .066 .019	.130 .065 .019	.125 .064 .019	.121 .063 .018	.117 .062 .018
.123 .061 .017	.121 .060 .017	.117 .059 .017	.113 .058 .017	.110 .057 .017
.115 .056 .016	.113 .055 .016	.109 .055 .016	.106 .054 .015	.103 .053 .015
.107 .052 .014	.106 .051 .014	.102 .051 .014	.099 .050 .014	.097 .049 .014
.101 .048 .013	.099 .048 .013	.096 .047 .013	.094 .046 .013	.091 .046 .013
.335 .191 .060	.323 .184 .058	.299 .172 .055	.277 .160 .051	.257 .149 .048
.293 .161 .049	.282 .155 .048	.260 .145 .045	.241 .135 .042	.222 .126 .040
.262 .139 .042	.251 .135 .040	.232 .126 .038	.214 .117 .036	.197 .109 .034
.236 .123 .036	.226 .119 .035	.209 .111 .033	.192 .103 .031	.177 .096 .029
.215 .109 .032	.206 .106 .031	.190 .099 .029	.175 .092 .027	.161 .086 .026
.197 .099 .028	.189 .095 .027	.174 .089 .026	.160 .083 .024	.147 .078 .023
.181 .090 .025	.174 .087 .025	.161 .081 .023	.148 .076 .022	.136 .071 .021
.168 .082 .023	.162 .080 .022	.149 .075 .021	.137 .070 .020	.126 .065 .019
.157 .076 .021	.151 .073 .021	.139 .069 .020	.128 .064 .018	.118 .060 .017
.147 .070 .020	.141 .068 .019	.130 .064 .018	.120 .060 .017	.111 .056 .016

Ceiling Cavity Exitance Coefficients for 20 Per Cent Floor Cavity Reflectance (ρFC = 20)

ρCC = 80 (50 / 30 / 10)	70 (50 / 30 / 10)	50 (50 / 30 / 10)	30 (50 / 30 / 10)	10 (50 / 30 / 10)
.125 .125 .125	.107 .107 .107	.073 .073 .073	.042 .042 .042	.013 .013 .013
.115 .103 .092	.098 .088 .079	.067 .061 .055	.039 .035 .032	.012 .011 .010
.108 .087 .069	.092 .075 .060	.063 .052 .042	.037 .030 .024	.012 .010 .008
.102 .075 .053	.087 .065 .046	.060 .045 .032	.035 .026 .019	.011 .008 .006
.097 .066 .042	.083 .057 .036	.057 .040 .026	.033 .023 .015	.011 .008 .005
.092 .059 .034	.079 .051 .029	.055 .035 .021	.032 .021 .012	.010 .007 .004
.088 .053 .028	.076 .046 .024	.052 .032 .017	.030 .019 .010	.010 .006 .003
.084 .048 .024	.072 .042 .021	.050 .029 .015	.029 .017 .009	.009 .006 .003
.080 .045 .020	.069 .039 .018	.048 .027 .013	.028 .016 .007	.009 .005 .002
.077 .041 .018	.066 .036 .015	.046 .025 .011	.027 .015 .006	.009 .005 .002
.073 .039 .016	.063 .034 .014	.044 .024 .010	.026 .014 .006	.008 .005 .002
.114 .114 .114	.097 .097 .097	.066 .066 .066	.038 .038 .038	.012 .012 .012
.105 .094 .084	.090 .080 .072	.061 .055 .050	.035 .032 .029	.011 .010 .009
.099 .079 .063	.085 .068 .054	.058 .047 .038	.033 .027 .022	.011 .009 .007
.094 .068 .048	.080 .059 .042	.055 .041 .029	.032 .024 .017	.010 .008 .006
.089 .060 .038	.077 .052 .033	.053 .036 .023	.030 .021 .014	.010 .007 .004
.085 .054 .030	.073 .046 .026	.050 .032 .019	.029 .019 .011	.009 .006 .004
.082 .049 .025	.070 .042 .022	.048 .029 .015	.028 .017 .009	.009 .006 .003
.078 .044 .021	.067 .039 .018	.046 .027 .013	.027 .016 .008	.009 .005 .003
.075 .041 .018	.064 .036 .016	.044 .025 .011	.026 .015 .007	.008 .005 .002
.072 .037 .016	.062 .033 .014	.042 .023 .010	.025 .014 .006	.008 .005 .002
.068 .036 .014	.059 .031 .012	.041 .022 .009	.024 .013 .005	.008 .004 .002
.092 .092 .092	.078 .078 .078	.053 .053 .053	.031 .031 .031	.010 .010 .010
.087 .075 .064	.074 .064 .055	.051 .044 .038	.029 .026 .022	.009 .008 .007
.084 .063 .047	.072 .055 .040	.049 .038 .028	.028 .022 .016	.009 .007 .005
.080 .055 .035	.069 .048 .030	.047 .033 .021	.027 .019 .013	.009 .006 .004
.077 .049 .027	.066 .042 .024	.045 .029 .017	.026 .017 .010	.008 .006 .003
.073 .044 .022	.063 .038 .019	.043 .026 .014	.025 .016 .008	.008 .005 .003
.070 .040 .018	.060 .035 .016	.042 .024 .011	.024 .014 .007	.008 .005 .002
.067 .037 .015	.057 .032 .013	.040 .022 .010	.023 .013 .006	.007 .004 .002
.064 .034 .013	.055 .029 .012	.038 .021 .008	.022 .012 .005	.007 .004 .002
.061 .031 .011	.052 .027 .010	.036 .019 .007	.021 .011 .004	.007 .004 .001
.058 .029 .010	.050 .025 .009	.035 .018 .006	.020 .011 .004	.007 .003 .001
.236 .236 .236	.201 .201 .201	.138 .138 .138	.079 .079 .079	.025 .025 .025
.230 .210 .193	.196 .181 .166	.134 .124 .115	.077 .072 .062	.025 .023 .022
.224 .193 .167	.192 .166 .144	.131 .115 .101	.076 .067 .059	.024 .022 .019
.218 .180 .150	.187 .155 .130	.128 .108 .091	.074 .063 .054	.024 .020 .018
.213 .170 .138	.182 .147 .120	.125 .103 .084	.073 .060 .050	.023 .020 .016
.207 .163 .130	.178 .141 .113	.123 .098 .080	.071 .058 .047	.023 .019 .016
.202 .157 .124	.174 .136 .108	.120 .095 .076	.069 .056 .045	.022 .018 .015
.197 .152 .120	.169 .131 .104	.117 .092 .074	.068 .054 .044	.022 .018 .015
.192 .147 .117	.166 .128 .102	.115 .090 .072	.067 .053 .043	.022 .017 .014
.188 .144 .114	.162 .125 .100	.112 .088 .071	.065 .052 .042	.021 .017 .014
.184 .141 .112	.158 .122 .098	.110 .086 .070	.064 .051 .041	.021 .017 .014
.114 .114 .114	.097 .097 .097	.066 .066 .066	.038 .038 .038	.012 .012 .012
.108 .093 .080	.089 .078 .069	.063 .055 .047	.036 .032 .028	.012 .010 .009
.103 .079 .058	.089 .068 .050	.061 .047 .035	.035 .027 .020	.011 .009 .007
.099 .068 .043	.085 .059 .038	.058 .041 .026	.034 .024 .016	.011 .008 .005
.094 .060 .033	.082 .052 .029	.056 .036 .021	.032 .021 .013	.010 .007 .004
.091 .054 .027	.078 .046 .023	.054 .032 .017	.031 .019 .010	.010 .007 .003
.086 .048 .021	.074 .042 .019	.051 .029 .014	.030 .017 .008	.010 .006 .003
.083 .043 .018	.071 .038 .016	.049 .027 .012	.029 .016 .007	.009 .006 .002
.079 .040 .015	.068 .035 .013	.047 .025 .010	.027 .014 .006	.009 .005 .002
.076 .037 .013	.065 .032 .011	.043 .024 .009	.026 .014 .005	.009 .005 .002
.071 .036 .013	.061 .032 .011	.043 .022 .008	.025 .013 .005	.008 .004 .002
.325 .325 .325	.278 .278 .278	.189 .189 .189	.109 .109 .109	.035 .035 .035
.321 .295 .272	.275 .253 .234	.188 .174 .162	.108 .101 .094	.035 .032 .030
.316 .275 .241	.270 .237 .208	.185 .164 .145	.107 .095 .085	.034 .031 .028
.309 .261 .222	.265 .225 .192	.182 .156 .135	.105 .091 .080	.034 .030 .026
.303 .250 .209	.260 .216 .180	.179 .150 .128	.103 .088 .076	.033 .029 .025
.296 .241 .201	.254 .209 .175	.175 .145 .123	.101 .086 .073	.033 .028 .024
.289 .234 .195	.249 .203 .169	.172 .142 .120	.100 .084 .071	.032 .027 .024
.283 .228 .190	.244 .198 .166	.168 .139 .117	.098 .082 .070	.032 .027 .023
.277 .224 .187	.239 .194 .163	.165 .136 .115	.096 .080 .069	.031 .026 .023
.272 .219 .184	.234 .190 .160	.162 .134 .114	.095 .079 .068	.031 .026 .022
.267 .216 .182	.230 .187 .159	.160 .132 .113	.093 .078 .067	.030 .026 .022

Fig. 61 Continued (See page 119 for instructions and notes)

Typical Luminaire	Maint. Cat.	SC	RCR ↓	80 pw→ 50	30	10	70 pw→ 50	30	10	50 pw→ 50	30	10	30 pw→ 50	30	10	10 pw→ 50	30	10	0 pw→ 0	WDRC	RCR ↓
				Coefficients of Utilization for 20 Per Cent Effective Floor Cavity Reflectance (pFC = 20)																	
49 2-lamp fluorescent strip unit with 235° reflector fluorescent lamps	I	1.4/1.2	0	1.13	1.13	1.13	1.09	1.09	1.09	1.01	1.01	1.01	.94	.94	.94	.88	.88	.88	.85		0
			1	.95	.90	.86	.92	.87	.83	.85	.82	.78	.79	.76	.74	.74	.72	.69	.66	.464	1
			2	.82	.74	.68	.79	.72	.66	.73	.68	.63	.68	.64	.60	.63	.60	.56	.53	.394	2
			3	.71	.62	.55	.69	.61	.54	.64	.57	.52	.59	.54	.49	.55	.51	.47	.44	.342	3
			4	.62	.53	.46	.60	.52	.45	.56	.49	.43	.52	.46	.41	.49	.44	.40	.37	.300	4
			5	.55	.46	.39	.54	.45	.39	.50	.43	.37	.47	.40	.36	.44	.38	.34	.32	.267	5
			6	.50	.41	.34	.48	.40	.33	.45	.38	.32	.42	.36	.31	.39	.34	.30	.27	.240	6
			7	.45	.36	.30	.43	.35	.29	.41	.34	.28	.38	.32	.27	.36	.30	.26	.24	.218	7
			8	.41	.32	.26	.40	.32	.26	.37	.30	.25	.35	.29	.24	.33	.27	.23	.21	.199	8
			9	.37	.29	.24	.36	.28	.23	.34	.27	.22	.32	.26	.22	.30	.25	.21	.19	.183	9
			10	.34	.26	.21	.33	.26	.21	.32	.25	.20	.30	.24	.20	.28	.23	.19	.17	.170	10

(Luminaire 49 markings: 12½ , 85)

Typical Luminaires	RCR ↓	80 pw→ 50	30	10	70 pw→ 50	30	10	50 pw→ 50	30	10	30 pw→ 50	30	10	10 pw→ 50	30	10	0 pw→ 0
		Coefficients of utilization for 20 Per Cent Effective Floor Cavity Reflectance, pFC															
50 Single row fluorescent lamp cove without reflector, mult. by 0.93 for 2 rows and by 0.85 for 3 rows.	1	.42	.40	.39	.36	.35	.33	.25	.24	.23	Coves are not recommended for lighting areas having low reflectances.						
	2	.37	.34	.32	.32	.29	.27	.22	.20	.19							
	3	.32	.29	.26	.28	.25	.23	.19	.17	.16							
	4	.29	.25	.22	.25	.22	.19	.17	.15	.13							
	5	.25	.21	.18	.22	.19	.16	.15	.13	.11							
	6	.23	.19	.16	.20	.16	.14	.14	.12	.10							
	7	.20	.17	.14	.17	.14	.12	.12	.10	.09							
	8	.18	.15	.12	.16	.13	.10	.11	.09	.08							
	9	.17	.13	.10	.15	.11	.09	.10	.08	.07							
	10	.15	.12	.09	.13	.10	.08	.09	.07	.06							
51 pcc from below ~65% Diffusing plastic or glass 1) Ceiling efficiency ~60%; diffuser transmittance ~50%; diffuser reflectance ~40%. Cavity with minimum obstructions and painted with 80% reflectance paint—use pc = 70. 2) For lower reflectance paint or obstructions—use pc = 50.	1				.60	.58	.56	.58	.56	.54							
	2				.53	.49	.45	.51	.47	.43							
	3				.47	.42	.37	.45	.41	.36							
	4				.41	.36	.32	.39	.35	.31							
	5				.37	.31	.27	.35	.30	.26							
	6				.33	.27	.23	.31	.26	.23							
	7				.29	.24	.20	.28	.23	.20							
	8				.26	.21	.18	.25	.20	.17							
	9				.23	.19	.15	.23	.18	.15							
	10				.21	.17	.13	.21	.16	.13							
52 pcc from below ~60% Prismatic plastic or glass. 1) Ceiling efficiency ~67%; prismatic transmittance ~72%; prismatic reflectance ~18%. Cavity with minimum obstructions and painted with 80% reflectance paint—use pc = 70. 2) For lower reflectance paint or obstructions—use pc = 50.	1				.71	.68	.66	.67	.66	.65	.65	.64	.62				
	2				.63	.60	.57	.61	.58	.55	.59	.56	.54				
	3				.57	.53	.49	.55	.52	.48	.54	.50	.47				
	4				.52	.47	.43	.50	.45	.42	.48	.44	.42				
	5				.46	.41	.37	.44	.40	.37	.43	.40	.36				
	6				.42	.37	.33	.41	.36	.32	.40	.35	.32				
	7				.38	.32	.29	.37	.31	.28	.36	.31	.28				
	8				.34	.28	.25	.33	.28	.25	.32	.28	.25				
	9				.30	.25	.22	.30	.25	.21	.29	.25	.21				
	10				.27	.23	.19	.27	.22	.19	.26	.22	.19				
53 pcc from below ~45% Louvered ceiling. 1) Ceiling efficiency ~50%; 45° shielding opaque louvers of 80% reflectance. Cavity with minimum obstructions and painted with 80% reflectance paint—use pc = 50. 2) For other conditions refer to Fig. 6–18. *	1							.51	.49	.48				.47	.46	.45	
	2							.46	.44	.42				.43	.42	.40	
	3							.42	.39	.37				.39	.38	.36	
	4							.38	.35	.33				.36	.34	.32	
	5							.35	.32	.29				.33	.31	.29	
	6							.32	.28	.26				.30	.28	.26	
	7							.29	.26	.23				.28	.25	.23	
	8							.27	.23	.21				.26	.23	.21	
	9							.24	.21	.19				.24	.21	.19	
	10							.22	.19	.17				.22	.19	.17	

*IES Lighting Handbook, *1984 Reference Volume*

Wall Exitance Coefficients for 20 Per Cent Effective Floor Cavity Reflectance ($\rho_{FC} = 20$)

80			70			50			30			10		
50	30	10	50	30	10	50	30	10	50	30	10	50	30	10
.357	.203	.064	.346	.197	.063	.326	.187	.060	.307	.177	.057	.290	.168	.054
.316	.173	.053	.306	.169	.052	.288	.160	.050	.271	.152	.048	.256	.145	.046
.284	.151	.045	.275	.147	.044	.259	.140	.043	.243	.133	.041	.229	.127	.039
.257	.134	.039	.249	.130	.039	.234	.124	.037	.221	.118	.036	.208	.113	.034
.235	.120	.035	.227	.117	.034	.214	.111	.033	.202	.106	.032	.190	.101	.030
.215	.108	.031	.209	.106	.030	.197	.101	.029	.185	.096	.028	.175	.092	.027
.199	.098	.028	.193	.096	.027	.182	.092	.027	.172	.088	.026	.162	.084	.025
.185	.090	.025	.180	.088	.025	.169	.085	.024	.160	.081	.023	.151	.077	.023
.172	.083	.023	.168	.082	.023	.158	.078	.022	.150	.075	.021	.141	.072	.021
.162	.077	.022	.157	.076	.021	.149	.073	.020	.140	.070	.020	.133	.067	.019

Ceiling Cavity Exitance Coefficients for 20 Per Cent Floor Cavity Reflectance ($\rho_{FC} = 20$)

80			70			50			30			10		
50	30	10	50	30	10	50	30	10	50	30	10	50	30	10
.280	.280	.280	.239	.239	.239	.163	.163	.163	.094	.094	.094	.030	.030	.030
.275	.247	.222	.235	.212	.191	.161	.146	.132	.093	.084	.077	.030	.027	.025
.268	.224	.188	.230	.193	.162	.157	.134	.113	.091	.078	.066	.029	.025	.022
.261	.208	.166	.224	.180	.144	.154	.125	.101	.089	.073	.060	.028	.024	.019
.253	.196	.152	.217	.169	.132	.150	.118	.093	.086	.069	.055	.028	.022	.018
.246	.186	.142	.211	.161	.123	.145	.112	.087	.084	.066	.052	.027	.022	.017
.239	.178	.135	.205	.154	.117	.142	.108	.083	.082	.064	.049	.027	.021	.016
.232	.172	.130	.199	.149	.113	.138	.104	.080	.080	.061	.048	.026	.020	.016
.225	.166	.126	.194	.144	.109	.134	.101	.078	.078	.060	.046	.025	.020	.015
.219	.161	.122	.189	.140	.107	.131	.099	.076	.076	.058	.045	.025	.019	.015
.214	.157	.120	.184	.137	.105	.128	.096	.074	.074	.057	.044	.024	0.19	.015

54

910 mm x 910 mm (3′ x 3′) fluorescent troffer with 1220 mm (48″) lamps mounted along diagonals—use units 40, 41 or 42 as appropriate

55

610 mm x 610 mm (2′ x 2′) fluorescent troffer with two "U" lamps—use units 40, 41 or 42 as appropriate

Tabulation of Luminous Intensities Used to Compute Above Coefficients
Normalized Average Intensity (Candelas per 1000 lumens)

Angle ↓	Luminaire No.													
	1	2	3	4	5	6	7	8	9	10	11	12	13	14
5	72.5	6.5	256.0	238.0	808.0	1320.0	695.0	374.0	2680.0	2610.0	208.0	152.0	190.0	316.0
15	72.5	8.0	246.0	264.0	671.0	1010.0	630.0	357.0	1150.0	1200.0	220.0	148.0	196.0	311.0
25	72.5	9.5	238.0	248.0	494.0	584.0	286.0	305.0	209.0	411.0	254.0	141.0	199.0	301.0
35	72.5	10.0	238.0	191.0	340.0	236.0	88.0	212.0	13.5	97.0	220.0	125.0	212.0	271.0
45	72.5	8.0	203.0	122.0	203.0	22.0	5.0	81.0	0	15.0	130.0	106.0	206.0	156.0
55	72.0	6.5	168.0	62.5	91.0	0	0	40.5	0	0	59.0	87.5	125.0	63.0
65	71.5	4.5	130.0	45.5	33.0	0	0	20.5	0	0	26.0	69.5	68.5	31.5
75	70.5	2.5	34.0	38.0	12.5	0	0	9.5	0	0	11.0	47.0	41.5	17.5
85	70.0	2.0	7.0	32.0	4.0	0	0	2.5	0	0	3.5	23.5	26.0	4.0
95	67.0	15.0	0	28.0	0	0	0	0	0	0	0	9.5	12.5	0
105	62.5	147.0	0	28.0	0	0	0	0	0	0	0	4.5	6.0	0
115	58.0	170.0	0	41.0	0	0	0	0	0	0	0	1.0	3.5	0
125	54.5	168.0	0	42.5	0	0	0	0	0	0	0	0	1.5	0
135	51.0	183.0	0	33.0	0	0	0	0	0	0	0	0	0	0
145	48.0	159.0	0	22.5	0	0	0	0	0	0	0	0	0	0
155	46.5	139.0	0	9.0	0	0	0	0	0	0	0	0	0	0
165	45.0	94.5	0	3.0	0	0	0	0	0	0	0	0	0	0
175	44.0	50.5	0	1.0	0	0	0	0	0	0	0	0	0	0

Angle ↓	Luminaire No.													
	15	16	17	18	19	20	21	22	23	24	25	26	27	28
5	288.0	999.0	470.0	294.0	576.0	274.0	203.0	136.0	155.0	0	263.0	246.0	284.0	244.0
15	321.0	775.0	384.0	282.0	519.0	302.0	192.0	151.0	169.0	0	258.0	260.0	262.0	248.0
25	331.0	475.0	344.0	294.0	426.0	344.0	194.0	171.0	185.0	0	236.0	264.0	226.0	242.0
35	260.0	188.0	290.0	294.0	274.0	321.0	252.0	175.0	188.0	0	210.0	248.0	187.0	218.0
45	202.0	90.5	210.0	246.0	127.0	209.0	230.0	182.0	162.0	0	163.0	192.0	145.0	152.0
55	114.0	32.0	86.5	137.0	69.5	45.5	119.0	158.0	119.0	0	98.0	98.0	83.0	70.0
65	13.5	8.5	18.0	26.0	20.0	8.0	52.5	90.0	57.0	0	55.5	32.5	36.5	26.0

Fig. 61 Continued (See page 119 for instructions and notes)

Angle ↓	Luminaire No.													
	15	16	17	18	19	20	21	22	23	24	25	26	27	28
75	6.0	6.0	5.0	6.5	2.5	3.0	21.0	41.0	4.5	0	29.5	12.5	18.5	10.0
85	2.0	1.0	1.0	1.0	1.5	2.5	3.5	17.0	0	0	11.0	4.0	5.5	3.0
95	1.0	0.5	0.5	0.5	0.5	3.5	0	8.0	0	19.0	8.0	3.5	3.5	2.5
105	0	0.5	0.5	0.5	0.5	8.0	0	7.0	0	64.0	14.5	6.5	11.0	6.0
115	0	0.5	0.5	0.5	4.5	15.5	0	7.0	0	212.0	21.5	12.0	21.0	13.0
125	0	1.0	0.5	0.5	10.5	22.5	0	5.0	0	205.0	31.0	21.5	34.5	24.0
135	0	1.5	1.0	0.5	16.5	29.0	0	0	0	160.0	47.0	33.5	51.5	36.0
145	0	8.0	3.0	1.5	20.5	33.5	0	0	0	128.0	59.5	50.0	71.5	49.5
155	0	8.5	8.0	7.5	32.0	42.0	0	0	0	115.0	82.5	70.5	92.0	70.0
165	0	0.5	0.5	0.5	33.0	27.5	0	0	0	106.0	105.0	92.0	109.0	88.5
175	0	0.5	0.5	0.5	16.5	2.5	0	0	0	102.0	111.0	102.0	115.0	95.5

Angle ↓	Luminaire No.																				
	29	30	31	32	33	34	35	36	37	38	39	40	41	42	43	44	45	46	47	48	49
5	189.0	270.0	218.0	199.0	41.5	194.0	210.0	206.0	107.0	272.0	312.0	218.0	206.0	253.0	288.0	197.0	90.0	132.0	135.0	157.0	238.0
15	176.0	249.0	220.0	194.0	38.5	192.0	211.0	199.0	104.0	244.0	268.0	207.0	202.0	249.0	284.0	196.0	104.0	144.0	142.0	156.0	232.0
25	147.0	200.0	224.0	184.0	35.5	187.0	212.0	185.0	98.5	202.0	213.0	187.0	183.0	236.0	271.0	199.0	125.0	181.0	167.0	153.0	218.0
35	110.0	144.0	222.0	170.0	32.5	169.0	204.0	158.0	90.0	156.0	148.0	164.0	162.0	214.0	246.0	235.0	140.0	202.0	171.0	147.0	200.0
45	64.0	86.5	187.0	154.0	29.0	123.0	164.0	108.0	79.5	106.0	87.0	135.0	133.0	172.0	190.0	223.0	131.0	173.0	151.0	137.0	176.0
55	34.5	53.5	99.0	137.0	22.0	77.5	78.5	51.5	66.5	68.0	51.0	106.0	104.0	95.5	97.0	99.5	104.0	113.0	120.0	122.0	149.0
65	20.5	34.0	15.5	117.0	14.5	37.5	36.5	35.5	52.0	42.0	30.0	74.0	70.5	45.0	25.0	18.5	65.5	63.0	82.0	104.0	119.0
75	10.0	20.5	3.5	88.5	7.0	18.5	26.0	34.5	36.0	21.5	15.5	42.5	36.5	19.0	6.0	3.0	27.5	42.5	41.5	79.0	86.5
85	2.5	10.0	1.0	59.0	2.0	10.5	17.5	32.0	21.5	6.0	4.0	15.5	7.0	7.0	2.5	0.5	8.0	27.5	7.5	52.5	50.5
95	4.0	7.0	0	49.5	11.0	14.5	15.5	32.5	14.5	0	0	5.5	0	0	0	0	0	23.0	0	45.0	32.5
105	19.0	8.5	0	32.5	49.5	40.0	22.0	49.0	14.5	0	0	2.5	0	0	0	0	0	31.0	0	43.5	27.5
115	40.5	9.5	0	6.5	96.0	57.0	27.0	49.0	14.0	0	0	0	0	0	0	0	0	30.0	0	38.5	22.0
125	67.0	10.0	0	0	130.0	68.5	23.0	44.5	13.0	0	0	0	0	0	0	0	0	19.5	0	33.0	17.5
135	93.0	11.0	0	0	155.0	71.5	18.5	36.0	12.0	0	0	0	0	0	0	0	0	10.0	0	27.0	13.5
145	117.0	11.0	0	0	172.0	67.5	12.0	28.5	10.0	0	0	0	0	0	0	0	0	7.5	0	20.0	10.5
155	136.0	11.5	0	0	183.0	65.0	7.5	24.0	8.5	0	0	0	0	0	0	0	0	4.5	0	13.0	7.5
165	151.0	12.0	0	0	189.0	67.5	4.5	21.0	6.5	0	0	0	0	0	0	0	0	1.5	0	7.0	5.0
175	155.0	13.0	0	0	201.0	73.5	4.0	18.0	5.5	0	0	0	0	0	0	0	0	0	0	2.5	2.5

Fig. 62. Procedure for
Determining Luminaire Maintenance Categories

To assist in determining Luminaire Dirt Depreciation (LDD) factors, luminaires are separated into six maintenance categories (I through VI). To arrive at categories, luminaires are arbitrarily divided into sections, a *Top Enclosure* and a *Bottom Enclosure*, by drawing a horizontal line through the light center of the lamp or lamps. The characteristics listed for the enclosures are then selected as best describing the luminaire. Only one characteristic for the top enclosure and one for the bottom enclosure should be used in determining the category of a luminaire. Percentage of uplight is based on 100 per cent for the luminaire.

The maintenance category is determined when there are characteristics in both enclosure columns. If a luminaire falls into more than one category, the lower numbered category is used.

Mainte-nance Cat-egory	Top Enclosure	Bottom Enclosure
I	1. None.	1. None
II	1. None 2. Transparent with 15 per cent or more uplight through apertures. 3. Translucent with 15 per cent or more uplight through apertures. 4. Opaque with 15 per cent or more uplight through apertures.	1. None 2. Louvers or baffles
III	1. Transparent with less than 15 per cent upward light through apertures. 2. Translucent with less than 15 per cent upward light through apertures. 3. Opaque with less than 15 per cent uplight through apertures.	1. None 2. Louvers or baffles
IV	1. Transparent unapertured. 2. Translucent unapertured. 3. Opaque unapertured.	1. None 2. Louvers
V	1. Transparent unapertured. 2. Translucent unapertured. 3. Opaque unapertured.	1. Transparent unapertured 2. Translucent unapertured
VI	1. None 2. Transparent unapertured. 3. Translucent unapertured. 4. Opaque unapertured.	1. Transparent unapertured 2. Translucent unapertured 3. Opaque unapertured

Fig. 63. Evaluation of Operating Atmosphere
Factors for Use in Table Below

1 = Cleanest conditions imaginable
2 = Clean, but not the cleanest
3 = Average

4 = Dirty, but not the dirtiest
5 = Dirtiest conditions imaginable

Type of Dirt	Area Adjacent to Task Area			Filter Factor (per cent of dirt passed)	Area Surrounding Task			Sub Total
	Intermittent Dirt	Constant Dirt	Total		From Adjacent	Intermittent Dirt	Constant Dirt	
Adhesive Dirt		+	=	×	=	+	+	=
Attracted Dirt		+	=	×	=	+	+	=
Inert Dirt		+	=	×	=	+	+	=

Total of Dirt Factors

0–12 = Very Clean	13–24 = Clean	25–36 = Medium	37–48 = Dirty	49–60 = Very Dirty

Fig. 64. Five Degrees of Dirt Conditions

	Very Clean	Clean	Medium	Dirty	Very Dirty
Generated Dirt	None	Very little	Noticeable but not heavy	Accumulates rapidly	Constant accumulation
Ambient Dirt	None (or none enters area)	Some (almost none enters)	Some enters area	Large amount enters area	Almost none excluded
Removal or Filtration	Excellent	Better than average	Poorer than average	Only fans or blowers if any	None
Adhesion	None	Slight	Enough to be visible after some months	High—probably due to oil, humidity or static	High
Examples	High grade offices, not near production; laboratories; clean rooms	Offices in older buildings or near production; light assembly; inspection	Mill offices; paper processing; light machining	Heat treating; high speed printing; rubber processing	Similar to Dirty but luminaires within immediate area of contamination

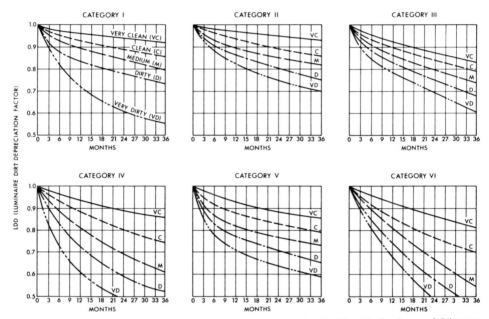

Fig. 65 Luminaire Dirt Depreciation factors (LDD) for six luminaire categories (I to VI) and for five degrees of dirtiness as determined from either Fig. 9–8 or 9–9.

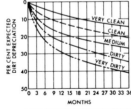

Fig. 66 Room Surface Dirt Depreciation Factors

	Luminaire Distribution Type																			
	Direct				Semi-Direct				Direct-Indirect				Semi-Indirect				Indirect			
Per Cent Expected Dirt Depreciation	10	20	30	40	10	20	30	40	10	20	30	40	10	20	30	40	10	20	30	40
Room Cavity Ratio																				
1	.98	.96	.94	.92	.97	.92	.89	.84	.94	.87	.80	.76	.94	.87	.80	.73	.90	.80	.70	.60
2	.98	.96	.94	.92	.96	.92	.88	.83	.94	.87	.80	.75	.94	.87	.79	.72	.90	.80	.69	.59
3	.98	.95	.93	.90	.96	.91	.87	.82	.94	.86	.79	.74	.94	.86	.78	.71	.90	.79	.68	.58
4	.97	.95	.92	.90	.95	.90	.85	.80	.94	.86	.79	.73	.94	.86	.78	.70	.89	.78	.67	.56
5	.97	.94	.91	.89	.94	.90	.84	.79	.93	.86	.78	.72	.93	.86	.77	.69	.89	.78	.66	.55
6	.97	.94	.91	.88	.94	.89	.83	.78	.93	.85	.78	.71	.93	.85	.76	.68	.89	.77	.66	.54
7	.97	.94	.90	.87	.93	.88	.82	.77	.93	.84	.77	.70	.93	.84	.76	.68	.89	.76	.65	.53
8	.96	.93	.89	.86	.93	.87	.81	.75	.93	.84	.76	.69	.93	.84	.76	.68	.88	.76	.64	.52
9	.96	.92	.88	.85	.93	.87	.80	.74	.93	.84	.76	.68	.93	.84	.75	.67	.88	.75	.63	.51
10	.96	.92	.87	.83	.93	.86	.79	.72	.93	.84	.75	.67	.92	.83	.75	.67	.88	.75	.62	.50

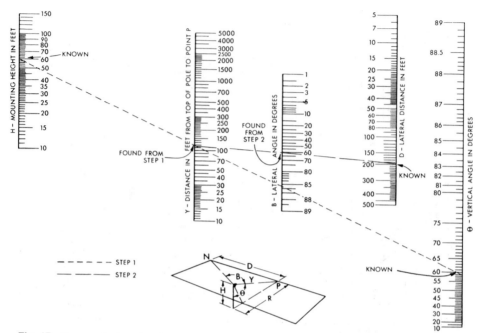

Fig. 67 . Nomogram for determining lateral angle *B* in terms of vertical angle *θ*, mounting height *H* and distances *Y* and *D*.

Considerations and Ideas for Lighting Energy Management

(From IES Lighting Handbook—*1987 Application Volume*)

Fig. 68. Lighting Considerations and Ideas for Energy Management

What to Consider	What Can Be Done in New Construction	What Can Be Done in Existing Spaces	Resource and Comments
1. The lighting needs—for productivity, safety and esthetics. a. Seeing tasks b. Seeing task locations c. Purposes of nonseeing task areas d. Illumination recommendations e. Uniform lighting f. Nonuniform lighting	Seeing tasks and their locations should be identified so that recommended illuminances can be provided for those tasks with less in surrounding non-critical seeing areas. Where there are no tasks, there is no visual need for task levels. Then safety and esthetics are of prime consideration. When tasks and locations can be identified, it may be possible to use a nonuniform lighting system, such as a nonuniform pattern of luminaires arranged to light work stations or built-in or supplementary lighting at work stations, coordinated with a general lighting system. When tasks and their locations cannot be identified, an illuminance can be selected for the expected tasks and a uniform pattern of luminaires installed with controls for lowering the level at specific points when no tasks are present.	Seeing tasks and their locations should be identified so that recommended illuminances can be provided for those tasks with less in surrounding non-critical seeing areas. Where there are no tasks, there is no visual need for task levels. Then safety and esthetics are of prime consideration. Since tasks and their locations can be identified, a lighting survey would show where illuminances are in excess of recommended maintained values. Lighting then can be adjusted to meet the recommendations. Careful consideration should be given at task locations that any change in the lighting system will not produce veiling reflections in the tasks.	Other sections of this handbook *should be consulted for specific lighting needs.
2. Luminaires a. Effectiveness for task lighting b. Effectiveness for non-task lighting c. Efficiency d. Heat transfer capability e. Cleaning capabilities	In selecting a luminaire for task lighting, consideration should be given to its effectiveness in providing high task contrast (minimum veiling reflections) and sufficiently high visual comfort (VCP). Luminaire light distribution and appearance are also important, particularly for esthetics, but consideration also should be given to efficiency. Luminaires with heat transfer capabilities should be considered so that the lighting heat can be utilized or removed and coordinated with the building thermal design and total building energy use. See Section 2.* Luminaires that can be cleaned easily and those with low dirt accumulation will reduce maintenance needs and cost.	Review luminaire effectiveness for task lighting and efficiency, and if ineffective or inefficient consider luminaire or component replacement. Check to see if all components are in good working condition. Transmitting or diffusing media should be examined and badly discolored and depreciated media replaced to improve efficiency (without producing excessive brightness and unwanted visual discomfort).	Luminaire manufacturers' data are useful guides in determining luminaire effectiveness in terms of comfort (VCP), efficiency and ability to provide high contrast rendering. Energy use will be affected by luminaire light source and distribution characteristics as shown in one study (see Fig. 2–16).*That study shows that for a medium size room, indirect, incandescent luminaires can consume over 5½ times the energy compared with a direct, fluorescent system, for the same hours of use and average illuminance (but not the same lighting quality).
3. Light sources (lamps and ballasts) a. Efficacy (lumens per watt) b. Color (chromaticity) c. Color rendering d. Lumen maintenance	Highest efficacy lamps and lamp ballast combinations should be used that are compatible with the desired light source color, color rendering capabilities, source size, life and light output depreciation. Compare sources being considered on basis of life cycle cost and energy use through life.	Where inefficient sources are used, consider relighting with more efficient sources, compatible with desired light source color and color rendering capabilities, based on life cycle costing. Consider reduced wattage fluorescent lamps in existing luminaires where 10 to 20 per cent reduction in illuminance can	Light source efficacy (lumens per watt) varies with lamp type and within types. Section 8 of the 1984 Reference Volume and Lamp manufacturers' catalogs will provide lumen output, lumens per watt, life and lumen maintenance data as well as cost for comparison purposes.

*In IES Lighting Handbook, *1981 Application Volume.*

Fig. 68 *Continued*

Consider multi-level ballasts for flexibility in achieving nonuniform lighting as needed in hour-to-hour operations throughout each day or with changing work layouts over the years.

Consider reduced current ballasts where module size and/or luminaire spacing permit needed illumination to be achieved.

be tolerated. (With improved maintenance procedures—periodic cleanings, group relamping—as much light can be obtained as with standard lamps.)

Consider multi-level ballasts for flexibility in varying lighting during occupied hours and cleaning periods, and reduced current ballasts where a reduction in illuminance can be tolerated.

4. Daylighting
a. Availability of daylight
b. Fenestration (windows, sky-lights)
c. Controls

Evaluate the daylighting potential—the levels and hours of availability—keeping in mind that glare from fenestration should be controlled to the same degree as from luminaires and that the heat gained or lost through fenestration needs to be coordinated with the building thermal system. Coordinate the electric lighting design with the daylighting so that glare, heat and illumination are controlled.

Consider the use of "high performance" heat reflecting insulating glass in windows to minimize heat gain in summer and heat loss in winter, while permitting a view of the exterior.

If daylighting can be used to replace some of the electric lighting during substantial periods of the day, lighting in those areas should be dimmed or switched off. If control is not provided, consider adding controls based on life-cycle-costing. Evaluate the effectiveness of the existing fenestration shading controls (interior and exterior) for possible replacements or additions.

Section 7 of the 1984 Reference Volume provides useful information on availability of daylight, daylight control systems, and design and evaluation methods. Recent data from manufacturers of fenestration materials and controls should be consulted.

5. Room surfaces

Work with the interior designer toward the specification of room surface and equipment reflectances at the higher end of recommended reflectances, not forgetting the importance of proper color schemes for esthetics. (Reflectances higher than those recommended may produce excessive luminance ratios and glare.)

Where the reflectances of room surfaces are lower than, or at the lower end of the recommended reflectance range, consider repainting using matte paints with reflectances toward the upper end of the range. When equipment is replaced, select light color finishes. (Reflectances higher than those recommended may produce excessive luminance ratios and glare.)

The use of higher reflectance finishes saves energy. Below is a chart showing the decreased energy needs for equal general illumination in a medium size room with improved reflectances. Reflectances are:

	Ceiling	Walls	Floor
A	50	30	10—Not recommended
B	80	40	20
C	90	60	40

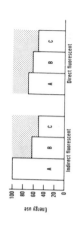

The grayed area shows per cent energy savings.

6. Maintenance procedures

Carefully consider a planned lighting maintenance program early in the design stage to allow for desired maintained levels using less equipment and less installed lighting wattage. Also consider the daylighting (fenestration and controls) maintenance program when planning on the daylight contribution to the desired illuminance levels.

The owner must be committed to the maintenance program used in the lighting system design. If not, the lighting will be less than planned and will be an energy waster.

Reevaluate the present lighting maintenance program and revise it as necessary to provide desired maintained illuminances. This may allow some reductions in lighting energy.

The diagram below shows the effect of maintenance procedures on energy use.

In system A, luminaires are cleaned and re-lamped every three years. In system B, luminaires are cleaned every year and one-third of the lamps are replaced every year.

For equal maintained illuminance over a 12-year period, the more frequent cleanings and re-lampings in system B saves about 15 per cent energy use due to the need for less equipment initially.

7. Operating procedures
a. During working hours
b. During building cleaning periods.

In spaces with tasks, consider switching arrangements so that only general surround lighting can be used when tasks are not performed; however, it is important that an adequate level of illumination be maintained for building cleaning. Allow for switching and switching ease to encourage turning lights off when not needed.

Prepare a suggested lighting switching scheme, based on the design, to aid operations.

Analyze the lighting use during working and building cleaning periods. Institute an educational program to have workers turn off lights when they are not needed. Also have cleaners' schedules adjusted to minimize the lighting use, such as by cleaning fewer spaces at the same time, turning off lights in unoccupied areas.

Where large floor or room areas are controlled by a single switch, consider adding more switching flexibility to turn off lights in areas when and where not needed.

8. Space utilization

Where seeing task locations have not been specified, work with the office space planner to show locations where higher levels will occur to best take advantage of the lighting design and where daylighting can be used effectively. The use of open plan offices, when practical, increases room size and improves utilization of light.

If new space is sparsely populated, consider locating employees with related work close together to efficiently provide the illumination needed for their tasks, and the remaining open area can be lighted to the lower level values for surrounding space and circulation areas.

Where it is found that workers are sparsely distributed consideration might be given to moving workers closer together and closing off unused space (with minimum heating, cooling and lighting). Also, an analysis of the existing lighting can show where tasks may be located to take advantage of the existing illumination provided and where ESI values are highest.

Fig. 69   Checklist of Energy Saving Lighting Ideas

A. Lighting Needs...Tasks, Task and Luminaire Location, Illumination Requirements and Utilization of Space.

1. *Identify seeing tasks and locations so recommended illuminances can be provided for tasks with less in surrounding areas.*
2. *Identify seeing tasks where maintained illuminance is greater than recommended and modify to meet the recommendations.*
3. *Consider replacing seeing tasks with those of higher contrast which call for lower illuminance requirements.*
4. *Where there are no visual tasks, task illumination is not needed. Review lighting requirements then, to satisfy safety and esthetics.*
5. *Group tasks having the same illuminance requirements or widely separated work stations, and close off unused space (with minimum heating, cooling and lighting).*
6. *When practical, have persons working after-hours work in close proximity to lessen all energy requirements.*
7. *Coordinate layout of luminaires and tasks for high contrast rendition rather than uniform space geometry. Analyze existing lighting to show where tasks may be relocated to provide better contrast rendition. Use caution when relocating tasks to minimize direct and reflected glare and veiling reflections in the tasks.*
8. *Relocate lighting from over tops of stacked materials.*
9. *Consider lowering the mounting height of luminaires if it will improve illumination, or reduce connected lighting power required to maintain adequate task lighting.*
10. *Consider illuminating tasks with luminaires properly located in or on furniture with less light in aisles.*
11. *Consider wall lighting luminaires, and lighting for plants, paintings and murals, to maintain proper luminance ratios in place of general overhead lighting.*
12. *Consider high efficacy light sources for required flood-lighting and display lighting.*
13. *Consider the use of open-plan spaces versus partitioned spaces. Where partitions are tall or stacked equipment can be eliminated, the general illumination may increase, and the lighting system connected power may be reduced.*
14. *Consider the use of light colors for walls, floors, ceilings and furniture to increase utilization of light, and reduce connected lighting power required to achieve needed light. Avoid glossy finishes on room and work surfaces.*

B. Lighting Equipment...Lamps and Luminaires.

15. *Establish washing cycles for lamps and luminaires.*
16. *Select a group lamp replacement time interval for all light sources.*
17. *Install lamps with higher efficacy (lumens per input watts) compatible with desired light source color and color-rendering capabilities.*
18. *In installations where low wattage incandescent lamps are used in luminaires, investigate the possibility of using fewer higher wattage (more efficient) lamps to get the needed light. Lamp wattages must not exceed luminaire rating.*
19. *Evaluate the use of R, PAR or ER lamps to get the needed light with lower watts depending on luminaire types or application.*
20. *Evaluate use of reduced wattage lamps when the illuminance is above task requirements, and whenever luminaire location must be maintained.*

B. Lighting Equipment *continued*

21. *Consider reduced wattage fluorescent lamps in existing luminaires along with improved maintenance procedures. CAUTION: Not recommended where ambient space temperature may fall below 16 °C (60 °F).*
22. *Check luminaire effectiveness for task lighting and efficiency, and if ineffective or inefficient, consider luminaire and component replacement or relocation for greater effectiveness.*
23. *Consider reduced-current ballasts where a reduction in illuminance can be tolerated.*
24. *Consider the use of ballasts which can accomodate high pressure sodium or metal halide lamps interchangeably with other lamps.*
25. *Consider multi-level ballasts where a reduction in illuminance can be tolerated.*
26. *Consider substituting interchangeable-type metal halide lamps on compatible ballasts in existing mercury lighting systems. Two options: Upgrade sub-standard lighting in a mercury system with no increase in lighting power, or reduce lighting power by removing luminaires that may increase lighting levels above task lighting requirements.*
27. *Consider substituting interchangeable high pressure sodium lamps on retrofit ballasts in existing mercury lighting systems. Results: reduced connected lighting power with lamp substitution and more light.*
28. *Consider using heat removal luminaires whenever possible to improve lamp performance and reduce heat gain to space.*
29. *Select luminaires which do not collect dirt rapidly and which can be easily cleaned.*

C. Daylighting.

30. *If daylighting can be used to replace some of the electric lighting near the windows during substantial periods of the day, lighting in those areas should be dimmed or switched off.*
31. *Maximize the effectiveness of existing fenestration-shading controls (interior and exterior) or replace with proper devices or media.*
32. *Use daylighting effectively by locating work stations requiring the most illumination nearest the windows.*
33. *Daylighting, whenever it can be effectively used, should be considered in areas when a net energy conservation gain is possible, considering total energy for lighting, heating, and cooling.*

D. Controls and Distribution Systems.

34. *Install switching for selective control of illumination.*
35. *Evaluate the use of low-voltage (24 volts or lower) switching systems to obtain maximum switching capability.*
36. *Install switching or dimmer controls to provide flexibility when spaces are used for multiple purposes and require different amounts of illumination for various activities.*
37. *Consider a solid state dimmer system as a functional means for variable lighting requirements of high intensity discharge lamps.*
38. *Consider photocells and/or timeclocks for turning exterior lights on and off.*
39. *Install selective switching on luminaires according to grouping of working tasks at different working hours, and when not needed.*
40. *Consider plug-in electrical wiring to allow for flexibility in moving/removing/adding luminaires to suit changing furniture layouts.*

Fig. 69 *Continued*

| D. Controls and Distribution System *continued* | F. Operating Schedules *continued* |

D. Controls and Distribution System *continued*

41. *Consider coding on light control panels and switches according to a predetermined schedule of when lights should be turned off and on.*

E. Lighting Maintenance Procedures.

42. *Evaluate the present lighting maintenance program, and revise it as necessary to provide the most efficient use of the lighting system.*
43. *Clean luminaires and replace lamps on a regular maintenance schedule.*
44. *Check to see if all components are in good working condition.* Transmitting or diffusing media should be examined, and badly discolored or deteriorated media replaced, to improve efficiency (without producing excessive brightness and unwanted visual discomfort).
45. *Replace outdated or damaged luminaires with modern luminaires* which have good cleaning capabilities, and which use lamps with higher efficacies and good lumen maintenance characteristics.
46. *Trim trees and bushes that may be obstructing luminaire distribution and creating unwanted shadows.*

F. Operating Schedules.

47. *Analyze lighting used during working and building cleaning periods* and institute an education program

F. Operating Schedules *continued*

to have workers turn off lights when they are not needed. Inform and encourage personnel to turn off light sources such as: (a) Incandescent—promptly when space is not in use; (b) Fluorescent—if the space will not be used for five minutes or longer; (c) High Intensity Discharge Lamps—(Mercury, Metal Halide, High Pressure Sodium)—if space will not be used for 30 minutes or longer.

48. *Light building for occupied periods only,* and when required for security purposes.
49. *Restrict parking to specific lots* so lighting can be reduced to minimum security requirements in unused parking areas.
50. *Try to schedule routine building cleaning during occupied hours.*
51. *Reduce illuminance levels during building cleaning periods.*
52. *Adjust cleaning schedules to minimize the lighting use,* such as by concentrating cleaning activities in fewer spaces at the same time, and by turning off lights in unoccupied areas.

G. Post Instruction Covering Lighting Operation and Maintenance Procedures in All Management and General Work Areas.

Lighting Cost Analysis

From IES Lighting Handbook—1987 Application Volume

the results must be normalized by dividing the various cost factors such as total capital expense per year (item 24), total operating and maintenance expense per year (item 46), and total lighting expense per year (item 49) by the maintained illuminance (item 14).

Either individual (spot) or group replacement of lamps can be handled by the model using the following equations:

Individual replacement

$$= \frac{B}{R}(c + i) \text{ dollars per socket annually.} \quad (1)$$

Group replacement (early burnouts

$$\text{replaced)} = \frac{B}{A}(c + g + KL + Ki) \quad (2)$$

dollars per socket annually.

Group replacement (no replacement

$$\text{for early burnouts)} = \frac{B}{A}(c + g) \quad (3)$$

dollars per socket annually.

where

B = burning hours per year.
R = rated average lamp life in hours.
A = burning time between replacements in hours.
c = net cost of lamps in dollars.
i = cost per lamp for replacing lamps individually in dollars.
g = cost per lamp for replacing lamps in a group in dollars.
K = proportion of lamps failing before group replacement (from mortality curve).
L = net cost of replacement lamps in dollars.

No general rule can be given for the use of group replacements; each installation should be considered separately. In general, group replacement should be given consideration when individual replacement cost i is greater than half the cost c and when group replacement cost g is small compared to i.

An iteration of equations (2) or (3) using various burning times between replacements (A) will indicate whether or not group relamping is economical and the best (lowest annual cost per socket) relamping interval.

The choice of a periodic relamping or cleaning interval will have a direct effect on the light loss factor (item 12). An economic analysis which includes a study of variable maintenance econom-

LIGHTING ECONOMICS

One way to put the various costs associated with the design, purchase, ownership and operation of a lighting system into perspective is to subject that system to a thorough economic analysis. Focusing on the initial cost of equipment, maintenance or energy costs may improperly bias a purchasing decision or subject the owner of a system to higher than necessary operating costs, and the potential value built into the design of the system may never be realized.

Lighting economic analyses have other rationales as well. For either new or existing systems they might be used for:

1. Comparing alternative systems as part of a decision making process.
2. The evaluation of maintenance techniques and procedures.
3. Determining the impact of lighting on other building systems.
4. Budgeting and cash flow planning.
5. Simplifying and reducing complex lighting system characteristics to a generally understandable common measure—cost.
6. Helping to determine the benefit of lighting (cost/benefit analysis).

Lighting Cost Comparisons

Fig. 70 is an example of an annual cost model in which both initial and recurring costs are put on a "per year" basis. For comparisons among systems to be valid, however, each lighting method must provide the same illuminance or

Fig. 70 Lighting Cost Comparison (Annual Cost Model)

	Item	Lighting Method #1	Lighting Method #2
A. Installation Data			
Type of installation (office, industrial, etc.)	1	. . . .	. . . .
Luminaires per row	2	. . . .	. . . .
Number of rows	3	. . . .	. . . .
Total luminaires	4	. . . .	. . . .
Lamps per luminaire	5	. . . .	. . . .
Lamp type	6	. . . .	. . . .
Lumens per lamp	7	. . . .	. . . .
Watts per luminaire (including accessories)	8	. . . .	. . . .
Hours per start	9	. . . .	. . . .
Burning hours per year	10	. . . .	. . . .
Group relamping interval or rated life	11	. . . .	. . . .
Light loss factor	12	. . . .	. . . .
Coefficient of utilization	13	. . . .	. . . .
Illuminance, maintained	14	. . . .	. . . .
B. Capital Expenses			
Net cost per luminaire	15	. . . .	. . . .
Installation labor and wiring cost per luminaire	16	. . . .	. . . .
Cost per luminaire (luminaire plus labor and wiring)	17	. . . .	. . . .
Total cost of luminaires	18	. . . .	. . . .
Assumed years of luminaire life	19	. . . .	. . . .
Total cost per year of life	20	. . . .	. . . .
Interest on investment (per year)	21	. . . .	. . . .
Taxes (per year)	22	. . . .	. . . .
Insurance (per year)	23	. . . .	. . . .
Total capital expense per year	24	. . . .	. . . .
C. Annual Operating and Maintenance Expenses			
Energy expense			
Total watts	25	. . . .	. . . .
Average cost per kWh (including demand charges)	26	. . . .	. . . .
Total energy cost per year*	27	. . . .	. . . .
Lamp renewal expense			
Net cost per lamp	28	. . . .	. . . .
Labor cost each individual relamp	29	. . . .	. . . .
Labor cost each group relamp	30	. . . .	. . . .
Per cent lamps that fail before group relamp	31	. . . .	. . . .
Renewal cost per lamp socket per year**	32	. . . .	. . . .
Total number of lamps	33	. . . .	. . . .
Total lamp renewal expense per year	34	. . . .	. . . .
Cleaning expense			
Number of washings per year	35	. . . .	. . . .
Man-hours each (est.)	36	. . . .	. . . .
Man-hours for washing	37	. . . .	. . . .
Number of dustings per year	38	. . . .	. . . .
Man-hours per dusting each	39	. . . .	. . . .
Man-hours for dustings	40	. . . .	. . . .
Total man-hours	41	. . . .	. . . .
Expense per man-hour	42	. . . .	. . . .
Total cleaning expense per year	43	. . . .	. . . .
Repair expenses			
Repairs (based on experience, repairman's time, etc.)	44	. . . .	. . . .
Estimated total repair expense per year	45	. . . .	. . . .
Total operating and maintenance expense per year	46	. . . .	. . . .
D. Recapitulation			
Total capital expense per year	47	. . . .	. . . .
Total operating and maintenance expense per year	48	. . . .	. . . .
Total lighting expense per year	49	. . . .	. . . .

* Total energy cost per year = $\dfrac{\text{Total watts} \times \text{burning hours per year} \times \text{cost per kWh}}{1000}$

** See formulas (2), (3) and (4) in the text. They can be used to determine the most economical replacement method.

Note: Items 38, 39 and 40 may be eliminated and "washings" in items 35 and 37 changed to "cleanings".

ics can, therefore, be used in the design phase of a project as a way to minimize system light output depreciation over time. As a result, for general lighting systems, fewer luminaires may be needed for a given maintained illuminance level and both initial as well as owning and operating costs can be reduced for the entire system over its life.

Life Cycle Costs

A second type of mathematical model known as the present worth or net present value analysis is becoming more widely used for lighting economic studies because of its power as a decision making tool. In contrast to the annual cost model in which all expenditures are put into a cost per year form, the present worth model is set up to take into account expenditures at the time they actually take place. It is, therefore, somewhat easier to consider all of the costs which are expected to occur over the life time or life cycle of the lighting system up to and including any salvage value.

All expenditures, whether they are periodic repetitive costs, such as for electric energy, cleaning, or lamp replacement, or not periodic costs such as damage repair or ballast replacement are factored in at the future time the expenditures are expected to occur. All costs are then transferred to the present time using relationships which take into account the time value of money via an interest factor or "opportunity rate." Tables [1,2] are available to simplify the arithmetic, or the following seven basic equations may be used (terms used in these equations are: P = present or first cost; F = future worth; A = uniform annual cost; i = opportunity rate or interest; and y = number of years):

1. Uniform compound amount factor.

$$F = A \left[\frac{(1 + i)^y - 1}{i} \right]$$

2. Uniform present worth factor.

$$P = A \left[\frac{(1 + i)^y - 1}{i(1 + i)^y} \right]$$

3. Uniform sinking fund factor.

$$A = F \left[\frac{i}{(1 + i)^y - 1} \right]$$

4. Single present worth factor.

$$P = F \left[\frac{1}{(1 + i)^y} \right]$$

5. Uniform capital recovery factor.

$$A = P \left[\frac{i(1 + i)^y}{(1 + i)^y - 1} \right]$$

6. Single compound amount factor.

$$F = P \left[(1 + i)^y \right]$$

7. The equations below give the present worth of an escalating annual cost (r = rate of escalation and A = annual amount before escalation).

(a) $$P = \sum_{k=1}^{y_n} A \left[\frac{(1 + r)^k}{(1 + i)^k} \right]$$

(b) $$P = A \left[\frac{(1 + r) [(1 + i)^{y_n} - (1 + r)^{y_n}]}{(i - r) (1 + i)^{y_n}} \right] \text{if } i \neq r$$

(c) $$P = A [y_n] \qquad if\ i = r$$

System Comparison Based on Present Worth. In the process of comparing several alternative lighting systems, all costs might be converted into a single number representing the present worth of each system over its life. The system with the lowest overall cost would then be readily apparent.

Public bodies have adopted this approach particularly for projects involving energy conservation and a number of calculational aids including a computer program in the public domain are available. [2]

Using the lighting system cost comparison outlined in Fig. 71, a comparison of lighting systems can be made based on present worth as follows: [3]

	System A	System B
First Costs	A1	B1
Power and maintenance costs	Ap	Bp
Salvage value	As	Bs
Terms = y years		
Opportunity rate = i (e.g., i = .09 means 9 percent)		

$$A_{T1} = A1 + Ap \cdot \left[\frac{(1 + i)^y - 1}{i(1 + i)^y} \right] - As \cdot \left[\left(\frac{1}{1 + i} \right)^y \right]$$

$$B_{T1} = B1 + Bp \cdot \left[\frac{(1 + i)^y - 1}{i(1 + i)^y} \right] - Bs \cdot \left[\left(\frac{1}{1 + i} \right)^y \right]$$

This present worth method is valid only if Systems A and B have equal lives.

In the event it is required to determine the number of years (y) at a given opportunity rate for a payout of one system over another the following formulas are given:

Let X_1 = difference between (present cost salvage value) of the two systems ($A_1 - As$).

Let X_2 = difference between annual operating costs of the two systems (Ap).

Let i = opportunity rate.

Find $X_3 = \dfrac{X_2}{(i \cdot X_1)}$

Let $a = 1 + i$

$$b = \frac{X_3}{(X_3 - 1)}$$

Then $y = \dfrac{\ln b}{\ln a}$

Note: The above formulas are correct if $X_2/X_1 > i$; however, if $X_2/X_1 = i$, then both systems are equal; and if $X_2/X_1 < i$, then the increase in present cost (X_1) will never recover through the annual benefits (X_2).

In comparing a number of different lighting systems, a life cycle cost study is not relevant unless all systems are functionally acceptable. There is no monetary expression that says one alternative doesn't provide enough light, or that another is a system with a great deal of discomfort glare. Therefore, the analysis assumes that all alternatives are functionally equivalent.

Notes for Fig. 71.
II. 1. The number of burning hours and cost/kWh will depend upon occupancy and local power rates.
II. 2. Only that portion needed to remove lighting heat.
II. 3. Reduction in cost for fuel for heating equipment because of increased heat obtained from lighting system.
II. 8. Percentage of first cost (I.1.) (1 to 1.5 percent) (compensating).
II. 9. Percentage of first cost (I.1.) (4 to 6 percent) (compensating).
*Based on reference 3.

Fig. 71 Lighting System Cost Comparison (Life Cycle Costs—Present Worth Model)*

Life cycle cost analysis for _____
Area _____ m² (ft²) Luminaire _____ Luminaire _____
Layout _____ Layout _____

I. Lighting and Air Conditioning Installed Costs (initial)

1. Luminaire installed costs: luminaire, lamps, material and labor:	$ _____	$ _____
2. Total kW lighting:	____ kW	____ kW
3. Tons of air conditioning required for lighting:	____tons	____tons
4. First cost of air-conditioning machinery:	$ _____	$ _____
5. Variation in first cost of heating equipment:	$ _____	$ _____
6. Other differential costs:	$ _____	$ _____
7. Subtotal mechanical and electrical installed cost:	$ _____	$ _____
8. Initial taxes:	$ _____	$ _____
9. Total costs:	$ __(A1)	$ __(B1)
10. Installed cost per square meter (foot)	$ _____	$ _____
11. Watts per square meter (foot) of lighting:	____ watts	____ watts
12. Salvage (at y years):	$ __(As)	$ __(Bs)

II. Annual Power and Maintenance Costs

1. Lamps: burning hours × kW × $/kWh	$ _____	$ _____
2. Air conditioning:	$ _____	$ _____
3. Reduction in heating costs fuel used: ____	$ _____	$ _____
4. Other differential costs (i.e., air-conditioning, maintenance cost):	$ _____	$ _____
5. Cost of lamps: (No. of lamps ____ @ $____/lamp per N) (Group relamping every one, two or three years, depending on burning schedule.)	$ _____	$ _____
6. Cost of ballast replacement: (No. of ballast ____ @ $____/ballast per n) (n = number of years of ballast life.)	$ _____	$ _____
7. Luminaire cleaning cost: No. of luminaires ____ @ $____ each. (Cost to clean one luminaire includes cost to replace or clean lamps.)	$ _____	$ _____
8. Annual insurance cost:	$ _____	$ _____
9. Annual property tax cost:	$ _____	$ _____
10. Total annual power and maintenance cost:	$ __(Ap)	$ __(Bp)
11. Cost per square meter (foot):	$ _____	$ _____

References

1. *Life Cycle Cost-Benefit Analysis*, K.G. Associates, Box 7596, Inwood Station, Dallas, TX 75209.

2. Ruegg, R.T., *et al* : *A Guide for Selecting Energy Conservation Projects for Public Buildings*, NBS BSS113, National Bureau of Standards, U.S. Dept. of Commerce, Washington, D.C., 1978.

3. Design Practice Committee of the IES: "Life Cycle Cost Analysis of Electric Lighting Systems," *Light. Des. Appl.*, Vol. 10, No. 5, p. 43, May, 1980.

4. DeLaney, W.B.: "How Much Does a Lighting System Really Cost?" *Light Des. Appl.*, Vol. 3, No. 1, p. 22, January 1973. Clear, R. and Berman: "Cost-Effective Visibility-Based Design Procedures for General Office Lighting." *J. Illum. Eng. Soc.*, Vol. 10, No. 45, p. 228, July, 1981.

Interior Lighting Survey Procedures

(From IES Lighting Handbook—*1984 Reference Volume*)

FIELD MEASUREMENTS—INTERIORS

In evaluating an actual lighting installation in the field it is necessary to measure or survey the quality and quantity of lighting in the particular environment. To help do this, the IES has developed a uniform survey method of measuring and reporting the necessary data.* The results of these uniform surveys can be used alone or with other surveys for comparison purposes, can be used to determine compliance with specifications, and can be used to reveal the need for maintenance, modification or replacement.

Field measurements apply only to the conditions that exist during the survey. Recognizing this it is very important to record a complete detailed description of the surveyed area and all factors that might affect results, such as: interior surface reflectances, lamp type and age, voltage, and instruments used in the survey.

In measuring illuminance, cell type instruments should be used which are cosine and color corrected. They should be used at a temperature above 15 °C (60 °F) and below 50 °C (120 °F), if possible. Before taking readings, the cells should be exposed to the approximate illuminance to be measured until they become stabilized. This usually requires 5 to 15 minutes. Casting a shadow on the light sensitive cell should be avoided while reading the instrument. A high intensity discharge or fluorescent system must be lighted for at least one hour before measurements are taken to be sure that normal operating output has been attained. In relatively new lamp installations, at least 100 hours of operation of a gaseous source should elapse before measurements are taken. With incandescent lamps, seasoning is accomplished in a shorter time (20 hours or less for common sizes).

Illuminance Measurements—Average

Determination of Average Illuminance on a Horizontal Plane from General Lighting Only. The use of this method in the types of areas described should result in values of average illuminance within 10 per cent of the values that would be obtained by dividing the area into 0.6-meter (2-foot) squares, taking a reading in each square and averaging.

The measuring instrument should be positioned so that when readings are taken, the surface of the light sensitive cell is in a horizontal plane and 760 millimeters (30 inches) above the floor. This can be facilitated by means of a small portable stand of wood or other material that will support the cell at the correct height and in the proper plane. Daylight may be excluded during illuminance measurements. Readings can be taken at night or with shades, blinds or other opaque covering on the fenestration.

Regular Area With Symmetrically Spaced Luminaires in Two or More Rows. (See Fig. 72a.) (1) Take readings at stations r-1, r-2, r-3, and r-4 for a typical inner bay. Repeat at stations r-5, r-6, r-7 and r-8 for a typical centrally located bay. Average the 8 readings. This is R in the equation directly below. (2) Take readings at stations q-1, q-2, q-3 and q-4 in two typical half bays on each side of room. Average the 4 readings. This is Q in the equation below. (3) Take readings at stations t-1, t-2, t-3 and t-4 in two typical half bays at each end of room. Average the 4 readings. This is T in the equation. (4) Take readings at stations p-1 and p-2 in two typical corner quarter bays. Average the 2 readings. This is P in the equation. (5) Determine the average illuminance in the area by solving the equation:

*Illum. Eng., Feb. 1963, p. 87.

$$\text{Average Illuminance} = \frac{R(N-1)(M-1) + Q(N-1) + T(M-1) + P}{NM}$$

where

N = number of luminaires per row;
M = number of rows.

Regular Area With Symmetrically Located Single Luminaire. (See Fig. 72b.) (1)

Take readings at stations p-1, p-2, p-3 and p-4 in all 4 quarter bays. Average the 4 readings. This is P, the average illuminance in the area.

Regular Area With Single Row of Individual Luminaires. (See Fig. 72c.) (1) Take readings at stations q-1 thru q-8 in 4 typical half bays located two on each side of the area. Average the 8 readings. This is Q in the equation directly below. (2) Take readings at stations p-1 and p-2 for two typical corner quarter bays. Average the

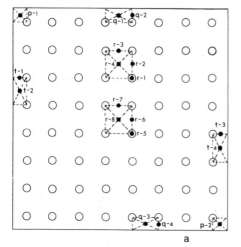

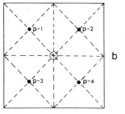

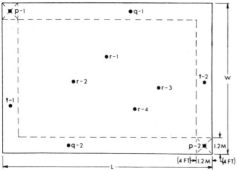

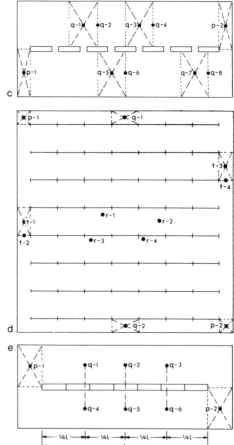

Fig. 72. Location of illuminance measurement stations in: (a) regular area with symmetrically spaced luminaires in 2 or more rows; (b) regular area with symmetrically located single luminaire; (c) regular area with single row of individual luminaires; (d) regular area with 2 or more continuous rows of luminaires; (e) regular area with single row of continuous luminaires; (f) regular area with luminous or louverall ceiling.

2 readings. This is P in the equation. (3) Determine the average illuminance in the area by solving the equation:

$$\text{Average Illuminance} = \frac{Q(N-1) + P}{N}$$

where N = number of luminaires.

Regular Area With Two or More Continuous Rows of Luminaires. (See Fig. 72d.) (1) Take readings at stations r-1 thru r-4 located near the center of the area. Average the 4 readings. This is R in the equation directly below. (2) Take readings at stations q-1 and q-2 located at each midside of the room and midway between the outside row of luminaires and the wall. Average the 2 readings. This is Q in the equation below. (3) Take readings at stations t-1 thru t-4 at each end of the room. Average the 4 readings. This is T in the equation. (4) Take readings at stations p-1 and p-2 in two typical corners. Average the 2 readings. This is P in the equation. (5) Determine the average illuminance in the area by solving the equation:

$$\text{Average Illuminance} = \frac{RN(M-1) + QN + T(M-1) + P}{M(N+1)}$$

where

N = number of luminaires per row;
M = number of rows.

Regular Area With Single Row of Continuous Luminaires. (See Fig. 72e.) (1) Take readings at stations q-1 thru q-6. Average the 6 readings. This is Q in the equation directly below. (2) Take readings at stations p-1 and p-2 in typical corners. Average the 2 readings. This is P in the equation below. (3) Determine the average illuminance in the area by solving the equation:

$$\text{Average Illuminance} = \frac{QN + P}{N+1}$$

where N = number of luminaires.

Regular Area With Luminous or Louverall Ceiling. (See Fig. 72f.) (1) Take readings at stations r-1 thru r-4 located at random in the central portion of the area. Average the 4 readings. This is R in the equation directly below. (2) Take readings at stations q-1 and q-2 located 0.6 meter (2 feet) from the long walls at random lengthwise of the room. Average the 2 readings. This is Q in the equation below. (3) Take read-

ings at stations t-1 and t-2 located 0.6 meter (2 feet) from the short walls at random crosswise of the room. Average the 2 readings. This is T in the equation below. (4) Take readings at stations p-1 and p-2 located at diagonally opposite corners 0.6 meter (2 feet) from each wall. Average the 2 readings. This is P in the equation. (5) Determine the average illuminance in the area by solving the equation:

$$\text{Average Illuminance} = \frac{R(L-8)(W-8) + 8Q(L-8) + 8T(W-8) + 64P}{WL}$$

where

W = width of room;
L = length of room.

Illuminance Measurements—Point

With task, general and supplementary lighting in use, the illuminance at the point of work should be measured with the worker in his normal working position. The measuring instrument should be located so that when readings are taken, the surface of the light sensitive cell is in the plane of the work or of that portion of the work on which the critical visual task is performed—horizontal, vertical or inclined. Readings as shown in Fig. 73 should be recorded.

Luminance Measurements

Luminance surveys may be made under actual working conditions and from a specified work point location with the combinations of daylight and electric lighting facilities available. Consideration should be given to sun position and weather conditions, both of which may have marked effect on luminance distribution. All lighting in the area, task, general and supple-

Fig. 73. Form for Tabulation of Point Illuminance Measurements

Work Point	Description of Work Point	Height Above Floor	Plane (horizontal, vertical, or inclined)	Illuminance	
				Total (general + supplementary)	General Only
1—(max.)					
2—(min.)					
3—					
4—					
5—					

Fig. 74. Form for Tabulation of Luminance Measurements

Work Point Location*	Luminance					
	A	B	C	D	E	F
Luminaire at 45° above eye level						
Luminaire at 30° above eye level						
Luminaire at 15° above eye level						
Ceiling, above luminaire						
Ceiling, between luminaires						
Upper wall or ceiling adjacent to a luminaire						
Upper wall between two luminaires						
Wall at eye level						
Dado						
Floor						
Shades and blinds						
Windows						
Task						
Immediate surroundings of task						
Peripheral surroundings of task						
Highest luminance in field of view						

* Describe locations A thru F.

mentary, should be in normal use. Work areas used only in the daytime should be surveyed in the daytime; work areas used both daytime and nighttime should preferably have two luminance surveys made under the two sets of conditions, as the luminance distribution and the possible comfort or discomfort will differ markedly at these times. Nighttime surveys should be made with shades drawn. Daytime surveys should be made with shades adjusted for best control of daylight.

On a floor plan sketch of the area, an indication should be made of which exterior wall or walls, if any, were exposed to direct sunlight during the time of the survey by writing the word "Sun" in the appropriate location. Readings should be taken, successively, from the worker's position at each work point location A, B, C, etc. and luminance readings from each location recorded as shown in Fig. 74.

ILLUMINATING ENGINEERING SOCIETY OF NORTH AMERICA (IESNA)

The IES is the recognized technical authority for the illumination field. For over 80 years, its objective has been to communicate information on all aspects of good lighting. The IES serves its members, the lighting community, and consumers through a variety of programs, publications and services. Its strength is in its diversified membership which includes engineers, architects, designers, educators, students, contractors, distributors, utility personnel, manufacturers, and scientists—all *contributing to* and *benefiting from* the Society.

The IES is a forum for exchange of ideas and information, and a vehicle for its members' professional development and recognition. Through its technical committees, with hundreds of qualified members from the lighting community, the IES correlates vast amounts of research, investigations and discussions to guide lighting experts and laymen on research- and consensus-based lighting recommendations. Complete lists of current and available recommendations also may be obtained by writing to the IES Publications Office.

The IES publishes the *IES Lighting Handbook*, upon which this Ready Reference is based, and *Lighting Design + Application (LD+A)* and the *Journal of the Illuminating Engineering Society* as the official magazines of the Society. LD+A, a popular application-oriented monthly magazine, contains special feature articles and news of practical and innovative lighting layouts, systems, equipment and economics, and news of the industry and its people. The *Journal*, a technical publication, contains technically-oriented papers, articles, American National Standards, IES recommended practices and other committee reports.

The Society also publishes nearly 100 varied publications (including education courses; educational video tapes; IES technical committee reports covering many specific lighting applications; forms and guides used for measuring and reporting lighting values, lighting calculations, performance of light sources and luminaires and energy management). The *IES Lighting Library*, a complete reference package encompassing all essential IES documents, is also available along with a yearly updating service.

The Society has two types of membership: individual and sustaining. Applications and current dues schedules are available upon request from the Membership Department.

IES local, regional and transnational meetings, conferences, symposiums, seminars, workshops and lighting exhibitions provide current information on the latest developments in illumination. Basic and advanced lighting courses are offered by local IES Sections.

The IES established the Lighting Research Institute (LRI), a not-for-profit corporation, in 1982. LRI promotes and sponsors basic and applied research applicable to all aspects of lighting phenomena; the fundamental life and behavioral sciences; bases for practice of lighting design; and needs of consumers and users of lighting.

Hand-in-hand with the research program of the LRI, an equally comprehensive educational program has been set in motion by expanding the role of the IES in education. As recommended by the IES, a two-fold approach to education was implemented—*academic* (teaching materials, faculty development, continuing education of teachers, scholarships and fellowships); and *public* (career development for the lighting community, continuing education for the opinion makers and the general public).

Heterick Memorial Library
Ohio Northern University

DUE	RETURNED	DUE	RETURNED
SEP 07 1993 1.	8	13.	
11/12/0 DEC 13 2001 2.		14.	
3.		15.	
4.		16.	
5.		17.	
6.		18.	
7.		19.	
8.		20.	
9.		21.	
10.		22.	
11.		23.	
12.		24.	

Index